Advances in Experimental Medicine and Biology

For further volumes:
http://www.springer.com/series/5584

Johng S. Rhim • Richard Kremer
Editors

Human Cell Transformation

Role of Stem Cells and the Microenvironment

Springer

Editors
Johng S. Rhim
Department of Surgery
Uniformed Services
University of the Health Sciences
Bethesda, MD, USA
jrhim@verizon.net

Richard Kremer
Department of Medicine
McGill University Health Centre
Montreal, QC, Canada
richard.kremer@mcgill.ca

ISBN 978-1-4614-0253-4 e-ISBN 978-1-4614-0254-1
DOI 10.1007/978-1-4614-0254-1
Springer New York Dordrecht Heidelberg London

Library of Congress Control Number: 2011935228

Printed on acid-free paper

Springer is part of Springer Science+Business Media (www.springer.com)

Preface

The role of carcinogenic agents in the development of human cancers is being defined using a variety of human cells as experimental model systems. Thus, the study of human cell transformation in culture by carcinogenic agents is of particular importance for understanding the cellular and molecular mechanisms underlying human carcinogenesis. A conference on "Human Cell Transformation: Role of Stem Cells and the Environment" was held at the McGill University Health Center, Montreal, Canada, on October 20–21, 2010. The aims of the conference were to present the state-of-the-art in the transformation of human cells in culture, as well as to provide insight into the molecular and cellular changes involved in the conversion of normal cells to a neoplastic state of growth (Fig. 1).

The conference encompassed the most recent developments in human cell transformation including selected in vitro models, tumor xenografts, and transgenic preclinical models. The meeting provided a unique forum for the exchange of information in this important research field. Some of the novel topics to be addressed included: what is the relationship between normal and malignant phenotypes? what is the role of stem cells and the microenvironment in this process? what are the current approaches to identify and target the cancer cells at the different stages of tumor progression?

The following topics were closely related to the theme of the conference:

1. Derivation of in vitro model systems (epithelial, fibroblastic, and hematopoietic).
2. Usefulness of defined in vitro model systems for viral, chemical, and radiation carcinogenesis.
3. The role of stem cells in tumorigenesis.
4. Genes involved in multistep carcinogenesis.

The conference was organized by Johng S. Rhim and Richard Kremer. There were 20 speakers, 20 poster presentations, and about 100 participants.

The conference was well received and was perhaps the latest one devoted solely to the subject of human cell transformation systems *in vitro*. It is our privilege to have an opportunity to edit these proceedings and also on behalf of all the contributors to thank everyone who have helped us produce this book.

Bethesda, MD — Johng S. Rhim
Montreal, QC — Richard Kremer

Group speakers photograph of the 2010 conference of human cell transformation: role of stem cells and the mircoenvironment held on McGill university faculty club, Montreal, Canada on October 20–21, 2010.

Front row left to right: Tom Hei, Johng Rhim, Richard Kremer, and Hsiang Kung.

Back row left to right: Shafaat Rabbani, Vyomesh Patel, Leon Nesti, Jerry Shay, Mira Jung, Anatoly Dritschilo, Justin McCormick, Clayton Yates, Vimla Band, Norman Maitland, Jose Russo, Fionna Frame, Catalin Mihalcioiu, David Lim, and Sang Park

Chronology of the Human Cell Transformation Meeting

1991 *Neoplastic transformation in human cell culture: mechanisms of carcinogenesis*. Georgetown University Medical Center, Washington, DC, USA. Organized by J.S. Rhim and A. Dritschilo.

1993 *International symposium on molecular mechanisms of radiation and chemical-induced cell Transformation*. Mackinac Island, MI, USA. Organized by K. Chadwick, A. Karaoglou, M. Frazin, V.M. Mahr, and J.J. McCormick.

1995 *Neoplastic transformation in human cell culture systems in cultures: mechanisms of Carcinogenesis*. Chicago, IL, USA. Organized by R.R. Weischenbaum, J.S. Rhim, A. Drischilo, T.C. Yang, and J.C. Barrett.

1999 *In vitro transformation*. Cork, Ireland. A satellite meeting of the 11th international Congress of Radiation Research, Dublin, Ireland. Organized by J.S. Rhim and C. Mothersill.

2002 *Human cell culture 2002*, Oxford, UK. Joint meeting of European Tissue Culture Society, Scandinavian Society for Cell Toxicology and British Prostate Group, Oxford, UK. Organized by N.J. Maitland et al.

2010 *Human cell transformation: role of stem cells and the microenvironment*. McGill University Health Center, Montreal, QC, Canada. Organized by J.S. Rhim and Richard Kremer.

Acknowledgments

This conference was sponsored by: Musculoskeletal Axis of the McGill University Health Center, Montreal, QC H3A1A1, Canada, IRSC (Instituts de Recherche en Sante du Canada) and CIHR (Canadian Institutes of Health Research), Amgen, Reseau de recherche en santé buccodentaire et osseuse, The Centre for Bone and Periodontal Research.

Contents

Contributors

Orna Amrani Clinical and Experimental Orthopaedics Group, National Institutes of Arthritis and Musculoskeletal and Skin Diseases, National Institutes of Health, Bethesda, MD, USA

Hamid Band Department of Genetics, Cell Biology and Anatomy, University of Nebraska Medical Center, Omaha, NE, USA

Eppley Institute for Cancer and Allied Diseases and UNMC-Eppley Cancer Center, University of Nebraska Medical Center, Omaha, NE, USA

Vimla Band Department of Genetics, Cell Biology and Anatomy, University of Nebraska Medical Center, Nebraska Medical Center, Omaha, NE, USA

Eppley Institute for Cancer and Allied Diseases and UNMC-Eppley Cancer Center, University of Nebraska Medical Center, Nebraska Medical Center, Omaha, NE, USA

Ramiro Iglesias-Bartolome Oral and Pharyngeal Cancer Branch, National Institute of Dental and Craniofacial Research, National Institutes of Health, Bethesda, MD, USA

Aditya Bele Department of Genetics, Cell Biology and Anatomy, University of Nebraska Medical Center, Omaha, NE, USA

Anne Camirand Department of Medicine, McGill University Health Centre, Montreal, QC, Canada

Lady Davis Institute for Medical Research, Jewish General Hospital, Montreal, QC, Canada

Chu-Yan Chan Stanley Ho Centre for Emerging Infectious, The Chinese University of Hong Kong, Hong Kong, People's Republic of China

Department of Anatomy, The Chinese University of Hong Kong, Hong Kong, People's Republic of China

Amrita K. Cheema Department of Oncology, Lombardi Comprehensive Cancer Center, Georgetown University Medical Center, Washington, DC, USA

Flora Chik Department of Pharmacology and Therapeutics, McGill University, Montreal, QC, Canada

Gregory T. Christopherson Clinical and Experimental Orthopaedics Group, National Institutes of Arthritis and Musculoskeletal and Skin Diseases, National Institutes of Health, Bethesda, MD, USA

Anatoly Dritschilo Departments of Oncology and Radiation Medicine, Lombardi Comprehensive Cancer Center, Georgetown University Medical Center, Washington, DC, USA

Department of Radiation Medicine, The Lombardi Comprehensive Cancer Center, Georgetown University Medical Center, Washington, DC, USA

Fiona M. Frame Department of Biology, YCR Cancer Research Unit, University of York, Heslington, North Yorkshire, UK

Bungo Furusato Department of Surgery, Uniformed Services University of the Health Sciences, Bethesda, MD, USA

Department of Genitourinary Pathology, Armed Forces Institutes of Pathology, Washington, DC, USA

Channabasavaiah Basavaraju Gurumurthy Department of Genetics, Cell Biology and Anatomy, University of Nebraska Medical Center, Omaha, NE, USA

J. Silvio Gutkind Oral and Pharyngeal Cancer Branch, National Institute of Dental and Craniofacial Research, National Institutes of Health, Bethesda, MD, USA

Zhong Chao Han Institute of Hematology, Chinese Academy of Medical Sciences and Peking Union Medical College, Tianjin, People's Republic of China

Ming-Liang He Stanley Ho Centre for Emerging Infectious Diseases, The Chinese University of Hong Kong, Hong Kong, People's Republic of China

Tom K. Hei Department of Radiation Oncology, Center for Radiological Research, Columbia University Medical Center, New York, NY, USA

Nam-ho Huh Department of Cell Biology, Okayama University Graduate School of Medicine, Dentistry and Pharmaceutical Sciences, Kita-ku, Okayama, Japan

Vladimir Ivanov Department of Radiation Oncology, Center for Radiological Research, Columbia University Medical Center, New York, NY, USA

Wesley M. Jackson Clinical and Experimental Orthopaedics Laboratory, Department of Surgery, The Uniformed Services University, Bethesda, MD, USA

Youngmi Ji Clinical and Experimental Orthopaedics Group, National Institutes of Arthritis and Musculoskeletal and Skin Diseases, National Institutes of Health, Bethesda, MD, USA

Mira Jung Department of Radiation Medicine, The Lombardi Comprehensive Cancer Center, Georgetown University Medical Center, Washington, DC, USA

Andrew C. Karaplis Lady Davis Institute for Medical Research, Jewish General Hospital, Montreal, QC, Canada

Matthew W. Kluk Clinical and Experimental Orthopaedics Group, National Institutes of Arthritis and Musculoskeletal and Skin Diseases, National Institutes of Health, Bethesda, MD, USA

Integrated Department of Orthopaedics and Rehabilitation, Walter Reed Army Medical and National Naval Medical Center, Bethesda, MD, USA

Zacharoula Konsoula Department of Radiation Medicine, The Lombardi Comprehensive Cancer Center, Georgetown University Medical Center, Washington, DC, USA

Richard Kremer Department of Medicine, McGill University Health Centre, Montréal, QC, Canada

Hsiang-Fu Kung Stanley Ho Centre for Emerging Infectious Diseases, The Chinese University of Hong Kong, Hong Kong, People's Republic of China

State Key Laboratory in Oncology in South China, Cancer Center, Sun Yat-Sen University Cancer Center, Guangzhou, People's Republic of China

Kantima Leelahavanichkul Oral and Pharyngeal Cancer Branch, National Institute of Dental and Craniofacial Research, National Institutes of Health, Bethesda, MD, USA

Guo Li Stanley Ho Centre for Emerging Infectious Diseases, The Chinese University of Hong Kong, Hong Kong, People's Republic of China

Hongzhen Li Department of Surgery, Uniformed Services University of the Health Sciences, Bethesda, MD, USA

Jiarong Li Department of Medicine, McGill University Health Centre, Montreal, QC, Canada

Karen Li Li Ka Shing Institute of Health Sciences, The Chinese University of Hong Kong, Hong Kong, People's Republic of China

Department of Pediatrics, The Chinese University of Hong Kong, Hong Kong, People's Republic of China

David J. Lim Division of Clinical and Translational Research, House Ear Institute, Los Angeles, CA, USA

Department of Cell and Neurobiology, University of Southern California, Los Angeles, CA, USA

Marie Chia-Mi Lin Department of Chemistry, The University of Hong Kong, Hong Kong, People's Republic of China

Veronica M. Maher Department of Microbiology and Molecular Genetics and Department of Biochemistry and Molecular Biology, Carcinogenesis Laboratory – Food Safety and Toxicology, Michigan State University, East Lansing, MI, USA

Norman J. Maitland YCR Cancer Research Unit, Department of Biology, University of York, Heslington, North Yorkshire, UK

Gautam Malhotra Department of Genetics, Cell Biology and Anatomy, University of Nebraska Medical Center, Nebraska Medical Center, Omaha, NE, USA

Christina A. Marsh Oral and Pharyngeal Cancer Branch, National Institute of Dental and Craniofacial Research, National Institutes of Health, Bethesda, MD, USA

J. Justin McCormick Department of Microbiology and Molecular Genetics and Department of Biochemistry and Molecular Biology, Carcinogenesis Laboratory–Food Safety and Toxicology, Michigan State University, East Lansing, MI, USA

Chun-Ling Meng Stanley Ho Centre for Emerging Infectious Diseases, The Chinese University of Hong Kong, Hong Kong, People's Republic of China

Sameer Mirza Department of Genetics, Cell Biology and Anatomy, University of Nebraska Medical Center, Omaha, NE, USA

Shakur Mohibi Department of Genetics, Cell Biology and Anatomy, University of Nebraska Medical Center, Omaha, NE, USA

Alfredo A. Molinolo Oral and Pharyngeal Cancer Branch, National Institute of Dental and Craniofacial Research, National Institutes of Health, Bethesda, MD, USA

Sung K. Moon Division of Clinical and Translational Research, House Ear Institute, Los Angeles, CA, USA

Masayoshi Namba Niimi College, Niimi, Japan

Leon J. Nesti Clinical and Experimental Orthopaedics Group, National Institutes of Arthritis and Musculoskeletal and Skin Diseases, National Institutes of Health, Bethesda, MD, USA

Integrated Department of Orthopaedics and Rehabilitation, Walter Reed Army Medical and National Naval Medical Center, Bethesda, MD, USA

Sai-Ming Ngai Department of Biology, The Chinese University of Hong Kong, Hong Kong, People's Republic of China

Sang Chul Park Department of Biochemistry and Molecular Biology, Seoul National University Medical School, Chong No Ku, Seoul, South Korea

Lee Gil Ya Cancer and Diabetes Institute, Gachon University of Medicine and Science, Incheon, South Korea

Vyomesh Patel Oral and Pharyngeal Cancer Branch, National Institute of Dental and Craniofacial Research, National Institutes of Health, Bethesda, MD, USA

Meghan G. Quinn Department of Genetics, Cell Biology and Anatomy, University of Nebraska Medical Center, Omaha, NE, USA

Shafaat A. Rabbani Department of Medicine, McGill University Health Centre, Montreal, QC, Canada

Habtom Ressom Department of Oncology, Lombardi Comprehensive Cancer Center, Georgetown University Medical Center, Washington, DC, USA

Johng S. Rhim Department of Surgery, Uniformed Services University of the Health Sciences, Bethesda, MD, USA

Irma H. Russo Breast Cancer Research Laboratory, Women Cancer Program, Fox Chase Cancer Center, Philadelphia, PA, USA

Jose Russo Breast Cancer Research Laboratory, Women Cancer Program, Fox Chase Cancer Center, Philadelphia, PA, USA

Masakiyo Sakaguchi Department of Cell Biology, Okayama University Graduate School of Medicine, Dentistry and Pharmaceutical Sciences, Kita-ku, Okayama, Japan

Bradford Siegele Oral and Pharyngeal Cancer Branch, National Institute of Dental and Craniofacial Research, National Institutes of Health, Bethesda, MD, USA

Moshe Szyf Department of Pharmacology and Therapeutics, McGill University, Montreal, QC, Canada

Sackler Program for Epigenetics and Developmental Psychobiology, McGill University, Montreal, QC, Canada

Olga Timofeeva Department of Oncology, Lombardi Comprehensive Cancer Center, Georgetown University Medical Center, Washington, DC, USA

Sau-na Tsai Department of Biology, The Chinese University of Hong Kong, Hong Kong, People's Republic of China

Kam Sze Tsang Li Ka Shing Institute of Health Sciences and Department of Anatomical and Cellular Pathology, The Chinese University of Hong Kong, Hong Kong, People's Republic of China

Rency Varghese Department of Oncology, Lombardi Comprehensive Cancer Center, Georgetown University Medical Center, Washington, DC, USA

Alfredo Velena Department of Radiation Medicine, Lombardi Comprehensive Cancer Center, Georgetown University Medical Center, Washington, DC, USA

Hua Wang Stanley Ho Centre for Emerging Infectious Diseases, The Chinese University of Hong Kong, Hong Kong, People's Republic of China

Xin Wang Stanley Ho Centre for Emerging Infectious Diseases, The Chinese University of Hong Kong, Hong Kong, People's Republic of China

Clayton Yates Department of Biology and Center for Cancer Research, Tuskegee University, Tuskegee, AL, USA

David Tai Wai Yew Department of Anatomy, The Chinese University of Hong Kong, Hong Kong, People's Republic of China

Xiao-Ai Zhang Stanley Ho Centre for Emerging Infectious Diseases, The Chinese University of Hong Kong, Hong Kong, People's Republic of China

Xiangshan Zhao Department of Genetics, Cell Biology and Anatomy, University of Nebraska Medical Center, Omaha, NE, USA

Hongning Zhou Department of Radiation Oncology, Center for Radiological Research, Columbia University Medical Center, New York, NY, USA

Yongliang Zhou Department of Radiation Oncology, Center for Radiological Research, Columbia University Medical Center, New York, NY, USA

Part I

Multistep Models

1 Nuclear Barrier Hypothesis of Aging as Mechanism for Trade-Off Growth to Survival

Sang Chul Park

Abstract
When the aging-dependent cellular behaviors toward growth factors and toxic stress have been analyzed, the perinuclear accumulation of the activated signals, either mitogenic or apoptotic, has been observed, suggesting the aging-dependent inefficiency of the nucleocytoplasmic trafficking of the signals. Thereby, it would be natural to assume the operation of the functional nuclear barrier in aging-dependent manner, which would be designated as "Park and Lim's Barrier." And for the ultimate transcriptional factor for these aging-dependent changes of the functional nuclear barrier, Sp1 transcriptional factor has been suggested to be the most probable candidate. This novel mechanism of aging-dependent operation of the functional nuclear barrier is proposed as the ultimate checking mechanism for cellular protection against toxic environment and the general mechanism for the trade-off growth to survival in aging.

1.1 Introduction

When human diploid fibroblasts (HDFs) are senescent, several characteristic features are apparent such as distinct morphology, appearance of senescence-associated β-galactosidase activity, altered growth factor responses, apoptotic resistance against various stimuli, and broad changes in gene expression [18, 43, 47]. For the study on aging-related changes, the aging-dependent altered cellular responsiveness has been reanalyzed. The cellular responsiveness could be classified into three categories. The first category is the metabolic response toward the nutrient supply and utilization, which must be ensured to maintain the homeostasis regardless of young and old state. The second category is the stress response. The responses against a variety of toxic stresses are attenuated in general by a variety of mechanisms. The third category is the mitogenic response, which is distinctively reduced or even absent in the senescent cells. In this review, the molecular mechanism for the aging-dependent

S.C. Park (✉)
Department of Biochemistry and Molecular Biology, Seoul National University Medical School, 28 Yongon Dong, Chong No Ku

Lee Gil Ya Cancer and Diabetes Institute, Gachon University of Medicine and Science, Incheon, 406-799, South Korea
e-mail: scpark@snu.ac.kr

J.S. Rhim and R. Kremer (eds.), *Human Cell Transformation: Role of Stem Cells and the Microenvironment*, Advances in Experimental Medicine and Biology 720,
DOI 10.1007/978-1-4614-0254-1_1,

alterations in stress response as well as the growth factor response would be discussed with a special intention to elucidate the common and fundamental mechanism of the aging process.

1.2 Mechanism of Aging-Dependent Growth Arrest by the Membrane Barrier

Growth arrest of senescent cells is strongly associated with their attenuated responses to growth factors. For the study of cellular responsiveness, the signal transduction cascades should be assessed in a variety of levels, such as ligand supply, receptor status, ligand–receptor interaction, downstream signal cascade, intracellular signal networks, signal translocation, transcriptional status, chromatin remodeling, and posttranscriptional apparatus [47]. In case of epidermal growth factor (EGF), the senescent cells display downregulation of mitogenic response despite no significant changes in quantity of the signals and receptors [5, 43, 47, 50]. Functional recovery of the senescent cells, especially of mitogenic response, could be induced successfully simply by restoration of the receptor-mediated endocytosis [43]. In addition, not only the receptor-mediated tyrosine kinase system (RTK) but also G protein-coupled receptor system (GPCR) is altered during the aging process in a variety of organs [72]. And the major difference in GPCR system between young and old cells in response to mitogenic stimuli is found to be strongly associated with the cellular cAMP status [24–26]. Both of these signaling regulatory systems at the cellular membrane level may provide a primary barrier for the extrinsic signals either good or bad in the aging process.

Endocytosis is one of the major mechanisms for intercellular communication, maintenance of protective functions, removal of dead and dying cells, modulation of plasma membrane composition, or entry route to pathogens. The receptor-mediated endocytosis can be classified as the clathrin-dependent system and the caveolae-dependent system, both of which are downregulated in the senescent state [42, 47].

Caveolae are the sites of important dynamic and regulatory events at the plasma membrane, abundant in terminally differentiated cell types, i.e., adipocytes, endothelial cells, and muscle cells [2, 4, 19, 41]. Signaling molecules are enriched in caveolae structure, where their activities are regulated [55]. Therefore, caveolae may function as subcellular compartments for the storage of signaling molecules, regulation of activities, and adjustment of crosstalk between distinct signaling cascades.

The principal component of caveolae consists of caveolins, 21–24 kDa integral membrane proteins. The mammalian caveolin gene family consists of caveolin-1, -2, and -3. Caveolin-1 and -2 are co-expressed and form a hetero-oligomeric complex in many cell types, with particularly high levels in adipocytes, whereas expression of caveolin-3 is muscle specific [48, 56, 57, 61, 64].

Growth factors such as EGF and platelet-derived growth factor (PDGF) stimulate their specific cell surface receptors and the subsequent activation of the intrinsic tyrosine kinases. However, senescent cells do not respond efficiently to external growth factors, such as EGF and PDGF [50, 71].

Interaction of caveolin-1 with signaling molecules is mediated via a membrane-proximal region of caveolin-1, termed the "caveolin-scaffolding domain" (residues 82–101) [13,14,58]. Through this domain, caveolin-1 interacts with G protein alpha-subunits, H-Ras, Src-family tyrosine kinases, PKC isoforms, EGF-R, Neu, and eNOS, resulting in suppression of their activities [6, 13, 14, 37, 38, 61]. It is illustrated that the targeted downregulation of caveolin-1 is sufficient to drive transformation of cells and hyperactivate the Erk kinase cascade [19]. Moreover, co-expression of EGFR, Raf, MEK-1, or Erk-2 with caveolin-1 resulted in inhibition of signaling translocation from the cytoplasm to the nucleus in vivo [19]. Senescent HDFs show the increased level of caveolin-1 and its colocalization with EGFR. Overexpression of caveolin-1 in young HDF suppressed the activation of Erk-1/2 upon EGF stimulation, indicating the direct role of caveolin-1 in EGF signaling [47]. Moreover, overexpression of caveolin-1 induces premature cellular senescence in primary cultures of murine fibroblasts [68].

We investigated the modulation of caveolin-1 status in senescent cells to determine the effect of caveolin-1 on mitogenic signaling efficiency and cell cycling. Reduction of caveolin-1 expression can be induced by different kinds of methods, such as antisense-oligonucleotide (AS-ON) and small interfering RNA (siRNA). Interestingly, we found that the downregulation of caveolin-1 by AS-ON or siRNA treatment led to restoration of the basal p-Erk level and Erk activation upon EGF stimulation as well as the downstream activation of Elk phosphorylation in senescent HDFs. A simple reduction of the caveolin-1 level in senescent HDF cells can induce the restoration of the Erk signaling system upon EGF stimulation, not only in terms of its phosphorylation but also its translocation into and activation of transcriptional factors in the nuclei [12]. Recently, it has been suggested that the activation of Erk signaling is regulated by phosphorylation/dephosphorylation through PP1 and PP2A, whose activities are controlled by caveolin-1 binding [35].

Caveolin-1-mediated cell cycle arrest occurs through a p53/p21Waf1-dependent pathway [20]. Downregulation of caveolin-1 in senescent HDF reduced the level of the cell cycle inhibitors, p53 and p21. Therefore, the reentry of the senescent cells into the cell cycle was examined by monitoring BrdU incorporation in response to EGF stimuli after reduction of caveolin-1 status, confirming the restoration of cell cycle reentry [12]. These results indicate that a simple adjustment of the caveolin-1 level in senescent cells can profoundly influence the aging phenotype. In addition, microarray data showed that transport-related genes are strongly associated with the caveolin-1 gene, which was demonstrated by downregulation of caveolin-1 expression through siRNA in senescent cells. The reduction of caveolin-1 gene activity resulted in the restoration of intracellular transport proteins, suggesting another role of caveolin-1 in cellular trafficking system (Manuscript in preparation).

Moreover, evidences for a role of caveolin-1 in cancer cells have been well documented [51], wherein caveolin-1 was suggested to be a tumor suppressor gene and metastasis-related gene [36]. And caveolin-1 levels in most cancer tissues are significantly lower than those in normal tissues, suggesting the circumvention of signal suppression by caveolin-1. It is apparent, therefore, that caveolin-1 plays an important role in the regulation of both aging and cancer not only for signaling control but also for intracellular trafficking of the signaling.

Senescent cells show morphological changes, such as flat and large cell shape. Structural alteration of senescent cells is deeply related with increase of focal adhesion and Rb family proteins [9]. Interestingly, caveolin-1 is associated with the focal adhesion complex through integrin in the membrane [8, 69] and the activation of focal adhesion kinase (FAK) is significantly decreased by downregulation of caveolin-1 [65]. These results indicate that caveolin-1 can play an important role in focal adhesion and thereby participate in adhesion signal cascade. We have observed that the formation of focal adhesion and actin stress fiber is increased and anchored in membrane via interaction with caveolin-1 in senescent cells. Therefore, reduction of caveolin-1 in senescent cells can resume the morphological changes as small and spindle shape like young cells by adjusting focal adhesion complexes [11].

The conversion of the senescent phenotype into a functionally active and structurally normal state, simply by restoring the membrane signaling apparatus, confirms the significance of membrane signaling system in the aging phenomenon [45, 46]. Therefore, we assumed that the aging process could be initiated and regulated at the membrane by a membrane-associated signal switch system. These results led us to postulate the gate theory of aging, in which the fundamental role of the membrane switch on/off system for a variety of signals has been emphasized [72]. Several other molecules could be the tentative gate switches of aging at the membrane. Thus far, we have observed that the simple adjustment of caveolin-1 status in the senescent cells can efficiently restore physiological and morphological properties to the youthful state. Therefore, we suggest that caveolin-1 would be one of the prime gate switches of the cellular aging process.

This gate theory of aging strongly implicates the possibility of restoration of the senescent phenotype simply by alteration of signaling system at the membrane level. These data implicate the novel concept of aging which should be substituted from the conventional irreversible and inevitable characteristics of aging to the flexible and plastic nature of the aging [44]. In addition, we observed that the nuclear translocation of p-ERK1/2 in response to EGF is significantly inhibited in senescent HDFs, despite the increase of p-ERK1/2 and MEK activities in the senescent cells [47]. Many signaling molecules as well as transcription factors should be imported to the nucleus to exert their own function. For example, upon EGF stimulation, p-ERK should be imported to the nucleus to activate AP1 transcription factor and thereby triggers cell cycle progression [30]. Therefore, if the nucleocytoplasmic trafficking is defective, cells cannot properly respond to growth factor stimulation [32]. These findings suggest the operation of the double checking system for growth factor responses in the senescent cells, primarily at the cellular membrane level and ultimately at the nuclear membrane level.

1.3 Mechanism of Aging-Dependent Apoptosis Resistance by the Nuclear Barrier

We have reported the aging-dependent apoptosis resistance in vitro and in vivo [63, 70]. Apoptosis plays an important role in many physiological and pathological processes, including aging itself and age-related diseases. The MAPKs are a family of kinases that transduce signals from the plasma membrane to the nucleus in response to various stimuli not only of growth factor signaling but also of apoptotic stress [31, 32]. MAPKs are serine/threonine kinases that phosphorylate specific substrates. ERK1 and ERK2 are well-characterized MAPKs that are activated in response to growth stimuli, though also involved in stress response. JNKs and p38-MAPK are simultaneously activated in response to a variety of cellular and environmental stresses, including DNA damage, heat shock, ischemia, inflammatory cytokines, exposure to UV radiation, and oxidative stress [7, 16, 27, 52, 59, 60]. Phosphorylation of MAPKs, induced by mediators of apoptosis, was controlled in agonist-specific manner [53]. Hydrogen peroxide, a reactive oxygen species that causes apoptosis [21, 40], induced an increase in ERK phosphorylation. Staurosporine, a broad-spectrum kinase inhibitor that is a strong inducer of caspase 3-dependent apoptosis [66], strongly induced the phosphorylation of p38. In contrast, thapsigargin, which disrupts endoplasmic reticulum Ca^{2+} stores, leading to the release of Ca^{2+} and activates mitochondrial-dependent and mitochondrial-independent apoptotic pathways, did not significantly influence MAPK phosphorylation [17]. However, many signaling molecules can execute their specific functions only after their import into the nucleus. Thus, it is hypothesized that the apoptotic resistance of senescent cells might be related with the senescence-associated defects in nucleocytoplasmic trafficking of the apoptotic signaling.

Bcl-2 family proteins contribute to regulation of programmed cell death/apoptosis. In mammalian cells, relative levels of proapoptotic proteins, such as Bax, Bak BID, BAD, PUMA, and NOXA, and antiapoptotic Bcl-2 family members, such as Bcl-Xl and Bcl-2, are important for determining whether and when apoptosis is triggered [29, 49]. We previously reported that levels of Bcl-xL and Bax were comparable in young and senescent HDFs, but that Bak, Bok, Bik, and PUMA were present at lower levels in senescent HDFs than in young HDFs [53]. These observations suggest that decreased levels of proapoptotic gene expression probably by failure of signaling molecules to enter the nucleus might desensitize senescent cells to apoptotic stimuli. Moreover, antiapoptotic Bcl-2 decreases readily in young cells in response to apoptotic cells in contrast to the old cells, where Bcl-2 level remains intact despite the apoptotic stress. The mechanism of the stable expression of Bcl-2 in the senescent cells might be related with the maintenance of phosphorylation status of CREB due to inactivation of protein phophatase-2A (PP2A) as well as block of the apoptotic signaling to the nucleus [53]. These data implicate the presence of aging-dependent barrier at the nuclear membrane level. Moreover, in search of the senescence-associated resistant nature to apoptosis in vitro and in vivo [63, 70], we have observed the senescence-dependent nuclear localization of actin and gelsolin [1], and overexpression of major vault protein (MVP) [54], implicating the operation of aging-dependent signal localization mechanism. Thereby, we conjectured that these phenomena might be ultimately related with senescence-dependent defects in intracellular signaling, especially in the nucleocytoplasmic trafficking, which might provide a secondary or ultimate barrier for cellular protection against toxic stress.

1.4 Nuclear Barrier Hypothesis of Aging

Many signaling molecules as well as transcription factors should be imported to the nucleus to exert their own function. For example, upon EGF stimulation, p-ERK should be imported to the nucleus to activate AP1 transcription factor and thereby triggers cell cycle progression [10]. Actually, we have previously found that senescent HDFs accumulate p-ERK1/2 in the cytoplasm and failed to translocate them to the nucleus in response to EGF stimulation. In addition, we have reported a role of caveolin as a suppressor of mitogenic signaling, especially to EGF stimulation [47] and senescence-dependent nuclear accumulation of actin [39]. These data suggest the aging-dependent alteration in nucleocytoplasmic trafficking.

In addition to the hyporesponsiveness to growth factors, we have previously reported that senescent HDFs are resistant to apoptosis induced by various stimuli [70]. Apoptotic response of cells is a well-programmed process requiring intracellular signaling pathways. Therefore, the senescence-dependent hyporesponsiveness to apoptotic stress as well as to growth factors could be attributed to the efficiency of nucleocytoplasmic trafficking. The nucleocytoplasmic trafficking is a highly sophisticated process involving many specific proteins. Nucleoporins are major components of nuclear pore complexes (NPCs) and are often used as markers for NPCs [3, 62]. NPCs allow passive diffusion of ions and small molecules, and facilitate active transport of macromolecules. The cargo molecules usually have short sequence elements called nuclear localization sequences (NLSs) and nuclear export sequences (NESs). Karyopherin α binds to the NLSs of cargo molecules, but karyopherin α itself cannot bind to nucleoporins. Karyopherin β can bind to both karyopherin α and nucleoporins, by which karyopherin β provides the cargo-karyopherin α complex with an access to the NPCs [15, 22]. Ran plays a critical role in importing cargoes as well as in exporting cargoes. Ran exists in two distinct forms, that is, a GTP-bound form or a GDP-bound form with a high concentration of the GTP-bound form in the nucleus. After the cargo-karyopherin α-karyopherin β complex enters the nucleus, RanGTP binds to karyopherin β (importin β), which triggers the dissociation of the complex. Also, RanGTP binds to karyopherin β (exportin β) and facilitates export of cargo molecules. The nucleotide state of Ran is regulated by several factors such as Ran GTP–GDP exchange factor (RanGEF or RCC1) and Ran GTPase-activating protein (RanGAP). In the nucleus, RanGEF associates with Ran to maintain its state as RanGTP. In the cytoplasm, RanGAP stimulates the GTPase activity of Ran to maintain its state as RanGDP [23, 67].

1.4.1 Aging-Dependent Impaired Nuclear Translocation

The initial observation of aging-dependent nuclear accumulation of actin [39] and gelsolin [1] attracted our concern on the mechanism of the aging-dependent specific localization of the biomolecules. Furthermore, the aging-dependent perinuclear accumulation of many signaling molecules could be observed in a variety of apoptotic and mitotic stress conditions [31, 32, 53]. These data led us to assume that the nuclear translocation of activated signaling might be inhibited by a certain barrier function in aging-dependent manner, as shown in p-ERK1/2 and NF-kB p50 in response to growth stimuli or LPS in senescent fibroblasts [31]. However, the activation steps of these enzymes were not impaired, because the phosphorylation and activation of ERK1/2 occurs as efficiently in senescent cells as in presenescent cells. These findings suggested that ERK and NF-kB might be unable to redistribute properly to the nucleus upon activation in senescent conditions due to operation of the aging-dependent functional nuclear barrier.

Furthermore, in the young cells, this nucleocytoplasmic trafficking system can be regulated by interacting with growth signals and starvation signals. When the cellular energy state is lowered, AMPK pathway is activated, which drives the import of the biomolecules into the nucleus. In contrast, when cell receives the growth signal, PI3K signaling system is activated, which stimulate the export of the biomolecules from the nucleus, as demonstrated by GAPDH (glyceraldehydes 3 phosphate dehydrogenase) [34]. These data indicate the plastic nature of nuclear translocation either import or export in the ordinary maintenance of living state, but which is limited and its efficiency is lowered for the sake of survival in sacrifice of proliferation in the senescent state.

1.4.2 Aging-Dependent Overall Repression of Nucleocytoplasmic Trafficking Gene Expression

To investigate the mechanism by which senescent cells show hyporesponsiveness, we analyzed gene expression profiles of young and senescent HDFs by using Affymetrix GeneChip® Human Gene 1.0 ST oligonucleotide arrays. The microarray analysis on 28,869 genes showed that expression of most nucleocytoplasmic trafficking genes was downregulated in senescent HDFs as compared to young HDFs [31]. We have observed the vast repression in the expression levels of most nucleoporin genes such as Nup107, Nup155, Nup205, Nup43, and Nup85 and of transport receptor genes, including importin, karyopherin $\alpha 2$, karyopherin $\beta 1$, and exportin 1 as well as of Ran and Ran-regulating factors, such as RAN, Ran-binding protein 1, and Ran GTPase-activating protein 1. In order to verify the microarray data, we confirmed the expression levels of some selected genes by using semiquantitative reverse transcription-polymerase chain reaction (RT-PCR) method and protein levels of Nup50, Nup88, Nup107, Nup155, karyopherin α, karyopherin β, and Ran by western blotting with available antibodies.

Taken together, these results demonstrate that expression of nucleocytoplasmic trafficking genes is generally repressed in senescent HDFs, and suggest that the nucleocytoplasmic trafficking is severely defective in senescent HDFs, which might play the role of nuclear barrier.

1.4.3 The Tentative Ultimate Transcriptional Regulator for Operation of the Functional Nuclear Barrier

The senescence-dependent hyporesponsiveness with a general reduction in the trafficking gene expression was traced to its common mechanism through analysis of their upstream transcriptional signaling of those genes. With bioinformatics tools, 46 nucleocytoplasmic trafficking genes were subjected to promoter analysis and 131 transcriptional factors were identified to be related with those systems. Among them, Sp1 (specificity protein 1) has been found to be the most common and dominant transcription factor for genetic regulation of the trafficking-related genes (Manuscript in preparation). Interestingly, most of the promoter regions of those genes have the multiple Sp1 binding sites. Actually, it has been reported that Sp1 stability decreases in senescent HDFs as compared to young HDFs. In addition, the DNA-binding activity of Sp1 was observed to decrease in aged brain and liver tissue [28]. Therefore, it is worth to investigate the mechanism of how Sp1 expression or activity decreases in senescent cells and aged organisms.

The transcription factor Spl is a DNA-binding protein present in most types of mammalian cells. It has three zinc finger motifs to bind to the GC box in the genome, through which it activates a variety of target genes including growth factor signaling genes such as PDGF, PDGFR, EGFR, c-Src, Cyclin D1, Cyclin E, Cyclin B1, c-Myc, c-Jun, and c-Fos. In addition, Sp1 activates several Cdk inhibitor genes and proapoptotic genes such as p15, p16, p19, p21, p27, p57, FasL, DR5, and PUMA depending on the cellular context. Moreover, Sp1 activity is regulated by posttranslational modifications such as phosphorylation, glycosylation, acetylation, ubiquitination, or poly (ADP-ribosyl) ation. Therefore, it might be assumed that Sp1 could be downregulated in senescent HDFs through the posttranslational modifications by activation of certain senescence signaling pathways, which remains to be elucidated.

Taken together, these data suggest that the aging-dependent functional nuclear barrier might be responsible for bilateral resistance of the senescent cells to growth factors and apoptotic stress and the barrier would be controlled by the transcriptional control of the nucleocytoplasmic trafficking-related genes, for which certain general ultimate factor(s) could be traced.

1.5 Nomenclature of Aging-Dependent Functional Nuclear Barrier as "Park and Lim's Barrier"

The assumed aging-dependent operation of the functional nuclear barrier might contribute to explain one of the mysteries of aging process; that is, the bilateral resistances to growth factors and to the apoptotic stress, which would explain the mechanism of the senescent cells for trade-off growth to survival. In the original disposable soma theory of aging, the utilization of energy has been primarily emphasized for trade-off survival with reproduction in the level of organism [33]. But at the cellular level, it can be proposed that the operation of the functional nuclear barrier might be the mechanism for the fundamental nature of trade-off growth to survival in aging process. Therefore, this nuclear barrier of aging concept would be the biologically significant signal integrating system, operating on the basis of the crosstalk of signaling between growth and death. Furthermore, this barrier concept can be extended to cancer research as well, since cancer cell has the high availability of Sp1 and the property of sensitivity to growth factor signaling and apoptotic signaling as well. It might be natural to assume that the new therapeutic modality of cancer could be developed through adjusting the nuclear barrier either by modulation of Sp1 status with other transcriptional modes or stimulation of barrier formation and novel mechanism to solve the multidrug resistance of the cancers could be developed from this new concept of nuclear barrier.

In the development of this concept, the initial collaboration for aging research with Professor In Kyoung Lim of Ajou University Medical School has actually stimulated me to pay an attention on the significance of aging-dependent perinuclear accumulation of the signals as well as the aging-dependent nuclear accumulation of biomolecules, which led me to develop the concept of nuclear barrier hypothesis of aging [39]. Therefore, in honor of collaboration with her, I would like to name this aging-dependent functional nuclear barrier for growth and survival as "Park and Lim's Barrier."

1.6 Conclusion

Our study strongly suggests that senescent cells gain the feature of hyporesponsiveness either to growth factors or to apoptotic stress via forming functional nuclear barriers, preventing nucleocytoplasmic trafficking of signaling molecules (Fig. 1.1). And the nucleocytoplasmic trafficking-related genes are downregulated in senescent HDFs as a whole at the transcriptional level. We also recognize that the significance of this barrier would be valued for survival nature of aging in sacrifice of growth and could be the platform for checking the crosstalks between good and bad signals from extrinsic sources. These data provide

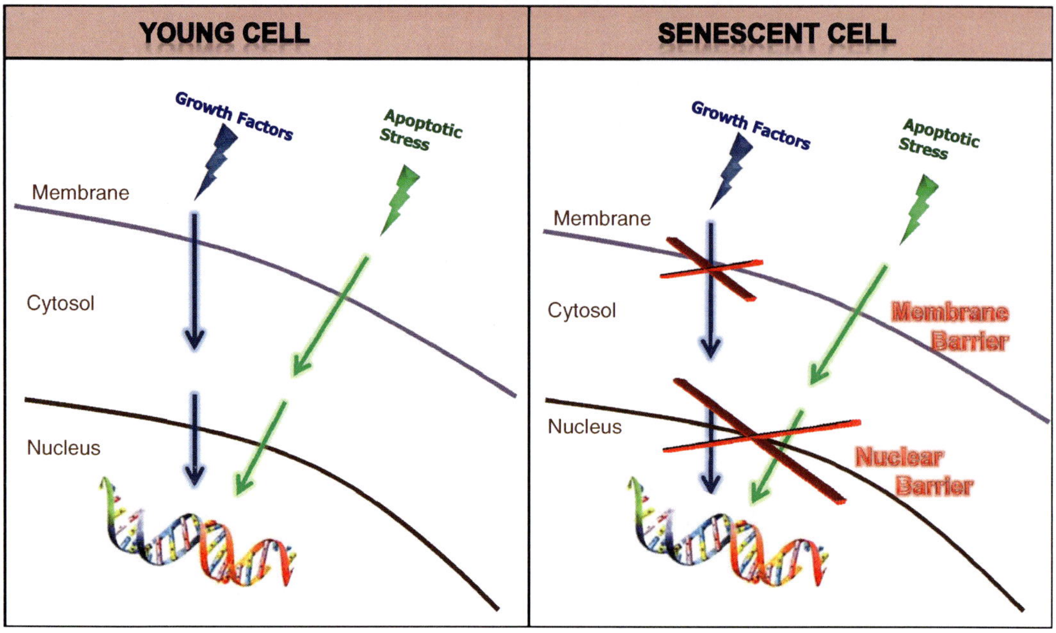

Fig. 1.1 The diagram of the aging-dependent functional barriers at the cellular membrane level and nuclear membrane level. Both of growth factor signaling and apoptotic signaling are limited in the senescent cells by primarily at the cellular membrane barrier and ultimately by the nuclear barrier, "Park and Lim's Barrier"

new insights into the mechanism of aging and address that more intensive studies are required to investigate the nature of aging for regulation of growth arrest and apoptosis resistance. Furthermore, this mechanism sheds a light for development of the new therapeutic modality of cancers as well.

Acknowledgements This work was supported by the National Research Foundation of Korea (NRF) grant funded by the Korean government (MEST) (2010-0029150) and also by the Korea Research Foundation for Health Science.

References

1. Ahn JS, Jang IS, Kim DI, Cho KA, Park YH, Kim KT, Kwak CS, Park SC (2003) Aging-associated increase of gelsolin for apoptosis resistance. Biochem Biophys Res Comun 312:1335–1341
2. Anderson RG (1998) The caveolae membrane system. Annu Rev Biochem 67:1996–2003
3. Beck M, Forster F, Ecke M, Plitzko JM, Melchior F, Gerisch G, Baumeister W, Medalia O (2004) Nuclear pore complex structure and dynamics revealed by cryoelectron tomography. Science 306:1387–1390
4. Bretscher MS, Whytock S (1977) Membrane-associated vesicles in fibroblasts. J Ultrastruct Res 6: 215–217
5. Carlin CR, Phillips PD, Knowles BB, Cristofalo VJ (1983) Diminished in vitro tyrosine kinase activity of the EGF receptor of senescent human fibroblasts. Nature 306(5943):617–620
6. Carman CV, Lisanti MP, Benovic JL (1999) Regulation of G protein-coupled receptor kinases by caveolin. J Biol Chem 274:8858–8864
7. Chang L, Karin M (2001) Mammalian MAP kinase signalling cascades. Nature 410:37–40
8. Chapman HA, Wei Y, Simon DI, Waltz DA (1999) Role of urokinase receptor and caveolin in regulation of integrin signaling. Thromb Haemost 82:291–297
9. Chen QM, Tu VC, Catania J, Burton M, Toussaint O, Dilley T (2000) Involvement of Rb family proteins, focal adhesion proteins and protein synthesis in senescent morphogenesis induced by hydrogen peroxide. J Cell Sci 113:4087–4097
10. Cho KA, Park SC (2005) Caveolin-1 as a prime modulator of aging: A new modality for phenotypic restoration? Mech Ageing Dev 126(1):105–110
11. Cho KA, Ryu SJ, Oh YS, Park JH, Lee JW, Kim HP, Kim KT, Jang IS, Park SC (2004) Morphological adjustment of senescent cells by modulating caveolin-1 status. J Biol Chem 279:42270–42278

12. Cho KA, Ryu SJ, Park JS, Is J, Ahn JS, Kim KT, Park SC (2003) Senescent pheonotype can be reversed by reduction of caveolin status. J Biol Chem 278(30): 27789–27795
13. Couet J, Sargiacomo M, Lisanti MP (1997) Interaction of a receptor tyrosine kinase, EGF-R, with caveolins: caveolin-binding negatively regulates tyrosine and serine/threonine kinase activities. J Biol Chem 272: 30429–30438
14. Couet J, Li S, Okamoto T, Ikezu T, Lisanti MP (1997) Identification of peptide and protein ligands for the caveolin-scaffolding domain. Implications for the interaction of caveolin with caveolae-associated proteins. J Biol Chem 272:6525–6533
15. D'Angelo MA, Hetzer MW (2008) Structure, dynamics and function of nuclear pore complexes. Trends Cell Biol 18:456–466
16. Dent P, Yacoub A, Fisher PB, Hagan MP, Grant S (2003) MAPK pathways in radiation responses. Oncogene 22:5885–5896
17. Diaz-Horta O, Van Eylen F, Herchuelz A (2003) Na/Ca exchanger overexpression induces endoplasmic reticulum stress, caspase-12 release, and apoptosis. Ann N Y Acad Sci 1010:430–432
18. Dimri GP, Lee X, Basile G, Acosta M, Scott G, Roskelley C, Medrano EE, Linskens M, Rubelj I, Pereira-Smith O et al (1995) A biomarker that identifies senescent human cells in culture and in aging skin in vivo. Proc Natl Acad Sci USA 92: 9363–9367
19. Engelman JA, Chu C, Lin A, Jo H, Ikezu T, Okamoto T, Kohtz DS, Lisanti MP (1998) Caveolin-mediated regulation of signaling along the p42/44 MAP kinase cascade in vivo. A role for the caveolin-scaffolding domain. FEBS Lett 428:205–211
20. Galbiati F, Volonte D, Liu J, Capozza F, Frank PG, Zhu L, Pestell RG, Lisanti MP (2001) Caveolin-1 expression negatively regulates cell cycle progression by inducing G(0)/G(1) arrest via a p53/p21(WAF1/Cip1)-dependent mechanism. Mol Biol Cell 12(8):2229–2244
21. Hardman RA, Afshari CA, Barrett JC (2001) Involvement of mammalian MLH1 in the apoptotic response to peroxide-induced oxidative stress. Cancer Res 61(4):1392–1397
22. Hoelz A, Blobel G (2004) Cell biology: popping out of the nucleus. Nature 432:815–816
23. Izaurralde E, Kutay U, von Kobbe C, Mattaj IW, Gorlich D (1997) The asymmetric distribution of the constituents of the Ran system is essential for transport into and out of the nucleus. EMBO J 16:6535–6547
24. Jang IS, Yeo EJ, Park JA, Ahn JS, Cho KA, Juhnn YS, Park SC (2003) Altered cAMP signaling induced by lysophosphatidic acid in senescent human diploid fibroblasts. Biochem Biophys Res Comm 302(4): 778–784
25. Jang IS, Rhim JH, Park SC, Yeo EJ (2006) Downstream molecular events in the altered profiles of lysophosphatidic acid-induced cAMP in senescent human diploid fibroblasts. Exp Mol Med 38(2):134–143
26. Jang IS, Rhim JH, Kim KT, Cho KA, Yeo EJ, Park SC (2006) Lysophosphatidic acid-induced changes in cAMP profiles in young and senescent human fibroblasts as a clue to the ageing process. Mech Ageing Dev 127(5):481–92006
27. Johnson GL, Lapadat R (2002) Mitogen-activated protein kinase pathways mediated by ERK, JNK, and p38 protein kinases. Science 298(5600):1911–1912
28. Kang HT, Ju JW, Cho JW, Hwang ES (2003) Down-regulation of Sp1 activity through modulation of O-glycosylation by treatment with a low glucose mimetic, 2-deoxyglucose. J Biol Chem 278: 51223–51231
29. Kuwana T, Newmeyer DD (2003) Bcl-2-family proteins and the role of mitochondria in apoptosis. Curr Opin Cell Biol 15(6):691–699
30. Kim K, Nose K, Shibanuma M (2000) Significance of nuclear relocalization of ERK1/2 in reactivation of c-fos transcription and DNA synthesis in senescent fibroblasts. J Biol Chem 275:20685–20692
31. Kim SY, Ryu SJ, Ahn HJ, Choi HR, Kang HT, Park SC (2010) Senescence-related functional nuclear barrier by down-regulation of nucleo-cytoplasmic trafficking gene expression. Biochem Biophys Res Commun 391:28–32
32. Kim SY, Kang HT, Choi HR, Park SC (2010) Reduction of Nup107 attenuates the growth factor signaling in the senescent cells. Biochem Biophys Res Commun 401:131–136
33. Kirkwood TB, Holliday R (1979) The evolution of ageing and longevity. Proc R Soc Lond B Biol Sci 205(1161):531–546
34. Kwon HJ, Rhim JH, Jang IS, Jun G, Park SC, Yeo EJ (2010) Activation of AMP-activated protein kinase stimulates the nuclear localization of glyceraldehyde 3-phosphate dehydrogenase in human diploid fibroblasts. Exp Mol Med 42(4):254–269
35. Li L, Ren CH, Tahir SA, Ren C, Thompson TC (2003) Caveolin-1 maintains activated Akt in prostate cancer cells through scaffolding domain binding site interaction with and inhibition of serine/threonine protein phosphatases PP1 and PP2A. Mol Cell Biol 23: 9389–9404
36. Li L, Yang G, Ebara S, Satoh T, Nasu Y, Timme TL, Ren C, Wang J, Tahir SA, Thonpson TC (2001) Caveloin-1 mediates testosterone-stimulated survival/clonal growth and promotes metastatic activities in prostate cancer cells. Cancer Res 61:4386–4392
37. Li S, Couet J, Lisanti MP (1996) Src tyrosine kinases, G alpha subunits and H-Ras share a common membrane-anchored scaffolding protein, caveolin. Caveolin binding negatively regulates the auto-activation of Src tyrosine kinases. J Biol Chem 271:29182–29190
38. Li S, Okamoto T, Chun M, Sargiacomo M, Casanova JE, Hansen SH, Nishimoto I, Lisanti MP (1995) Evidence for a regulated interaction between heterotrimeric G proteins and caveolin. J Biol Chem 270:15693–15701
39. Lim IK, Hong K, Kwak IH, Yoon G, Park SC (2000) Cytoplasmic retention of p-Erk1/2 and nuclear

accumulation of actin proteins during cellular senescence in human diploid fibroblasts. Mech Ageing Dev 119:113–130
40. Martin D, Salinas M, Fujita N, Tsuruo T, Cuadrado A (2002) Ceramide and reactive oxygen species generated by H2O2 induce caspase-3-independent degradation of Akt/protein kinase B. J Biol Chem 277(45): 42943–42952
41. Okamoto T, Schlegel A, Schlegel PE, Schlegel MP (1998) Caveolins, a family of scaffolding proteins for organizing 'preassembled signaling complexes' at the plasma membrane. J Biol Chem 273:5419–5422
42. Park JS, Park WY, Cho KA, Kim DI, Jhun BH, Kim SR, Park SC (2001) Down-regulation of amphiphysin-1 is responsible for reduced receptor-mediated endocytosis in senescent cells. FASEB J 15:1625–1627
43. Park SC (2002) Functional recovery of senescent cells through restoration of receptor-mediated endocytosis. Mech Ageing Dev 123:917–926
44. Park SC (2006) New molecular target for modulation of aging process. Antioxid Redox Signal 8(3–4):620–627
45. Park SC (2004) Phenotypic adjustment of senescent cells: replace or restore. Geriatr Gerontol Int 4: 517–520
46. Park SC, Cho KA, Jang IS, Kim KT, Ryu SJ (2004) Functional efficiency of the senescent cells: replace or restore. Ann NY Acad Sci 1019:309–316
47. Park WY, Park JS, Cho KA, Kim DI, Ko YG, Seo JS, Park SC (2000) Up-regulation of caveolin attenuates epidermal growth factor signaling in senescent cells. J Biol Chem 275(27):20847–20852
48. Parton RG, Way M, Zorzi N, Stang E (1997) Caveolin-3 associates with developing T-tubules during muscle differentiation. J Cell Biol 136:137–154
49. Pepper C, Bentley P (2000) The role of the Bcl-2 family in the modulation of apoptosis. Symp Soc Exp Biol 52:43–53
50. Phillips PD, Kuhnle E, Cristofalo VJ (1983) EGF binding ability is stable throughout the replicative life-span of WI-38 cells. J Cell Physiol 114(3): 311–316
51. Razani B, Schlegel A, Liu J, Lisanti MP (2001) Caveolin-1, a putative tumour suppressor gene. Biochem Soc Trans 29:494–499
52. Royuela M, Arenas MI, Bethencourt FR, Sanchez-Chapado M, Fraile B, Paniagua R (2002) Regulation of proliferation/apoptosis equilibrium by mitogen-activated protein kinases in normal, hyperplastic, and carcinomatous human prostate. Hum Pathol 33(3): 299–306
53. Ryu SJ, Oh YS, Park SC (2007) Failure of stress-induced downregulation of Bcl-2 contributes to apoptosis resistance in senescent human diploid fibroblasts. Cell Death Differ 14(5):1020–1028
54. Ryu SJ, An HJ, Oh YS, Choi HR, Ha MK, Park SC (2008) On the role of major vault protein in the resistance of senescent human diploid fibroblast to apoptosis. Cell Death Differ 15(11): 1678–1680
55. Sargiacomo M, Scherer PE, Tang Z, Kubler E, Song KS, Sanders MC, Lisanti MP (1995) Oligomeric structure of caveolin: implications for caveolae membrane organization. Proc Natl Acad Sci USA 92: 9407–9411
56. Scherer PE, Lewis RY, Volonte D, Engelman JA, Galbiati F, Couet J, Kohtz DS, Donselaar E, Peters P, Lisanti MP (1997) Cell-type and tissue-specific expression of caveolin-2. Caveolins-1 and -2 co-localize and form a stable hetero-oligomeric complex in vivo. J Biol Chem 272:29337–29346
57. Scherer PE, Okamoto T, Chun M, Nishimoto I, Lodish HF, Lisanti MP (1996) Identification, sequence, and expression of caveolin-2 defines a caveolin gene family. Proc Natl Acad Sci USA 93:131–135
58. Schlegel A, Schwab RB, Scherer PE, Lisanti MP (1999) A role for the caveolin-scaffolding domain in mediating the membrane attachment of caveolin-1. The caveolin-scaffolding domain is both necessary and sufficient for membrane binding in vitro. J Biol Chem 274:22660–22667
59. Seger R, Krebs EG (1995) The MAPK signaling cascade. FASEB J 9(9):726–735
60. Seluanov A, Gorbunova V, Falcovitz A, Sigal A, Milyavsky M, Zurer I, Shohat G, Goldfinger N, Rotter V (2001) Change of the death pathway in senescent human fibroblasts in response to DNA damage is caused by an inability to stabilize p53. Mol Cell Biol 21:1552–1564
61. Song KS, Li S, Okamoto T, Quilliam LA, Sargiacomo M, Lisanti MP (1996) Copurification and direct interaction of Ras with caveolin, an integral membrane protein of caveolae microdomains. Detergent free purification of caveolae membranes. J Biol Chem 271:9690–9697
62. Stoffler D, Fahrenkrog B, Aebi U (1999) The nuclear pore complex: from molecular architecture to functional dynamics. Curr Opin Cell Biol 11:391–401
63. Suh Y, Lee KA, Kim WH, Han BG, Vijg J, Park SC (2002) Aging alters the apoptotic response to genotoxic stress. Nat Med 8:3–4
64. Tang Z, Scherer PE, Okamoto T, Song KC (1996) Molecular cloning of caveolin-3, a novel member of the caveolin gene family expressed predominantly in muscle. J Biol Chem 271:2255–2261
65. Teixeira A, Chaverot N, Schroder C, Strosberg AD, Couraud PO, Cazaubon S (1999) Requirement of caveolae microdomains in extracellular signal-regulated kinase and focal adhesion kinase activation induced by endothelin-1 in primary astrocytes. J. Neurochemistry 72(1):120–128
66. Tesauro M, Thompson WC, Moss J (2006) Effect of staurosporine-induced apoptosis on endothelial nitric oxide synthase in transfected COS-7 cells and primary endothelial cells. Cell Death Differ 13(4): 597–606
67. Turpin P, Ossareh-Nazari B, Dargemont C (1999) Nuclear transport and transcriptional regulation. FEBS Lett 452:82–86

68. Volonte D, Zhang K, Lisanti MP, Galbiati F (2002) Expression of caveolin-1 induces premature cellular senescence in primary cultures of murine fibroblasts. Mol Biol Cell 13(7):2502–2517
69. Wei Y, Yang X, Liu Q, Wilkins JA, Chapman HA (1999) A role for caveolin and the urokinase receptor in integrin-mediated adhesion and signaling. J Cell Biol 144(6):1285–1294
70. Yeo EJ, Hwang YC, Kang CM, Choy HE, Park SC (2000) Reduction of UV-induced cell death in the human senescent fibroblasts. Mol Cells 10: 415–422
71. Yeo EJ, Jang IS, Lim HK, Ha KS, Park SC (2002) Agonist-specific differential changes of cellular signal transduction pathways in senescent human diploid fibroblasts. Exp Gerontol 37(7):871–883
72. Yeo EJ, Park SC (2002) Age-dependent agonist-specific dysregulation of membrane-mediated signal transduction: Emergence of the gate theory of aging. Mech Age Devel 123:1563–1578

Establishment of Cell Lines from the Human Middle and Inner Ear Epithelial Cells

2

David J. Lim and Sung K. Moon

Abstract

The middle ear infection is the most common childhood infection. In order to elucidate the cell and molecular mechanisms involved in bacterial recognition and innate immune response, we have established a stable human middle ear cell line, which has contributed to the current knowledge concerning the molecular pathogenesis of the middle ear infection. The inner ear, a sensory organ responsible for hearing and balance, is filled with inner ear fluid, and disturbance of the fluid homeostasis results in dizziness and hearing impairment. It has been suggested that the endolymphatic sac (ES) may play a critical role in the fluid homeostasis of the inner ear. We have established a stable human ES cell line and are undertaking cell and molecular characterization of this cell line.

2.1 Introduction

The middle ear infection (otitis media) is the most common pediatric infectious disease and is the most common reason for physician's office visits and antibiotic prescriptions [1]. It was estimated to cost the US alone roughly five billion dollars per year for the management of otitis media (OM) in 1996 [2]. However, exact molecular mechanisms involved in the pathogenesis of OM are poorly understood [3]. In order to elucidate a mechanism involved in bacterial recognition and innate and adaptive immune response, we have established a stable human middle ear cell line, which contributed to the current knowledge concerning the cell signaling involved in the pathogenesis of the middle ear infection.

The inner ear, which contains unique sensory organs responsible for hearing and balance, is filled with lymph fluids such as perilymph and endolymph [4]. The perilymph is characterized by high Na^+ ion, whereas the endolymph is characterized by high K^+ ion [5]. Disturbance of fluid homeostasis results in hearing impairment and vertigo. Particularly, Meniere's disease (MD), characterized by fluctuating hearing loss, tinnitus, and dizziness, is considered to be an example of

D.J. Lim (✉)
Division of Clinical and Translational Research, House Ear Institute, 2100 West 3rd Street, Los Angeles, CA 90057, USA

Department of Cell and Neurobiology, University of Southern California, Los Angeles, CA, USA
e-mail: dlim@hei.org

J.S. Rhim and R. Kremer (eds.), *Human Cell Transformation: Role of Stem Cells and the Microenvironment*, Advances in Experimental Medicine and Biology 720,
DOI 10.1007/978-1-4614-0254-1_2,

inner ear fluid homeostasis disorders. While the underlying causes of MD are not yet known, it has been suggested that the endolymphatic sac (ES) may play a critical role in the pathogenesis of MD [6, 7]. It has been suggested that the ES may play an important physiologic function to regulate endolymph fluid homeostasis and also may function as an immune organ [8]. In order to elucidate the cell and molecular mechanisms involved in the physiology and pathophysiology of MD, it is critical to establish a stable human ES cell line. We report the progress thus far, which we have made in the characterization of this cell line.

2.2 Functional Morphology of the Ear

2.2.1 Middle Ear Epithelium

The middle ear cavity is connected to the oropharynx through the eustachian tube (E-tube), which provides pressure regulation and aeration of the tympanic cavity and the mastoid cavity (Fig. 2.1). The lining mucosa of the middle ear consists of ciliated cells, secretory cells, nonsecretory cells, and basal cells (Fig. 2.2) [9, 10]. The middle ear epithelial cells vary depending

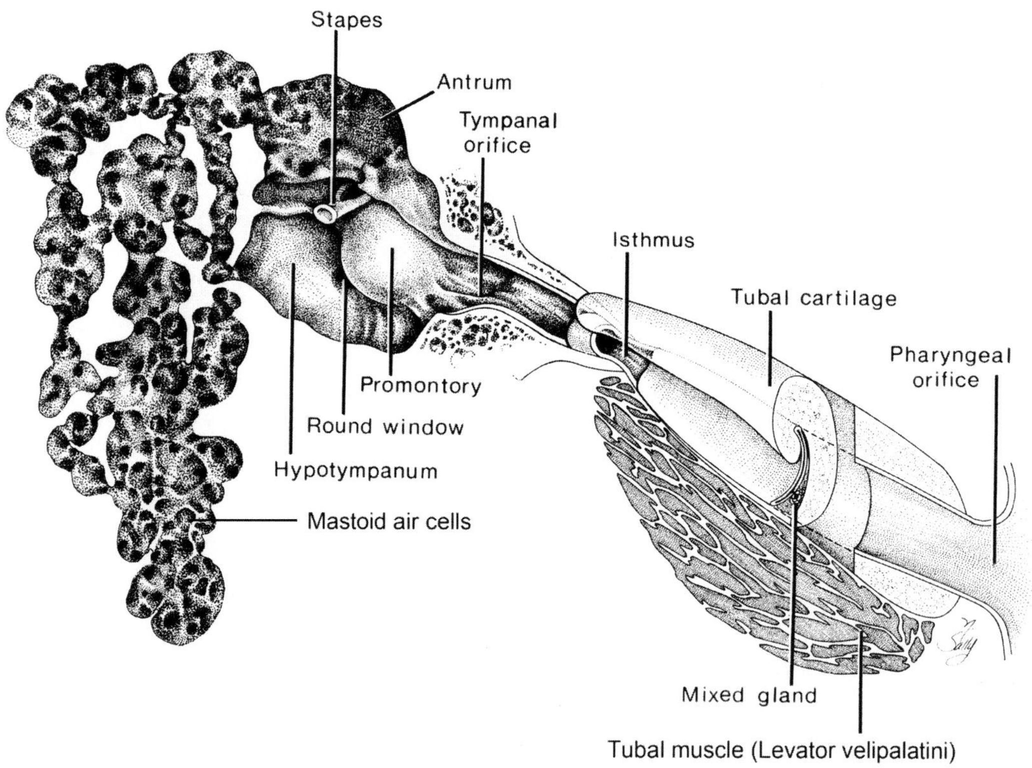

Fig. 2.1 Schematic diagram of mastoid-E-tube-middle ear complex. Mastoid cavity is continuous to the middle ear cavity, which is connected to eustachian tube which is consisted of bony part and cartilaginous part, where mixed glands are well developed. The cartilaginous E-tube, which is open to the pharynx, is closed by tubal cartilage and opens intermittently to aerate and equalize pressure between middle ear cavity and ambient pressure of the oropharynx. Poor tubal function is one of the major risk factors for otitis media susceptibility. The lining of the middle ear epithelium is covered by ciliated cells and secretory cells typical of respiratory epithelium. The mastoid cavity is largely covered by simple squamous mucosal epithelium. The eustachian tube is covered by tall columnar epithelial cells consists of ciliated and secretory cells. Adapted from Lim [13]

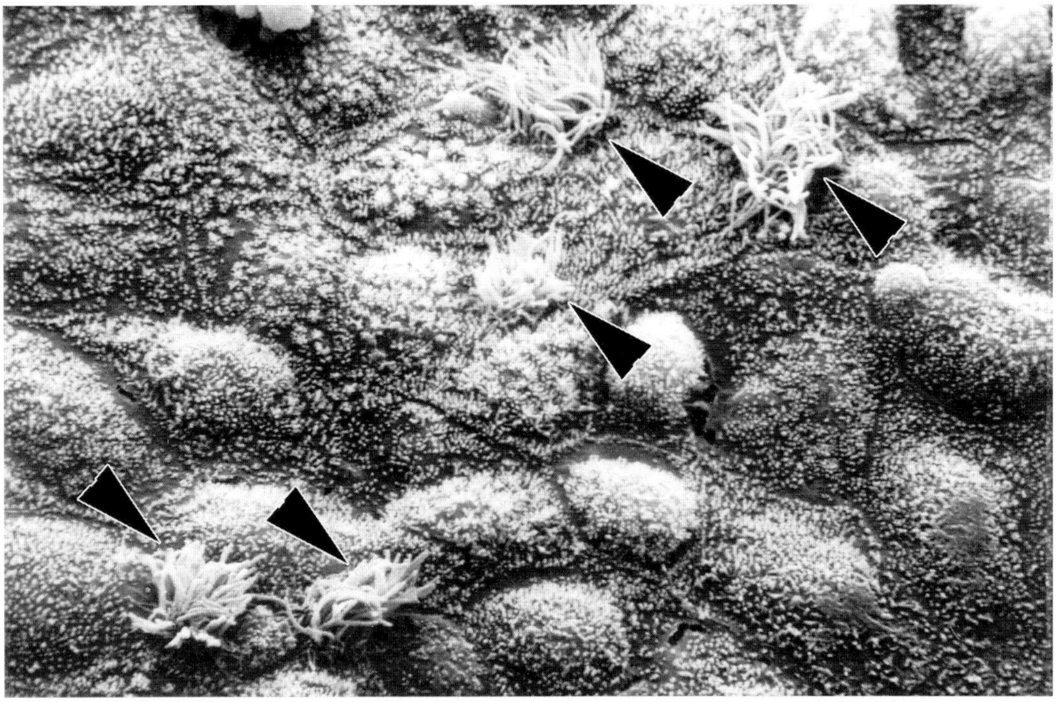

Fig. 2.2 Middle ear cavity is covered with respiratory epithelium consisting of ciliated cells (*arrowheads*), secretory cells, and nonsecretory cells. Viral upper respiratory infection destroys these respiratory epithelial cells impairing mucociliary transport system essential for mucosal protection

on the location. Near the eustachian tube orifice, these cells are columnar or cuboidal, and they are becoming flat epithelial cells toward the mastoid cavity. The number of secretory cells is proportional to the number of ciliated cells. In the promontory region, the epithelial cells are largely cuboidal. The eustachian tube consists of two parts: bony parts closer to the tympanic cavity and the cartilaginous part connected to the oropharyngeal opening. This latter part has rich mixed glands. Tubal cartilage is closed most of time, but opens intermittently to equate pressure and aeration. The lining epithelium is largely ciliated and secretory cells and its major function is to transport unwanted particles trapped in the mucous blanket by mucociliary transport system. The tubal dysfunction is believed to be involved with the risk factor for OM in children during the winter months when upper respiratory infections (URI) are prevalent [11]. URI is known to impair or destroy mucociliary epithelium, leading to the tubal dysfunction. The tubal dysfunction is believed to result in high negative pressure of the tympanic cavity. Therefore, the bolus of mucus containing pathogenic bacteria may enter into the tympanic cavity when the tube attempts to open.

The proportion of the ciliated and secretory cells reflects the history of inflammation in the middle ear, suggesting the middle ear epithelial cell type is in dynamic state [12]. The epithelial cells are also known to elaborate antimicrobial molecules, which constitute the innate immune defense mechanism [3].

2.2.2 Inner Ear

As to the functional morphology of the inner ear, the auditory sensory organ (organ of corti of the cochlea) is a unique sensory structure composed of one row of inner hair cells (IHC) and three (or four) rows of the outer hair cells in mammals

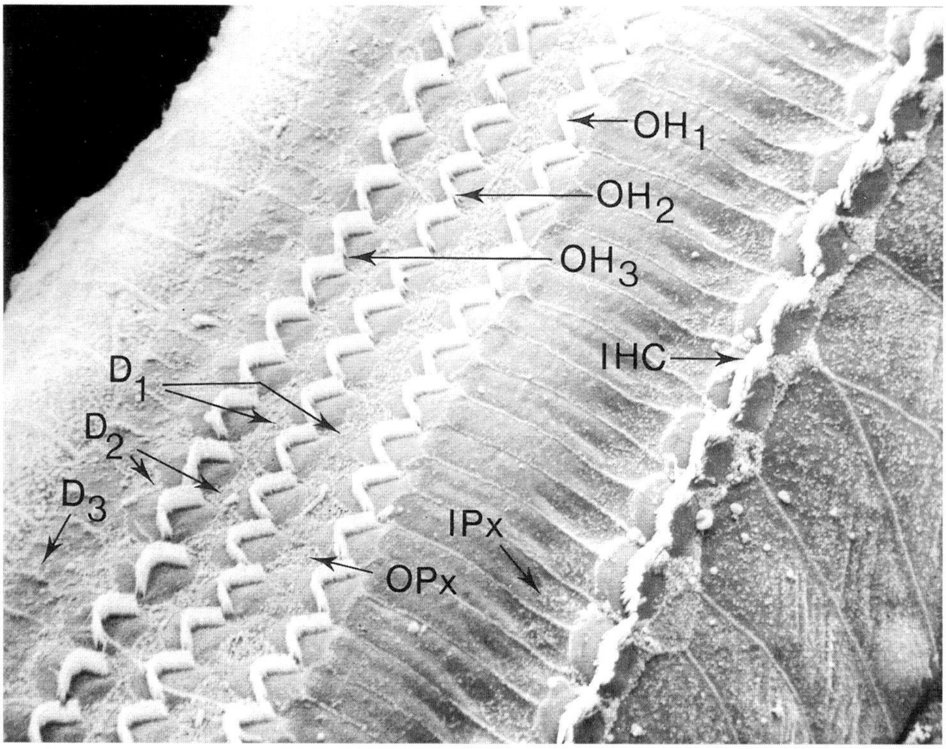

Fig. 2.3 A SEM photomicrograph of surface view of chinchilla organ of Corti showing complexity of cell structures including one row of inner hair cells (IHC) and three rows of outer hair cells (OH1–OH3), inner phalangeal cells (IPx), outer phalangeal cells (OPx), and three rows of Deiter cells (D1–D3). The sensory cell hair bundles (stereocilia), arranged in "W" formation, are exposed to the endolymph (scala media). Adapted from Lim [14]

(Fig. 2.3) [13–16]. The IHC are considered the primary sensory cells responsible for transmitting electrical signals to the brain, whereas the outer hair cells are responsible for fine tuning (frequency resolution) of the hearing through motile activity of the sensory cells, by regulating their lengths. Electrical stimulation of the outer hair cells induces motor activity through their unique motor proteins (known as Prestin) embedded in the membrane structure [17].

In addition, the sensory organ is bathed in different fluid compartments. The endolymph compartment (scala media) is filled with K^+ ion-rich fluid, whereas the perilymph compartments, composed of scala vestibuli and scala tympani, contain Na^+ ion-rich fluid. The perilymph space is communicating with the CSF through the cochlea aqueduct.

2.2.3 Endolymphatic Sac

The endolymphatic compartment is connected to the blind sac known as ES through endolymphatic duct (ED). The ES is composed of intraosseous and intradural parts. In human, the ES is composed of multiple interconnecting tubules [18], occasionally filled with dense PAS-positive materials [19], which is believed to be the osmotic agent required for fluid volume regulation of the endolymph system. Experimental evidences suggest that ES epithelial cells may secrete these osmotic agents. In addition, the ion transport activities of the ES involve mainly Na^+ and Cl^- ions. The ES epithelial cells express ion channels and cotransporters, such as ENaC, Na^+-K^+-2Cl-type 2 (NKCC-2) cotransporter in addition

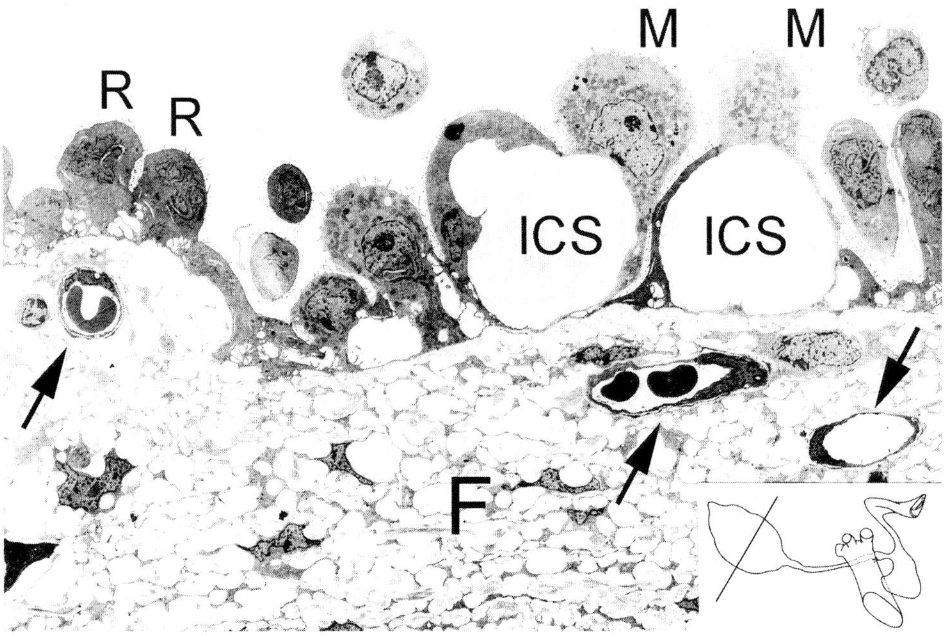

Fig. 2.4 Epithelial lining of the endolymphatic sac consists of mitochondria-rich light cell (M) and ribosome-rich dark cell (R). There are also considerable variations of luminal cell surface areas among different cells and extent of intercellular space (ICS) indicating various stages of fluid transport. This may represent dynamic states of the fluid transport (absorption) at any given time. Loose subepithelial connective tissue contains capillaries (*arrows*) and nerve fibers (not shown), and has extensive network of fibrocyte processes (F) contacting to each other. Insert is a diagram of the endolymphatic sac and a *line* indicates the plane of section. Adapted from Lim [13]

to SLC26A4, carbonic anhydrase, and Cl^-/HCO_3^- exchanger [5, 20, 21].

There are also strong evidences that the ES is the immune organ of the inner ear, containing resident macrophages and lymphocytes [6], and it is suggested to play a critical role in the immune response of the inner ear [22]. The luminal fluid often contains dense glycoprotein substances as well as free floating macrophages. In addition to macrophages, there are reports to indicate the presence of intraepithelial lymphocytes [23].

The ES consists of epithelial cells and loose subepithelial connective tissue, which is continuous with the perilympathic space of the ED. The ES epithelial cells are composed of relatively flat epithelial cells of two types (Fig. 2.4) [24]. Mitochondria-rich light cell is characterized by a large number of microvilli on its luminal surface, whereas mitochondria-poor ribosome-rich dark cell has a relatively smooth luminal cell surface (Fig. 2.5). Mitochondria-rich cell is also believed to express Pendrin and carbonic anhydrase based on the studies with immunolabeling [25]. Foxi-1 gene is upstream for Pendrin expression, and Foxi-1-deficient mice develop endolymphatic hydrops and the ES of these animals lack Pendrin-positive (mitochondria-rich) cells [26].

The subepithelial connective tissue contains a large number of blood vessels and accompanying nerve supply. Sympathetic (superior cervical ganglion), parasympathetic (pterygopalatine ganglion), and somatosensory (trigeminal ganglion) innervation is known to be largely involved in regulation of the blood flow and possibly in the fluid absorption of the ES. They are either myelinated or unmyelinated [6].

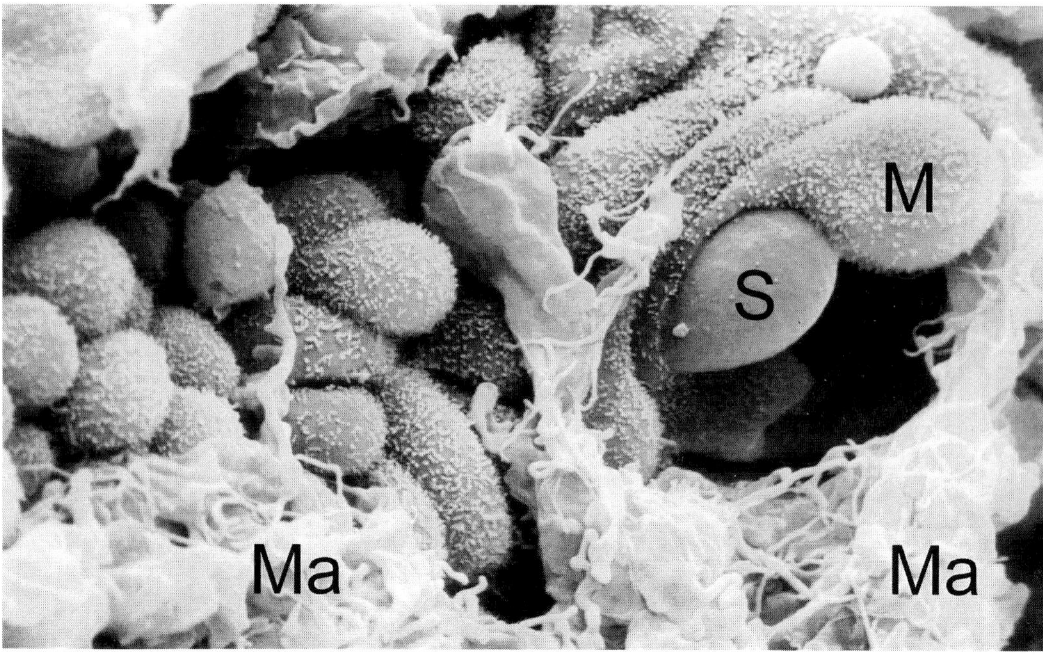

Fig. 2.5 SEM photomicrograph of ES surface showing two types of epithelial cells: cell with extensive microvilli (M) and cell with smooth surface (S), and free-floating macrophages (Ma) in the luminal surface. The cell with extensive microvilli is interpreted to be mitochondria-rich light cell and the cell with smooth surface is ribosome-rich dark cell. Adapted from Lim [13]

2.3 Human Middle Ear Cell Line

The middle ear epithelial cells are known to directly interact with the pathogens and pathogen-derived molecules. Since normal human middle ear epithelial cells are not easily obtainable, it is critical to establish a stable human middle ear epithelial cell line that express important genes/gene products involved in the bacteria–host interaction and resultant inflammatory response.

2.3.1 Immortalization

Small pieces of healthy human middle ear mucosa were harvested from the promontory area of the patient during translabyrinthine craniotomy. To induce proliferation of primary cells, explants of human middle ear mucosa were plated on 35-mm plastic culture dishes with a minimal volume of media, allowing them adhere to the bottom [27]. Cultures were maintained at 37°C in a humidified atmosphere of 5% CO_2. The culture medium used was a 1:1 mixture of bronchial epithelial growth media (BEGM; Clonetics, Walkersville, MD) and Dulbecco's Modified Eagle Medium (DMEM) (Gibco BRL, Gaithersburg, MD), containing hydrocortisone (0.5 μg/mL), insulin (5 μg/mL), transferrin (10 μg/mL), epinephrine (0.5 μg/mL), triiodothyronine (6.5 μg/mL), gentamycin (50 μg/mL), and amphotericin B (50 ng/mL), all supplied by Clonetics (Walkersville, MD), and further supplemented with EGF (25 ng/mL; Collaborative Research, Bedford, MA), all-trans retinoic acid (5 × 10¼ 8 M; Sigma, St. Louis, MO), bovine serum albumin (1.5 μg/mL; Sigma), and bovine pituitary extract (1% v/v; Pel Freez, Rogers, AR).

For immortalization, we infected cells using a retrovirus containing the E6/E7 genes of human papillomavirus type 16 [28]. Briefly, the PAS 17 amphotropic packaging cell line, stably transfected with a replication-defective retrovirus construct (pLXSN16E6-E7), coding for HPV type 16 transforming oncoproteins E6 and E7, was grown to 70% confluence, and supernatants were collected after 24 h. Primary human middle ear epithelial cells (passage 3 at 50% confluence) were infected with 1 mL of virus stock diluted in 9 mL of fresh BEGM for 48 h. The medium was replaced with fresh BEGM and the cells were allowed to proliferate. For selection, cells were reseeded in fresh BEGM containing 0.4 mg/mL of G418 (Gibco BRL) and were kept in the selection media for 14 days. Multiple colonies were isolated using cloning rings and each clone was expanded for further characterization. One of clones appeared to be stably immortalized and was designated as HMEEC-1. The average doubling time of the HMEEC-1 was 23.8 h and appeared to preserve characteristics of epithelial cells such as expression of pan-cytokeratin, dome formation, and anchorage dependency. In addition, subcutaneous injection of HMEEC-1 cells to the nude mice did not result in tumor formation. Karyotypic analysis confirmed the immortalized cells are derived from male humans and have no major abnormality of chromosomes.

2.3.2 Major Research Findings Resulting from the Human Middle Ear Cell Line

2.3.2.1 Secretion of Mucins

The sterility of the eustachian tube and the middle ear is maintained not only by the adaptive immune system, but also by the innate immune system such as mucociliary system and the antimicrobial molecules [3]. Mucins are high molecular weight glycoproteins that constitute the major component of mucus secretions in the eustachian tube and middle ear. The core proteins are encoded by different mucin genes (MUC genes); and we showed the expression of MUC genes such as MCU1, MUC2, MUC5AC, and MUC5B in the primary human middle ear epithelial cells [29]. Moreover, we demonstrated that HMEEC-1 cells up-regulate MUC5AC expression in response to cytoplasmic proteins of nontypeable *Haemophilus influenzae* via a p38 MAP kinase signaling pathway [30], which is negatively regulated via the transforming growth factor beta-Smad signaling pathway [31]. In addition, we showed that HMEEC-1 cells up-regulate MUC2 expression through the cooperation of transforming growth factor-beta-Smad signaling pathway and NF-kB activation [32].

2.3.2.2 Induction of Antimicrobial Agents

To defend against invading pathogens, the tubotympanum is known to secrete antimicrobial molecules including lysozyme, lactoferrin, and beta defensins [3]. The beta defensins are mainly produced by epithelial cells of the skin, kidneys, and respiratory lining of nearly all vertebrates [33]. The beta defensins are released upon microbial invasion and are located at the host–environment interfaces, such as mucosal surfaces and skin. Beta defensin 2, which is released from the epithelial cells in response to microorganisms or cytokines, exhibits potent antimicrobial activity against gram-negative bacteria and candida. HMEEC-1 cells were found to up-regulate beta defensin 2 in response to interleukin 1 alpha (IL-1α) via a Src-dependent Raf-MEK1/2-ERK signaling pathway [34]. We also found that nontypeable *H. influenzae*-induced up-regulation of beta defensin 2 requires a TLR2/MyD88-dependent p38 MAP kinase pathway [35]. As shown in Fig. 2.6, we demonstrated that IL-1α is secreted by middle ear epithelial cells upon exposure and that it can synergistically act with nontypeable *H. influenzae* molecules to up-regulate beta defensin 2 via the p38 MAP kinase pathway [36].

2.3.2.3 Expression of Toll-Like Receptors

Toll-like receptors (TLRs) have been suggested to play a critical role in the recognition of various bacterial components such as lipoprotein, peptidoglycan, lipoteichoic acid, and lipopolysaccharide [37]. We found that HMEEC-1 cells regulate TLR2 expression and that glucocorticoids synergistically enhance nontypeable *H. influenzae*-induced TLR2 expression via a negative cross-talk with p38 MAP kinase [38]. Moreover, we demonstrated that TLR2 is involved in nontypeable

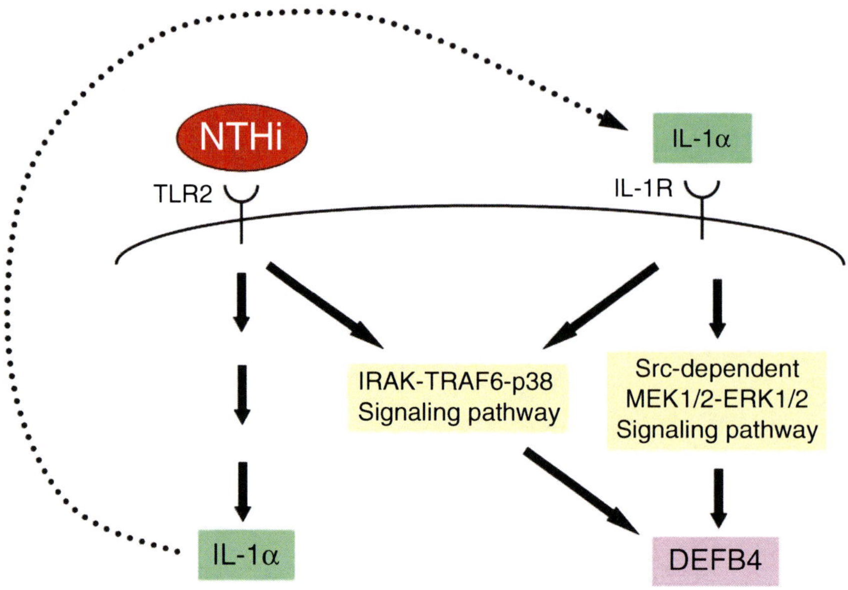

Fig. 2.6 Schematic illustration summarizes signaling pathways involved in the individual and synergistic effects of IL-1 alpha and nontypeable *H. influenzae* (NTHi) on beta defensin 2 (DEFB4) transcription in the middle ear epithelial cells

H. influenzae-induced NF-kB activation through the TAK1-dependent NIK-IKK-IKB and MKK3/6-p38 MAP kinase signaling pathways [39]. In addition, we found the involvement of TLR2 in the recognition of nontypeable *H. influenzae* molecules, resulting in induction of beta defensin 2 [35].

2.4 Human Endolymphatic Sac Cell Line

Because of the location of the ES deep inside the petrosal bone and the difficulty of obtaining epithelial cells, experiments using the human ES epithelial cells have been difficult to perform. Recently, we have successfully developed a human ES cell line preserving the characteristics of normal ES epithelial cells such as the expression of Pendrin.

2.4.1 Immortalization

Primary cell culture preparation of ES epithelial cell has been described [40]. Briefly, human ES were excised during translabyrinthine acoustic neuroma surgery at Uppsala University Hospital with an institutional approval. The intraosseous portion of the human ES was drilled out, leaving a thin, movable eggshell layer of bone on its anterior surface. The human ES was separated from the posterior bony surface with a mucosal knife and was cut with a pair of scissors at the external aperture of the vestibular aqueduct. Thus, only the intraosseous portion of the human ES was retained. The sample was then cut into small pieces (<1 mm^3) and incubated at 4°C in 0.17% trypsin in PBS for 15–18 h. The resulting suspension of single cells and small aggregates were transferred to a test tube, and an equal volume of growth medium was added. The cells were centrifuged for 5 min at 1,000 rpm, resuspended in growth medium, and seeded into a 35-mm diameter fibronectin and collagen coated tissue culture dish. The cells were grown in a 3:1 mixture of (DMEM) and Ham's F12 medium (Gibco BRL) supplemented with 10% fetal calf serum (FCS) (Hyclone), 0.4 μg/mL hydrocortisone (Sigma), 10 μg/mL human epidermal growth factor (EGF) (Austral Biologicals), 10^{-10} M cholera toxin

(Sigma), 5 μg/mL Zn-free insulin (Lilly Research Laboratories), 24 μg/mL adenine (Sigma), and 2×10^{-9} M 3,3,5-triiodo-L-thyronine (Sigma).

For immortalization, primary ES epithelial cells were infected using a retrovirus containing the E6/E7 genes of human papillomavirus type 16 as described [28]. Infected cells were selected using the G418-containing medium and multiple colonies were isolated. Each clone was expanded for the further characterization, and one clone showing stable immortalization was designated as HESC-1. HESC-1 cells appeared to preserve the characteristics of epithelial cells such as expression of ZO-1 and cytokeratin. Karyotypic analysis demonstrated that the immortalized cells have no major abnormality in chromosomes. Although HESC-1 cells were originated from a single cell, they appeared to differentiate into two distinct subtypes, resembling ribosome-rich dark cells and mitochondria-rich light cells of the ES epithelial cells. Immunolabeling and FACS analysis showed that there are two subpopulations according to the expression of Pendrin and richness of mitochondria.

2.4.2 Research Findings Resulting from the Human Endolymphatic Sac Cell Line

Secretory capacity of the ES has been postulated by some authors [41–43]. In about 30% of the cross-sections of the mouse ES, a small amount of homogeneous substance was observed. This homogeneous substance is believed to contain acidic protein-bound carbohydrates [44], proteoglycan [45], sulfur compounds [46], hyaluronan [47], and soluble megalin [48]. Although the function of this homogeneous substance is unclear, it is suggested to be involved in the attraction of water and small water soluble cations, which could influence the regulation of fluid volume of the inner ear endolymphatic compartment [45].

Unexpectedly, we found that our HESC-1 cells secrete viscous substance in response to EGF. Dot blot analysis showed that the supernatant of HESC-1 cells contains mucin core proteins and hyaluronan. qRT-PCR analysis demonstrated that HESC-1 cells up-regulate mucin core proteins and hyaluronic acid synthases in response to EGF. Further studies are needed to determine the regulatory mechanisms involved in secretion and biological functions of these molecules. We anticipate that our HESC-1 cells will provide an in vitro model for the studies of secretory function of the ES.

2.5 Conclusions

We have successfully immortalized epithelial cells from human middle ear and ES. The human middle epithelial cells have been used for elucidating cell signaling pathways involved in inflammatory responses including induction of beta defensins and mucin genes. The human ES cell line is in the process of further characterization. Preliminary results show the expression of genes involved in water and ion transportation, such as Pendrin and aquaporins. Particularly, the human ES cell line was found to secrete osmotically active substances, and further studies are necessary to identify these molecules and determine their function in the regulation of endolymphatic fluid volume.

Acknowledgement We would like to express our profound thanks for the guidance and assistance of Dr. Johng Rhim in establishing human middle ear and endolymphatic sac cell lines. This work was supported in part by grants DC5025 and DC6276 from the NIH, NIDCD.

References

1. Vergison A (2008) Microbiology of otitis media: a moving target. Vaccine 26(Suppl 7):G5–G10
2. Gates GA (1996) Cost-effectiveness considerations in otitis media treatment. Otolaryngol Head Neck Surg 114(4):525–530
3. Lim DJ, Chun YM, Lee HY, Moon SK, Chang KH, Li JD, Andalibi A (2000) Cell biology of tubotympanum in relation to pathogenesis of otitis media – a review. Vaccine 19(Suppl 1):S17–S25
4. Lim DJ (1981) Middle ear and inner ear structure and biological function. In: Bernstein JM, Ogra PL (eds) Immunology of the ear. Raven, New York, pp 1–38
5. Kim SH, Marcus DC (2010) Sodium homeostasis in the inner ear. In: Kim HN, Chung MH, Lee WS (eds) Current opinion on sensorineural hearing loss. Gunsa, Seoul, pp 41–63

6. Lim DJ, Moon SK (2010) The endolymphatic sac structure and function and their clinical implications: a review. In: Kim HN, Chung MH, Lee WS (eds) Current opinion on sensorineural hearing loss. Gunsa, Seoul, pp 65–84
7. Lim DJ (1999) Ultrastructure of the endolymphatic duct and sac in normal and Meniere's disease. In: Harris JP (ed) Meniere's disease. Kugler, Hague, pp 175–193
8. Tomiyama S, Harris JP (1987) The role of the endolymphatic sac in inner ear immunity. Acta Otolaryngol 103(3–4):182–188
9. Lim DJ, Paparella MM, Kimura RS (1967) Ultrastructure of the eustachian tube and middle ear mucosa in the guinea pig. Acta Otolaryngol 63(5): 425–444
10. Lim DJ (1974) Functional morphology of the lining membrane of the middle ear and eustachian tube: an overview. Ann Otol Rhinol Laryngol 83(Suppl 11): 15–26
11. Lim DJ, DeMaria TF, Bakaletz LO (1987) Functional morphology of the tubotympanum related to otitis media: a review. Am J Otol 8(5):385–389
12. Lim DJ, Birck H (1971) Ultrastructural pathology of the middle ear mucosa in serous otitis media. Ann Otol Rhinol Laryngol 80(6):838–853
13. Lim DJ (1980) Scanning electron microscopic morphology of the ear. In: Paparella MM, Shumrick CA (eds) Textbook of otolaryngology, vol 1. WB Saunders, Philadelphia, pp 439–469
14. Lim DJ (1986) Functional structure of the organ of corti: a review. Hear Res 22:117–146
15. Lim DJ, Lane WC (1969) Three-dimensional observation of the inner ear with the scanning electron microscope. Trans Am Acad Ophthalmol Otolaryngol 73(5):842–872
16. Lim DJ, Lane WC (1969) Cochlear sensory epithelium. A scanning electron microscopic observation. Ann Otol Rhinol Laryngol 78(4):827–841
17. Matsumoto N, Kalinec F (2005) Prestin-dependent and prestin-independent motility of guinea pig outer hair cells. Hear Res 208(1–2):1–13
18. Linthicum FH Jr, Tian Q, Milicic M (1995) Constituents of the endolymphatic tubules as demonstrated by three-dimensional morphometry. Acta Otolaryngol 115(2):246–250
19. Tian Q, Rask-Andersen H, Linthicum FH Jr (1994) Identification of substances in the endolymphatic sac. Acta Otolaryngol 114(6):632–636
20. Akiyama K, Miyashita T, Mori T, Inamoto R, Mori N (2008) Expression of thiazide-sensitive Na^+-Cl^- cotransporter in the rat endolymphatic sac. Biochem Biophys Res Commun 371(4):649–653
21. Akiyama K, Miyashita T, Mori T, Mori N (2007) Expression of the Na^+-K^+-$2Cl^-$ cotransporter in the rat endolymphatic sac. Biochem Biophys Res Commun 364(4):913–917
22. Wackym PA, Friberg U, Linthicum FH Jr, Bagger-Sjoback D, Bui HT, Hofman F, Rask-Andersen H (1987) Human endolymphatic sac: morphologic evidence of immunologic function. Ann Otol Rhinol Laryngol 96(3 Pt 1):276–281
23. Rask-Andersen H, Stahle J (1979) Lymphocyte-macrophage activity in the endolymphatic sac. An ultrastructural study of the rugose endolymphatic sac in the guinea pig. ORL J Otorhinol Relat Spec 41(4): 177–192
24. Barbara M, Rask-Andersen H, Bagger-Sjoback D (1987) Ultrastructure of the endolymphatic sac in the Mongolian gerbil. Arch Otorhinolaryngol 244(5): 284–287
25. Peters TA, Tonnaer EL, Kuijpers W, Cremers CW, Curfs JH (2002) Differences in endolymphatic sac mitochondria-rich cells indicate specific functions. Laryngoscope 112(3):534–541
26. Hulander M, Kiernan AE, Blomqvist SR, Carlsson P, Samuelsson EJ, Johansson BR, Steel KP, Enerback S (2003) Lack of pendrin expression leads to deafness and expansion of the endolymphatic compartment in inner ears of Foxi1 null mutant mice. Development 130(9):2013–2025
27. Moon SK, Lim DJ, Lee HK, Kim HN, Yoo JH (2000) Mucin gene expression in cultured human middle ear epithelial cells. Acta Otolaryngol 120(8):933–939
28. Chun YM, Moon SK, Lee HY, Webster P, Brackmann DE, Rhim JS, Lim DJ (2002) Immortalization of normal adult human middle ear epithelial cells using a retrovirus containing the E6/E7 genes of human papillomavirus type 16. Ann Otol Rhinol Laryngol 111(6):507–517
29. Moon SK, Yoo JH, Kim HN, Lim DJ, Chung MH (2000) Effects of retinoic acid, triiodothyronine and hydrocortisone on mucin and lysozyme expression in cultured human middle ear epithelial cells. Acta Otolaryngol 120(8):944–949
30. Wang B, Lim DJ, Han J, Kim YS, Basbaum CB, Li JD (2002) Novel cytoplasmic proteins of nontypeable *Haemophilus influenzae* up-regulate human MUC5AC mucin transcription via a positive p38 mitogen-activated protein kinase pathway and a negative phosphoinositide 3-kinase-Akt pathway. J Biol Chem 277(2): 949–957
31. Jono H, Xu H, Kai H, Lim DJ, Kim YS, Feng XH, Li JD (2003) Transforming growth factor-beta-Smad signaling pathway negatively regulates nontypeable *Haemophilus influenzae*-induced MUC5AC mucin transcription via mitogen-activated protein kinase (MAPK) phosphatase-1-dependent inhibition of p38 MAPK. J Biol Chem 278(30):27811–27819
32. Jono H, Shuto T, Xu H, Kai H, Lim DJ, Gum JR Jr, Kim YS, Yamaoka S, Feng XH, Li JD (2002) Transforming growth factor-beta-Smad signaling pathway cooperates with NF-kappa B to mediate nontypeable *Haemophilus influenzae*-induced MUC2 mucin transcription. J Biol Chem 277(47): 45547–45557
33. Ganz T, Selsted ME, Lehrer RI (1990) Defensins. Eur J Haematol 44(1):1–8
34. Moon SK, Lee HY, Li JD, Nagura M, Kang SH, Chun YM, Linthicum FH, Ganz T, Andalibi A, Lim DJ

(2002) Activation of a Src-dependent Raf-MEK1/2-ERK signaling pathway is required for IL-1alpha-induced upregulation of beta-defensin 2 in human middle ear epithelial cells. Biochim Biophys Acta 1590(1–3):41–51

35. Lee HY, Takeshita T, Shimada J, Akopyan A, Woo JI, Pan H, Moon SK, Andalibi A, Park RK, Kang SH, Kang SS, Gellibolian R, Lim DJ (2008) Induction of beta defensin 2 by NTHi requires TLR2 mediated MyD88 and IRAK-TRAF6-p38MAPK signaling pathway in human middle ear epithelial cells. BMC Infect Dis 8:87
36. Moon SK, Lee HY, Pan H, Takeshita T, Park R, Cha K, Andalibi A, Lim DJ (2006) Synergistic effect of interleukin 1 alpha on nontypeable *Haemophilus influenzae*-induced up-regulation of human beta-defensin 2 in middle ear epithelial cells. BMC Infect Dis 6:12
37. Medzhitov R, Janeway C Jr (2000) Innate immune recognition: mechanisms and pathways. Immunol Rev 173:89–97
38. Shuto T, Imasato A, Jono H, Sakai A, Xu H, Watanabe T, Rixter DD, Kai H, Andalibi A, Linthicum F, Guan YL, Han J, Cato AC, Lim DJ, Akira S, Li JD (2002) Glucocorticoids synergistically enhance nontypeable *Haemophilus influenzae*-induced Toll-like receptor 2 expression via a negative cross-talk with p38 MAP kinase. J Biol Chem 277(19):17263–17270
39. Shuto T, Xu H, Wang B, Han J, Kai H, Gu XX, Murphy TF, Lim DJ, Li JD (2001) Activation of NF-kappa B by nontypeable Hemophilus influenzae is mediated by toll-like receptor 2-TAK1-dependent NIK-IKK alpha /beta-I kappa B alpha and MKK3/6-p38 MAP kinase signaling pathways in epithelial cells. Proc Natl Acad Sci U S A 98(15):8774–8779
40. Linder B, Bostrom M, Gerdin B, Rask-Andersen H (2001) In vitro growth of human endolymphatic sac cells: a transmission electron microscopic and immunohistochemical study in patients with vestibular schwannoma and Meniere's disease. Otol Neurotol 22(6):938–943
41. Adlington P (1967) The ultrastructure and the functions of the saccus endolymphaticus and its decompression in Meniere's disease. J Laryngol Otol 81(7): 759–776
42. Adlington P (1984) The saccus endolymphaticus in the rabbit. Further studies. J Laryngol Otol 98(9): 857–877
43. Seymour JC (1954) Observations on the circulation in the cochlea. J Laryngol Otol 68(10):689–711
44. Porubsky ES, Marovitz WF, Arenberg IK (1972) Presence of acidic protein-bound carbohydrates in the endolymphatic sac and duct of fetal, neonatal and adult rats, and adult humans. Ann Otol Rhinol Laryngol 81(1):76–81
45. Friberg U, Wackym PA, Bagger-Sjoback D, Rask-Andersen H (1986) Effect of labyrinthectomy on the endolymphatic sac. A histological, ultrastructural and computer-aided morphometric investigation in the mouse. Acta Otolaryngol 101(3–4):172–182
46. Barbara M, Takumida M, Nilsson J, Bagger-Sjoback D, Rask-Andersen H (1989) Turnover of sulphur compounds in the endolymphatic sac: an autoradiographic study in the Mongolian gerbil. ORL J Otorhinol Relat Spec 51(1):1–7
47. Friberg U, Erwall C, Bagger-Sjoback D, Rask-Andersen H (1989) Hyaluronan content in human inner ear fluids. Acta Otolaryngol 108(1–2):62–67
48. Ishida T, Hatae T, Nishi N, Araki N (2006) Soluble megalin is accumulated in the lumen of the rat endolymphatic sac. Cell Struct Funct 31(2):77–85

Cellular Systems for Studying Human Oral Squamous Cell Carcinomas

3

Vyomesh Patel, Ramiro Iglesias-Bartolome, Bradford Siegele, Christina A. Marsh, Kantima Leelahavanichkul, Alfredo A. Molinolo, and J. Silvio Gutkind

Abstract

The human oral squamous epithelium plays an important role in maintaining a barrier function against mechanical, physical, and pathological injury. However, the self-renewing cells residing on the basement membrane of the epithelium can give rise to oral squamous cell carcinomas (OSCC), now the sixth most common cancer in the developed world, which is still associated with poor prognosis. This is due, in part, to the limited availability of well-defined culture systems for studying oral epithelial cell biology, which could advance our understanding of the molecular basis of OSCC. Here, we describe methods to successfully isolate large cultures of human oral epithelial cells and fibroblasts from small pieces of donor tissues for use in techniques such as three-dimensional cultures and animal grafts to validate genes suspected of playing a role in OSCC development and progression. Finally, the use of isolated oral epithelial cells in generating iPS cells is discussed which holds promise in the field of oral regenerative medicine.

3.1 Introduction

We now know that the normal skin and the oral mucosa, consist of the superficial epithelium, and the underlying lamina propria [1]. At this interface lies the basal lamina, or basement membrane, serving primarily as a supportive structural role to correctly direct migration and differentiation of the overlying epithelial cells up toward the epithelium surface [1]. Of note, resting on the basement membrane is a single layer of cuboidal cells known as the stratum basale, and includes a subpopulation of stem cells, which can undergo mitotic division to create new cells. Differentiation and upward migration of these new cells results in a population of larger polyhedral cells that form the stratum spinosum [2]. The cells of the stratum spinosum extend several layers thick and are held together by desmosomal adhesions visible by light microscopy [2]. The orderly differentiation of the cells of the squamous epithelium is completed with cells

V. Patel (✉)
Oral and Pharyngeal Cancer Branch, National Institute of Dental and Craniofacial Research, National Institutes of Health, 30 Convent Drive, Building 30, Room 211, Bethesda, MD 20892-4330, USA
e-mail: vpatel@mail.nih.gov

J.S. Rhim and R. Kremer (eds.), *Human Cell Transformation: Role of Stem Cells and the Microenvironment*, Advances in Experimental Medicine and Biology 720, DOI 10.1007/978-1-4614-0254-1_3, © Springer Science+Business Media, LLC 2011

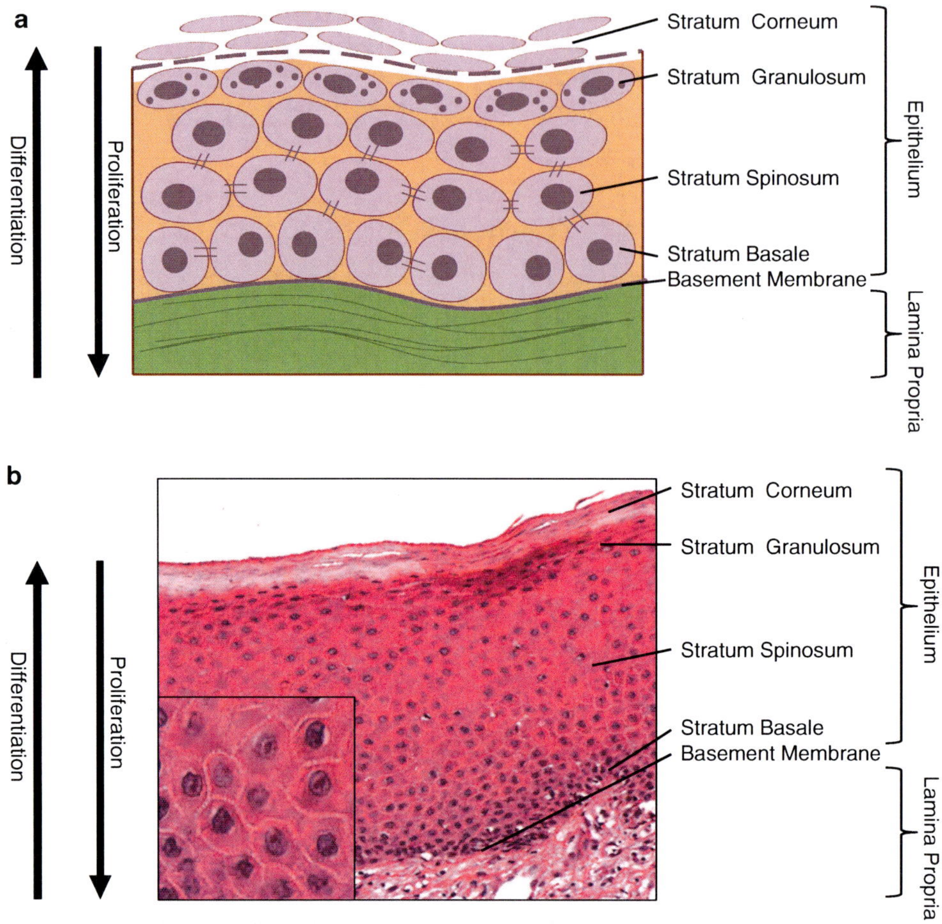

Fig. 3.1 (**a**) Schematic of normal oral squamous epithelium. Normal oral epithelial cell proliferation and differentiation are orderly and unidirectional, with keratinocytes originating in a relatively undifferentiated state in the stratum basale, progressing through the stratum spinosum and stratum granulosum layers, and reaching full differentiation at the surface stratum corneum. (**b**) H&E staining of normal oral squamous epithelium (20×). Normal oral squamous epithelial tissue is histologically bland, with hallmarks of orderly differentiation from the basement membrane to the epithelial surface. Less differentiated cells of the stratum basale are compact in size and round in shape, with prominently staining nuclei. Cells enlarge and nuclei grow less dense as they differentiate through the stratum spinosum, after which they flatten and eventually shed their nuclei in the stratum corneum

losing their nuclei and cytoplasmic organelles and becoming fully keratinized at the surface layers of the stratum corneum [3]. Collectively, the squamous epithelium represents a dynamic structure, whereby cells of the basement membrane work continuously in a coordinate manner to cease to proliferate, and migrate and differentiate up towards the surface culminating in the replenishment of the spinous, granular, and ultimately the cornified layer of the stratified epithelium, and in the process renewing and repairing defects if any, to ensure that integrity and correct barrier function is maintained [4]. The key components of a normal epithelium are shown in Fig. 3.1a, b. In this review, we describe features of the oral squamous epithelium, including differences between the skin equivalent, and how this crucial and very important structure can be exploited for the isolation of oral epithelial cell for use in investigating their biological properties, with emphasis on the understanding of the molecular basis of human oral squamous cell carcinoma (OSCC).

3.2 Oral Squamous Epithelium

Several differences exist between skin and oral squamous epitheliums [5]. For instance, the differentiation pattern of the superficial layers of the oral mucosa and the skin epidermis are very distinct. The oral epithelium can be classified into three types, masticatory, lining, and specialized, which include keratinized, parakeratinized, and nonkeratinized stratified epithelium [6]. In general, though, a parakeratinized stratified epithelium forms the lining of most of the oral mucosa, thus providing a barrier against mechanical, physical, and pathological injury [7, 8]. Specifically, the main morphological difference between the buccal mucosa and the skin epidermis is in the differentiation pattern found in the superficial layers of the epithelium. The oral mucosa lacks the granulous and cornified layers, but instead presents parakeratin as the last step in the differentiation process. In addition, emerging evidence suggests that the epithelial stem cells maintaining the homeostasis of the oral mucosa and skin might be distinct [9]. In the skin, both follicular and interfollicular epithelial (IF) stem cells can contribute to the replacement of the exfoliated cells and to regenerate the tissue upon injury [10]. In this regard, current information suggests that the IF stem cells and their derived transient amplifying cells may play a primary role in epidermal homeostasis and tissue repair, and that in some cases IF stem cells may be alone sufficient for the healing of cutaneous wounds [11, 12]. The oral mucosa, however, represents an epidermal anatomical site that presents similar architecture to that of the skin but devoid of any hair-follicular structures. Thus, the oral epithelium is expected to be maintained by a single type of adult epithelial stem cell possessing self-renewal capacity. While the distinct roles of these stem cells in tissue maintenance and their behavior during tissue regeneration represent areas of active investigation, we expect that the availability of primary human oral epithelial cell culture systems may aid in dissecting their multiple biological properties and underlying molecular mechanisms as well as help in our understanding of the molecular basis of human OSCC.

3.3 Human Oral Squamous Cell Carcinoma

A delicate balance exists between growth-promoting and growth-inhibiting physiological cues that are responsible for mediating normal homeostasis of stratum basale residing in the basal layer of the oral epithelium [13, 14]. Therefore, perturbation of this balance has the potential to contribute to unrestricted growth and the acquisition of a malignant phenotype [15]. Over time, these cells can progress into overt squamous carcinomas, resulting in invasion of nearby tissues and consequently to metastases, as depicted in Fig. 3.2a [16]. Clinically, OSCC arise predominantly in the floor of the mouth and on the lateral tongue but any aspect of the oral mucosa is susceptible [17]. Histologically, these lesions are characterized by the disruption of the process of orderly differentiation and the migration of oral epithelial cells from the stratum basale to the superficial mucosal layers [18]. The cellular proliferation is disordered, as manifested by the frequent presence of enlarged, irregularly shaped nuclei, multiple mitoses both within and beyond the stratum basale, and markedly abnormal mitotic structures [19]. Dysregulation of cellular differentiation is evidenced by the abnormal presence of nuclei in the stratum corneum, or parakeratosis, or with the abnormal keratinization of cells below the stratum granulosum or dyskeratosis [19]. Carcinoma cells in the stratum spinosum also frequently lose their adhesive capacities and detach from neighboring cells resulting in acantholysis. The histological hallmark of squamous cell carcinoma of the oral cavity remains the disordered invasive migration of epithelial cells downward through the basement membrane into the lamina propria, as shown in Fig. 3.2b. In stark reality, each year approximately 40,000 patients are diagnosed with OSCC predominately in the oral cavity, pharynx, or larynx, representing together nearly 5% of all diagnosed malignancies in the United States [20]. From these, close to

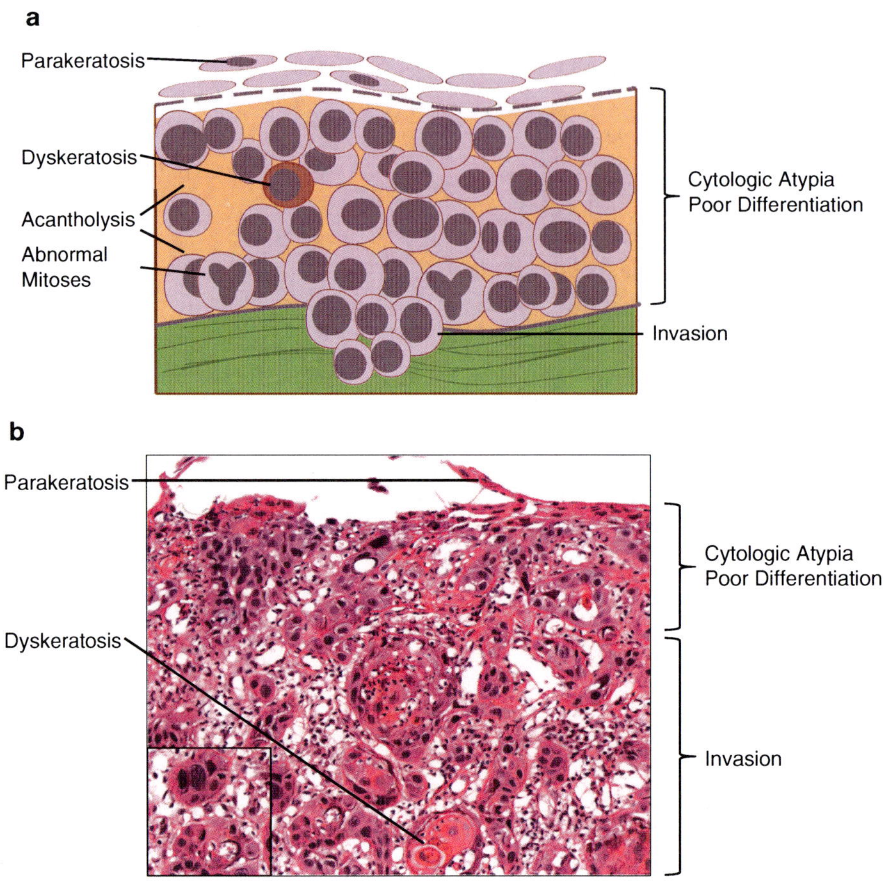

Fig. 3.2 (**a**) Schematic representing oral squamous cell carcinoma. Orderliness of proliferation and differentiation is lost in cancerous tissue of the squamous epithelium, with crowded cellular clustering common and less differentiated cells occurring in upper epithelial layers. Abnormal cellular features often occurring in cancers of the squamous epithelium include abnormal mitoses, acantholysis, dyskeratosis, and parakeratosis. The hallmark of squamous cell cancer, differentiating it from pre-neoplastic lesions, is invasion through the basement membrane into the lamina propria. (**b**) H&E staining to showing oral squamous cell carcinoma (20×). Oral squamous cell carcinoma tissue is strikingly disorderly. Cellular morphology often includes prominently stained nuclei and nucleoli, though morphologies vary widely across the cancer specimen and are frequently bizarre. Invasion here is inferred, as the basement membrane has been entirely obliterated

10,000 US patients die from oral cancer annually equating to one death every hour, and with more than 500,000 new cases per year worldwide, oral cancer is considered the sixth most common cancer in the developed world [20]. In spite of the recent advances in powerful technologies, to accelerate the identity of molecules causal to cancer development and progression, and with improving surgical and treatment options, the prognosis for oral cancer patients is still unfavorable [21]. This is due, in part, to the lack of reliable cellular systems needed to improve our understanding of oral epithelial cell biology and the molecular basis of how these can transform and develop into carcinomas, as well as to test new treatment strategies that can be translated to oral cancer patients for better management.

The identification of the oral epithelial stem cells may have also a broad impact in our understanding of oral cancer development [22, 23]. Emerging studies in our laboratory suggest that OSCC may result from the progressive accumulation of genetic and epigenetic alterations in the epithelial stem cell compartment or in cells that have regained self-renewal capacity [24]. Thus, it would be critical to investigate the mechanisms preserving the genomic integrity of the oral epithelial stem cells, and how their dysregulation, for example upon mutations caused by environmental and tobacco-related carcinogens, can contribute to the malignant progression of oral cancers. We believe that this effort may facilitate the identification of clinically relevant markers for the early diagnosis of oral malignancies, as well as novel molecular targets for the chemoprevention of this disease [22]. Because of the relative simplicity of the oral epithelium, the study of its stem cells may also afford a unique opportunity to understand the most basic principles governing the maintenance of the stem cell compartment, and how extracellular cues affect their cell fate decisions, particularly upon tissue injury. Indeed, although most of our current knowledge on epithelial stem cell biology is based on the study of multipotent epithelial stem cells residing in the bulge region within the hair follicle, it is still not even clear whether these follicle stem cells do in fact contribute to the interfollicular epidermis under physiological situations, or whether multiple lineage restricted stem cell populations give rise to each histologically defined dermal structure, the epidermis, hair follicles, and sebaceous glands. This situation is expected to be much simpler for oral epithelial stem cells, which only need to proliferate and differentiate into stratified epithelia thus enabling to ask fundamental questions in the stem cell research field in a well-defined and biologically relevant system [22, 25].

3.4 Oral Epithelial Stem Cells

The stratified epithelium lining the oral mucosa is continuously renewed by two classes of replicating cells: (1) stem cells that have the capacity for self-renewal, and extended or unlimited growth and (2) transiently amplifying cells that arise from stem cells and replicate a limited number of cycles before undergoing differentiation [26]. In particular, the existence of oral epithelial stem cells that harbor self-renewing capacity provides a likely mechanism whereby the integrity of this tissue can be maintained throughout the entire postnatal life, and wounds can heal rapidly upon injury. We have recently initiated a new concerted effort in the area of oral epithelial stem cell biology, which may facilitate the development of reliable procedures for the isolation and culture of self-renewing epithelial cells from the buccal mucosa, which can then be grafted into patients suffering from tissue injury and disabling epithelial defects in the oral cavity. These stem cells could also be stably transduced with retroviral and lentiviral vectors, therefore representing attractive targets for gene therapy for the sustained expression of therapeutic molecules in the oral mucosa. In addition, each year more than ten million patients worldwide are being treated for cancers with intensive chemotherapy alone or in combination with radiation therapy, which often result in damage to the oral mucosa and the depletion or death of its stem cells. The consequent injury to the mucosal barrier, manifesting as mucositis, leads to chronic pain, difficulty in swallowing, hyposalivation, morbidity, and in an overall reduction in the quality of life. Thus, learning how to protect the oral epithelial stem cells from the effects of anticancer treatment may facilitate the rapid recovery of cancer patients and reduce their suffering and financial burden [27, 28]. In this context, stem cell-therapy has great potential to allow prevention or treatment of radiation and/or chemotherapy-induced oral mucositis in cancer patients [29]. Therefore, advancement in our knowledge of epithelial cell biology with respect to their isolation and characterization as well as their response to damage, will greatly help towards our understanding of human OSCC, and its outcome.

3.5 Isolation of Human Oral Keratinocytes and Their Use in Experimental Systems

Initial attempts at deriving cultures of keratinocytes primarily involved the use of relatively large skin explants but these methods also encouraged the growth of host fibroblasts, which represented the most serious source of contamination of these cultures [30]. To counteract this, co-culture with a feeder layer of lethally irradiated mouse 3T3 mesenchymal cells, encouraged macroscopic colonies to form from single cells, while preventing the growth of host fibroblasts [31]. Also pertinent to this success, was the presence of hydrocortisone, cholera toxin, insulin, transferin, and EGF in the media, aiding to increase life span of the cultures while maintaining the clonogenic fraction and in the reduction of amounts of xenobiotic factors used (fetal bovine serum) [32].

Similarly, several different oral epithelial cell culture methods have been developed, and these range from explants outgrowth to the use of enzymatically digested tissues, relying on distinct media and supplements (described earlier), and coating of culture plates with several extracellular matrix components [33, 34]. However, the resulting cultures invariably grew with limited growth capacity and the cells not surviving beyond subculturing. As with the cultures of skin keratinocytes, the use of a 3T3 feeder layer encouraged the clonal growth of oral keratinocytes [35]. While this approach was the most widely used method with undoubted success, the presence of a mixed cell population and the unknown effects of 3T3 cells on human oral epithelial cells, as well as the presence of xenobiotic serum, has raised concerns, and key among these being the expression of nonhuman xeno-antigens on human cells (embryonic stem cells) dependent on a mouse feeder layer [36]. Consequently, current culture methods have been adapted to incorporate these doubts, as well as improving the efficiency of growth. In essence, gingival tissues from healthy donors are now generally used as a source of oral epithelial cells, where an initial incubation with dispase, an enzyme that facilitates the mechanical separation of the epithelium from the underlying stroma, and then the use of trypsin on the separated epithelium to obtain a single cell suspension and using defined media, are now commonly used ensuring the growth of a pure cell population with minimal contaminating fibroblasts [37]. The primary function of dispase is to selectively separate the surface epithelium from the underlying fibrous connective tissues, an important first step in reducing host fibroblast contamination. Also instrumental to this success, was the use of defined media with no serum, and low calcium concentration, allowing for the proliferation of oral epithelial cells that could be viably maintained for an extended period of time [38].

Our laboratory has essentially optimized a method using trypsin, commercially available defined keratinocyte serum-free medium (KSFM), and rat-tail collagen type I coated culture plates, for successfully obtaining a pure cell population of oral keratinocytes from the separated epithelium, while oral fibroblasts can be obtained if needed from the remaining dermis, in parallel cultures. The resulting fibroblast-free cultures of oral epithelial cells, as seen in Fig. 3.3a, b, can grow readily and are able to undergo propagating and repetitive passage. Additional factors for consideration in obtaining healthy and high yielding cultures are that tissues from younger donors are optimal, and based on a recent study, isolated epithelial cells grown under hypoxia (2%), which may mimic the microenvironment of the epithelial stem cell niche, can stimulate proliferation while inhibiting differentiation [39]. With the aforementioned said, the ability to culture normal oral squamous epithelial cells efficiently can provide an excellent opportunity to study basic oral keratinocyte biology as well as the processes of immortalization and malignant transformation, both of which are key steps in human OSCC.

As described earlier in this chapter, new techniques of culturing oral epithelial cells now makes its possible in obtaining large cultures from small pieces of donor tissues, and affords an excellent opportunity to explore their biological properties in greater detail, and to translate this information potentially to the field of oral regenerative medicine [22, 25]. However, before embarking on such an endeavor we must appreciate that monolayer cultures of cells are not accurate representations of tissue in vivo, primarily because they lack the types of interactions that cells have with other cells and with the extracellular matrix [40]. These interactions are known to influence gene expression as well as cell behavior. Organotypic

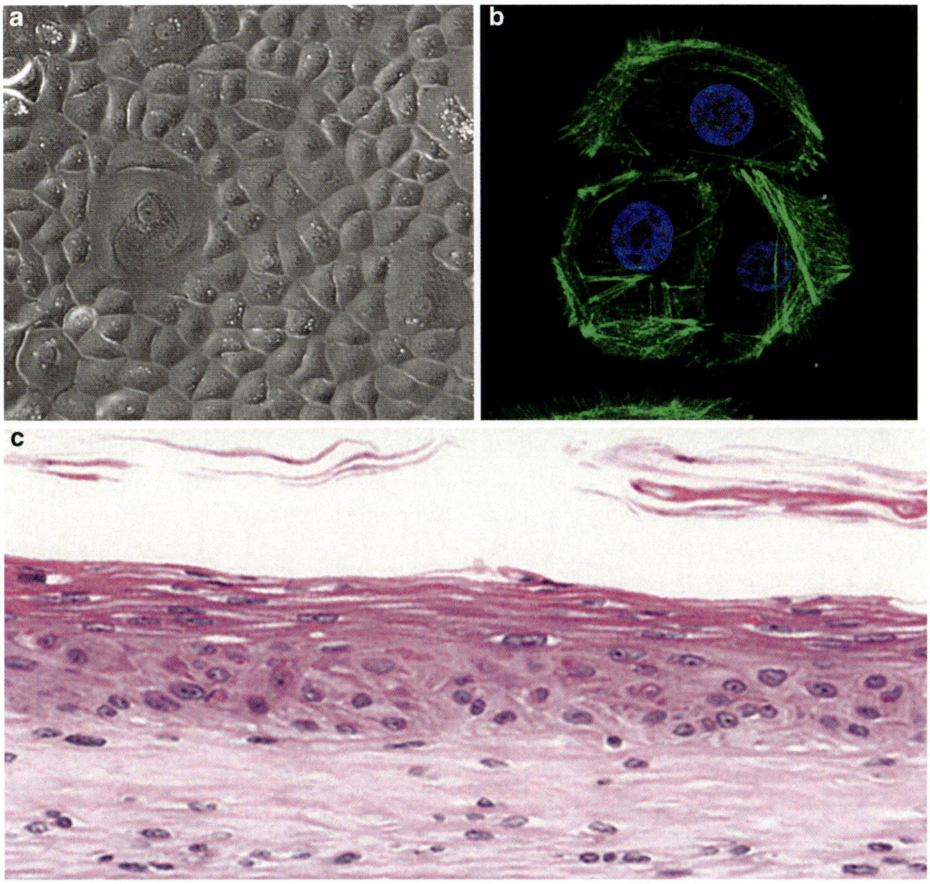

Fig. 3.3 (**a**) Bright field images of the primary culture of oral keratinocytes (40×). (**b**) Immunofluorescence staining of primary culture of oral keratinocytes showing cell epithelial cell morphology by Phalloiding-Alexa 488 actin staining and DAPI nuclear staining (63×). (**c**) Three-dimensional organotypic culture of primary normal oral keratinocytes and oral fibroblasts. H&E staining shows the different layers from the basal proliferative cells to the parakeratinized (differentiated) keratinocytes in the upper layers

three-dimensional (3D) cultures are tissue culture models that essentially mimic in vivo tissue architecture through manipulation of epithelial and stromal cells within and on top of an extracellular matrix [41]. It follows that 3D cultures can represent an ideal platform for studying oral epithelial cell biology, allowing the possibility of assessing multiple aspects of cellular behavior that might be intimately associated with carcinogens, such as tumor invasion which in essence the 3D system recapitulates the morphological features of human OSCC [42]. The 3D culture system essentially comprises of a collagen support that is embedded with human fibroblasts, which is then overlaid with isolated oral epithelial cells and after allowing for growth, these cultures are raised to achieve an air–liquid interface, and the exposed layer of oral epithelial cells soon after undergo organized differentiation and stratification, as seen in Fig. 3.3c, and thus better representing the normal mucosal epithelium [42]. However, a limitation associated with this method is the lack of standardization, whereby a large number of biopsies are usually needed to obtain sufficient number of primary keratinocytes; for efficient seeding of the collagen support, the cells from different donors may also compound differences in growth rates and life span. To circumvent this, the use of a single immortalized oral epithelial cell line would be deemed essential. Early attempts for immortalization involved the use of DNA virus such as SV40, but often the resulting cultures demonstrated chromosomal aneuploidy, and not ideal for use as research tool. Primary human oral keratinocytes have also been successfully immortalized with human papillomavirus type 16 [43, 44]. More recently, the observation that the expression of telomerase can maintain the telomeres at the ends of mammalian chromosomes has led to the use of exogenous human telomerase (hTERT) as an alternative way for the immortalization of oral keratinocytes [45]. Furthermore, using a modified method, the potential to grow sheets consisting of layers of oral keratinocytes for use in regenerating or reconstituting damaged tissues can be realized. This method essentially involves seeding the isolated epithelial cells onto a dermal matrix that is based on a fibrin glue gel containing live human fibroblast, and after 15 days in culture the cells grow into sheets consisting of three to four layers of cells with evidence of squamous differentiation displaying all the features of a stratified epithelium [46]. Conversely, the 3D cell culture system can be used to investigate the physiological roles of cancer-related genes and their validation, identified from high-throughput genomic analysis [14]. For instance, the epidermal growth factor receptor pathway substrate 8 gene (Eps8) was found to be overexpressed in OSCC, and by targeting this molecule for reduced expression, tumor cell invasion was inhibited in an organotypic (3D) model of oral cancer [47]. Another molecule that is frequently found to be up-regulated in OSCC is FOXM1, and to study the functional role of this transcription factor, normal oral epithelial cells with forced expression of this molecules were grown on a modified 3D cultures (organotypic epithelial tissue regeneration system), demonstrated loss of differentiation, while inducing hyperproliferatation [48, 49]. Noteworthy, while elevated mRNA levels of FOXM1 may be targeting genes involved in keratinocyte differentiation, for instance desmoglein 3 (dsg3), and recent data have suggested that reduced levels of dsg3 may serve as a marker of the stem/progenitor cell-like phenotype, suggesting that FOXM1 itself might play a role in early oncogenic pathways likely exposing the oral epithelial cells to tumorigenesis by expanding the stem/progenitor compartment and deregulating terminal differentiation [49].

The isolated human oral epithelial cells can be used for validating genes suspected in playing a role in OSCC development and progression [15]. As mentioned previously, squamous cell carcinomas originate from the proliferating basal layer of the epithelium, as a result of the accumulation of mutations in the genetic material caused by exposure to endogenous and exogenous genotoxic agents, or altered expression, in a limited set of genes. With recent technical advancements that now allow full genome sequencing with relative speed, are likely to play a pivotal role in our ability to survey genetic landscapes readily, and by comparing for instance tumor DNA with that of normal, we can potentially identify genes that may feature cancer-related characteristics such as somatic mutations, and potentially giving rise to a gain of function. Armed with this information, cell-based systems as well as animal models of cancers can be used for validation to assess whether these alterations are likely to play a role in driving the cancer. In this context, The International Cancer Genome Consortium (ICGC) has initiated to sequence and catalogue abnormalities in tumors in 50 different cancer types, and emerging information can potentially reveal oncogenic mutations causal for cancer progression as well as those that can identify, for instance, subsets that may influence cancer therapeutic responses and play an important role in personalized medicine [50]. A case study can be applied to prostate cancer, and using a whole-genome sequencing approach, whole segments of DNA were found to be altered, and with the loss of *Pten*, an important tumor suppressor gene involved in a pathway that regulates tumor growth by controlling cell growth and division, was identified as causal to development of lesions, but importantly, with minimal progression, and to better understand the molecular bases of this observation, additional transcriptome and pathway analysis comparing *Pten* ablated epithelium with corresponding normal, identified that levels of SMAD4, part of the TGF-β-BMP-SMAD4 signaling axis, cyclin D1 (CCND1) and secreted phosphoprotein 1 (SPP1) were instrumental in driving early stages of prostate cancer to progress to invade and metastasis in both human and mouse systems pertinent to prostate cancer [51]. In this context, the emerging information of gene signatures can now be translated to cell-based experimental systems for validation, to potentially reprogram normal prostate epithelia cells to undergo transformation, and convert to a fully malignant phenotype. Indeed, ongoing efforts in our laboratory are aimed at validating candidate genes causal to OSCC by overexpressing the altered gene products in isolated primary oral epithelial cultures, and by assessing and monitoring the specific contribution of each of the molecules to processes involved in transformation and malignant conversion. This strategy is being performed on oral epithelial cells transduced using lentiviral vectors for overexpression of target genes, either grown in 3D cultures as described earlier, as well as in vivo, whereby cultures of transduced oral epithelial cells can be grafted onto healthy recipient nude mice to study the ability of the gene(s) to cause tumors [52]. In this manner, functionally validated genes with specific roles in the development and progression of OSCC could potentially identify new drug targets that can be translated into effective treatment options, ultimately benefiting cancer patients.

Finally, oral epithelial cells hold great potential for use in the field of regenerative oral medicine [53]. As mentioned earlier, patients with oral cancer undergo very aggressive treatment that often results in damage to surrounding tissue usually with extensive and permanent tissue loss, resulting in severe facial disfigurement and permanent impairments in eating, swallowing, speaking, and a very poor quality of life [28]. In this regard, there is tremendous potential for generating induced pluripotent stem cells (iPS) from isolated oral, mature, and terminally differentiated epithelial cells, which may have use for regenerating damaged tissues in the oral cavity. The method for inducing pluripotency within a mature cell essentially requires global transcriptional and epigenetic reprogramming, as first reported by Takahashi et al. [54]. In this seminal study, the forced expression of four specific genes, Oct4, Sox2, Kif4, and c-Myc in mouse fibroblast, resulted in iPS cells that were functionally and molecularly indistinguishable from embryonic stem cells, possessing the capacity to self-renew and ability to give rise to all three germ layers [54, 55]. A worry though is that these genes required for reprogramming have oncogenic potential, and may induce tumors [56]. Thus, investigation has been ongoing to search for alternative methods for generating iPS cells, for instance, by using synthetic small molecules to alter transcription. To address this in part, the use of a defined mixture of small molecules, together with Oct4, was recently found to confer pluripotency in mouse embryonic fibroblasts [57]. A goal now remaining to make iPS cells is to replace Oct4 with an alternative molecule, and have a single chemical cocktail of defined small molecules capable of safe and efficient reprogramming. However, making and differentiating iPS cells is a very inefficient, and time-consuming procedure, with most protocols taking several weeks and a success rate at best, of 1 in 100 [58]. To address this in part, the use of mRNA transfection for reprogramming offers several advantages, key among these are that mRNAs are translated directly into proteins, and the small size allows them to be easily introduced into cells with precise control of the protein expression, sufficient to activate pluripotency target genes [58, 59]. We can further improve the reprogramming efficiency in part, by better understanding which of the pluripotency target genes is responsible for lineage specification, similar to how we now know, for instance, that the bone morphogenetic protein (BMP) can direct the reprogrammed cells to choose a mesodermal rather than an ectodermal pathway of differentiation [60]. Using this approach, ectopic expression of Oct4 in human dermal fibroblasts was found to activate hematopoietic transcription factors, giving rise to several different lineages of blood cells as assessed by the expression of the pan-leukocyte marker CD45, without the need for a middle pluripotent state, and the prospect of producing large quantities of human blood for use, for instance, in traumatic injuries [61]. Using a similar rational, isolated mature oral keratinocytes, may represent a promising source for iPS cell generation, particularly as these cultures also include cells with long-term self-renewal capacity, and experimental evidence suggests that the use of adult stem cells significantly increases iPS cell generation. Therefore, iPS cells generated from human oral keratinocytes, and pertinently from patients own cells isolated from normal tissue surrounding the tumor, hold a promising future for oral cancer patient survivors, whereby through induced differentiation of these cells regeneration of entire structures including salivary glands, a fully functional tongue comprising a mucosal epithelium including muscle, nerves, and taste receptors.

Acknowledgements This article was supported by the Intramural Research Program of the National Institutes of Health, National Institute of Dental and Craniofacial Research, project Z01DE00558.

References

1. Balogh K, Pantanowitz L (2007) Mouth, nose, and paranasal sinuses. In: Mills SE (ed) Histology for pathologists. Lipincott Williams and Wilkins, Philadelphia, pp 403–30
2. Strachan D, Avery JK (2001) Histology of the Oral Mucosa and Tonsils. In: Avery JK (ed) Oral Development and Histology. Thieme, New York, pp 243–62
3. Squier CA, Finkelstein MW (1989) Oral mucosa. In: Ten Cate AR (ed) Oral Histology, Development, Structure and Function. CV Mosby, St. Louis, pp 341–82
4. Swindle LD, Thomas SG, Freeman M, Delaney PM (2003) View of normal human skin in vivo as observed using fluorescent fiber-optic confocal microscopic imaging. J Invest Dermatol 121:706–12
5. Lindberg K, Rheinwald JG (1990) Three distinct keratinocyte subtypes identified in human oral epithelium by their patterns of keratin expression in culture and in xenografts. Differentiation 45:230–41
6. Nanci A, Ten Cate AR (2008) Ten Cate's oral histology: development, structure, and function, 7th edn. Mosby Elsevier, St. Louis, MO
7. Sawaf MH, Ouhayoun JP, Forest N (1991) Cytokeratin profiles in oral epithelial: a review and a new classification. J Biol Buccale 19:187–98
8. Wertz PW, Squier CA (1991) Cellular and molecular basis of barrier function in oral epithelium. Crit Rev Ther Drug Carrier Syst 8:237–69
9. Castilho RM, Squarize CH, Leelahavanichkul K, Zheng Y, Bugge T, Gutkind JS (2010) Rac1 is required for epithelial stem cell function during dermal and oral mucosal wound healing but not for tissue homeostasis in mice. PLoS One 5:e10503
10. Levy V, Lindon C, Zheng Y, Harfe BD, Morgan BA (2007) Epidermal stem cells arise from the hair follicle after wounding. FASEB J 21:1358–66
11. Ito M, Liu Y, Yang Z, Nguyen J, Liang F, Morris RJ, Cotsarelis G (2005) Stem cells in the hair follicle bulge contribute to wound repair but not to homeostasis of the epidermis. Nat Med 11:1351–54
12. Langton AK, Herrick SE, Headon DJ (2008) An extended epidermal response heals cutaneous wounds in the absence of a hair follicle stem cell contribution. J Invest Dermatol 128:1311–18
13. Debnath J, Brugge JS (2005) Modelling glandular epithelial cancers in three-dimensional cultures. Nat Rev Cancer 5:675–88
14. Ridky TW, Chow JM, Wong DJ, Khavari PA (2010) Invasive three-dimensional organotypic neoplasia from multiple normal human epithelia. Nat Med 16:1450–55
15. Ha PK, Benoit NE, Yochem R, Sciubba J, Zahurak M, Sidransky D, Pevsner J, Westra WH, Califano J (2003) A transcriptional progression model for head and neck cancer. Clin Cancer Res 9:3058–64
16. Mao L, Hong WK, Papadimitrakopoulou VA (2004) Focus on head and neck cancer. Cancer Cell 5:311–16
17. Scully C, Bagan J (2009) Oral squamous cell carcinoma: overview of current understanding of aetiopathogenesis and clinical implications. Oral Dis 15:388–99
18. Shi Z, Stack MS (2010) Molecules of cell adhesion and extracellular matrix proteolysis in oral squamous cell carcinoma. Histol Histopathol 25:917–32
19. Henley JD, Summerlin DJ (2006) Head and Neck. In: Cheng L, Bostwick DG (eds) Essentials of Anatomic Pathology. Humana, Totowa, NJ, pp 795–831
20. Jemal A, Siegel R, Xu J, Ward E (2010) Cancer statistics, 2010. CA Cancer J Clin 60:277–300
21. Sciubba JJ (2001) Oral cancer. The importance of early diagnosis and treatment. Am J Clin Dermatol 2:239–51
22. Izumi K, Tobita T, Feinberg SE (2007) Isolation of human oral keratinocyte progenitor/stem cells. J Dent Res 86:341–46
23. Calenic B, Ishkitiev N, Yaegaki K, Imai T, Kumazawa Y, Nasu M, Hirata T (2010) Magnetic separation and characterization of keratinocyte stem cells from human gingiva. J Periodontal Res 45:703–08
24. Chiou SH, Yu CC, Huang CY, Lin SC, Liu CJ, Tsai TH, Chou SH, Chien CS, Ku HH, Lo JF (2008)

Positive correlations of Oct-4 and Nanog in oral cancer stem-like cells and high-grade oral squamous cell carcinoma. Clin Cancer Res 14:4085–95
25. Izumi K, Inoki K, Fujimori Y, Marcelo CL, Feinberg SE (2009) Pharmacological retention of oral mucosa progenitor/stem cells. J Dent Res 88:1113–18
26. Blanpain C, Fuchs E (2009) Epidermal homeostasis: a balancing act of stem cells in the skin. Nat Rev Mol Cell Biol 10:207–17
27. Vissink A, Mitchell JB, Baum BJ, Limesand KH, Jensen SB, Fox PC, Elting LS, Langendijk JA, Coppes RP, Reyland ME (2010) Clinical management of salivary gland hypofunction and xerostomia in head-and-neck cancer patients: successes and barriers. Int J Radiat Oncol Biol Phys 78:983–91
28. Mosel D, Bauer R, Lynch D, Hwang S (2011) Oral complications in the treatment of cancer patients. Oral Dis. 2011 Jan 7. doi:10.1111/j.1601-0825.2011.01788.x. [Epub ahead of print]
29. Coppes RP, van der Goot A, Lombaert IM (2009) Stem cell therapy to reduce radiation-induced normal tissue damage. Semin Radiat Oncol 19:112–21
30. Rheinwald JG, Green H (1977) Epidermal growth factor and the multiplication of cultured human epidermal keratinocytes. Nature 265:421–24
31. Rheinwald JG, Green H (1975) Serial cultivation of strains of human epidermal keratinocytes: the formation of keratinizing colonies from single cells. Cell 6: 331–43
32. Parkinson KE, Yeudall WA (2002) The Epidermis. In: Freshney RI, Freshney MG (eds) Culture of epithelial cells, 2nd edn. Wiley-Liss, New York, pp 66–94
33. Mlinek A, Buchner A (1975) In vitro cultivation of adult human gingiva I. primary epithelial outgrowth from gingival explants. J Periodontal Res 10:73–8
34. Neiders ME, Weiss L (1970) Isolation of viable epithelial cells from human gingiva. J Periodontol 4:325–32
35. Prime SS, Nixon SV, Crane IJ, Stone A, Matthews JB, Maitland NJ, Remnant L, Powell SK, Game SM, Scully C (1990) The behaviour of human oral squamous cell carcinoma in cell culture. J Pathol 160:259–69
36. Martin MJ, Muotri A, Gage F, Varki A (2005) Human embryonic stem cells express an immunogenic non-human sialic acid. Nat Med 11:228–32
37. Kedjarune U, Pongprerachok S, Arpornmaeklong P, Ungkusonmongkhon K (2001) Culturing primary human gingival epithelial cells: comparison of two isolation techniques. J Craniomaxillofac Surg 29: 224–31
38. Oda D, Watson E (1990) Human oral epithelial cell culture I. Improved conditions for reproducible culture in serum-free medium. In Vitro Cell Dev Biol 26:589–95
39. Higa K, Shimazaki J (2008) Recent advances in cultivated epithelial transplantation. Cornea 27:S41–7
40. Hansson A, Bloor BK, Haig Y, Morgan PR, Ekstrand J, Grafström RC (2001) Expression of keratins in normal, immortalized and malignant oral epithelia in organotypic. Oral Oncol 37:419–30
41. Mackenzie IC (2004) Growth of malignant oral epithelial stem cells after seeding into organotypical cultures of normal mucosa. J Oral Pathol Med 33:71–8
42. Dongari-Bagtzoglou A, Kashleva H (2006) Development of a highly reproducible three-dimensional organotypic model of the oral mucosa. Nat Protoc 1:2012–18
43. Park NH, Min BM, Li SL, Huang MZ, Cherick HM, Doniger J (1991) Immortalization of normal human oral keratinocytes with type 16 human papillomavirus. Carcinogenesis 12:1627–31
44. Jeon GA, Lee JS, Patel V, Gutkind JS, Thorgeirsson SS, Kim EC, Chu IS, Amornphimoltham P, Park MH (2004) Global gene expression profiles of human head and neck squamous carcinoma cell lines. Int J Cancer 112:249–58
45. Dickson MA, Hahn WC, Ino Y, Ronfard V, Wu JY, Weinberg RA, Louis DN, Li FP, Rheinwald JG (2000) Human keratinocytes that express hTERT and also bypass a p16(INK4a)-enforced mechanism that limits life span become immortal yet retain normal growth and differentiation characteristics. Mol Cell Biol 20:1436–47
46. Kamolz LP, Luegmair M, Wick N, Eisenbock B, Burjak S, Koller R, Meissl G, Frey M (2005) The Viennese culture method: cultured human epithelium obtained on a dermal matrix based on fibroblast containing fibrin glue gels. Burns 31:25–9
47. Yap LF, Jenei V, Robinson CM, Moutasim K, Benn TM, Threadgold SP, Lopes V, Wei W, Thomas GJ, Paterson IC (2009) Upregulation of Eps8 in oral squamous cell carcinoma promotes cell migration and invasion through integrin-dependent Rac1 activation. Oncogene 28:2524–34
48. Gemenetzidis E, Elena-Costea D, Parkinson EK, Waseem A, Wan H, Teh MT (2010) Induction of human epithelial stem/progenitor expansion by FOXM1. Cancer Res 70:9515–26
49. Wan H, Yuan M, Simpson C, Allen K, Gavins FN, Ikram MS, Basu S, Baksh N, O'Toole EA, Hart IR (2007) Stem/progenitor cell-like properties of desmoglein 3dim cells in primary and immortalized keratinocyte lines. Stem Cells 25:1286–97
50. The International Cancer Genome Consortium (2010) International network of cancer genome projects. Nature 464:93–998
51. Ding Z, Wu CJ, Chu GC, Xiao Y, Ho D, Zhang J, Perry SR, Labrot ES, Wu X, Lis R, Hoshida Y, Hiller D, Hu B, Jiang S, Zheng H, Stegh AH, Scott KL, Signoretti S, Bardeesy N, Wang YA, Hill DE, Golub TR, Stampfer MJ, Wong WH, Loda M, Mucci L, Chin L, DePinho RA (2011) SMAD4-dependent barrier constrains prostate cancer growth and metastatic progression. Nature 470:269–73
52. Lichti U, Anders J, Yuspa SH (2008) Isolation and short-term culture of primary keratinocytes, hair follicle populations and dermal cells from newborn mice and keratinocytes from adult mice for in vitro analysis and for grafting to immunodeficient mice. Nat Protoc 3:799–810

53. Shortt AJ, Tuft SJ, Daniels JT (2010) Ex vivo cultured limbal epithelial transplantation. A clinical perspective. Ocul Surf 8:80–90
54. Takahashi K, Yamanaka S (2006) Induction of pluripotent stem cells from mouse embryonic and adult fibroblast cultures by defined factors. Cell 126:663–76
55. Okita K, Ichisaka T, Yamanaka S (2007) Generation of germline-competent induced pluripotent stem cells. Nature 448:313–17
56. Okita K, Yamanaka S (2010) Induction of pluripotency by defined factors. Exp Cell Res 316:2565–70
57. Yuan X, Wan H, Zhao X, Zhu S, Zhou Q, Ding S (2011) Combined chemical treatment enables Oct4-induced reprogramming from mouse embryonic fibroblasts. Stem Cells 29:549–553
58. Warren L, Manos PD, Ahfeldt T, Loh YH, Li H, Lau F, Ebina W, Mandal PK, Smith ZD, Meissner A, Daley GQ, Brack AS, Collins JJ, Cowan C, Schlaeger TM, Rossi DJ (2010) Highly efficient reprogramming to pluripotency and directed differentiation of human cells with synthetic modified mRNA. Cell Stem Cell 7:618–30
59. Plews JR, Li J, Jones M, Moore HD, Mason C, Andrews PW, Na J (2010) Activation of pluripotency genes in human fibroblast cells by a novel mRNA based approach. PLoS One 5:e14397
60. Doss MX, Chen S, Winkler J, Hippler-Altenburg R, Odenthal M, Wickenhauser C, Balaraman S, Schulz H, Hummel O, Hübner N, Ghosh-Choudhury N, Sotiriadou I, Hescheler J, Sachinidis A (2007) Transcriptomic and phenotypic analysis of murine embryonic stem cell derived BMP2+ lineage cells: an insight into mesodermal patterning. Genome Biol 8:R184
61. Szabo E, Rampalli S, Risueño RM, Schnerch A, Mitchell R, Fiebig-Comyn A, Levadoux-Martin M, Bhatia M (2010) Direct conversion of human fibroblasts to multilineage blood progenitors. Nature 468:521–26

Heterotopic Ossification Following Musculoskeletal Trauma: Modeling Stem and Progenitor Cells in Their Microenvironment

4

Youngmi Ji, Gregory T. Christopherson, Matthew W. Kluk, Orna Amrani, Wesley M. Jackson, and Leon J. Nesti

Abstract

Heterotopic ossification (HO), characterized by the formation of mature bone in the soft tissues, is a complication that can accompany musculoskeletal injury, and it is a frequent occurrence within the military population that has experienced orthopaedic combat trauma. The etiology of this disease is largely unknown. Our laboratory has developed strategies to investigate the cellular and molecular events leading to HO using clinical specimens that were obtained during irrigation and debridement of musculoskeletal injuries. Our approach enables to study (1) the cell types that are responsible for pathological transformation and ossification, (2) the cell- and tissue-level signaling that induces the pathologic transformation, and (3) the effect of extracellular matrix topography and force transduction on HO progression. In this review, we will report on our findings in each of these aspects of HO etiology and describe our efforts to recapitulate our findings in an animal model for traumatic HO.

4.1 Heterotopic Ossification: Clinical Prevalence and Military Relevance

Heterotopic ossification (HO) is a musculoskeletal disease that is characterized by the formation of mature bone in soft tissues, such as muscle tendon or fascia. It is a frequent complication that occurs following trauma orthopaedic that accompanies an injury or may be an iatrogenic artifact, as occurs during total hip arthroplasty. HO is prevalent in patients with severe time-of-war extremity wounds, and the incidence of HO increases by 57% in patients that sustain a polytraumatic blast injury [1].

L.J. Nesti (✉)
Clinical and Experimental Orthopaedics Group, National Institutes of Arthritis and Musculoskeletal and Skin Diseases, National Institutes of Health, Building 50, Room 1525, 50 South Drive, MSC 8022, Bethesda, MD 20892-8022, USA

Integrated Department of Orthopaedics and Rehabilitation, Walter Reed Army Medical and National Naval Medical Center, Bethesda, MD, USA
e-mail: nestil@mail.nih.gov

J.S. Rhim and R. Kremer (eds.), *Human Cell Transformation: Role of Stem Cells and the Microenvironment*, Advances in Experimental Medicine and Biology 720, DOI 10.1007/978-1-4614-0254-1_4, © Springer Science+Business Media, LLC 2011

This disease has emerged as a challenging consequence of battlefield injury that confounds the military's orthopaedic surgeons, as their efforts to return the injured soldiers to active duty are betrayed by their patients' own cells' pathological transformation into osteoprogenitors. This condition presents a host of problems, including skin and soft tissue breakdown over the residual limb, suboptimal prosthetic fitting, and impaired limb function [2]. Complicating this situation is the degree of difficulty in removing heterotopic bone after it has formed. This procedure requires a significant amount of soft tissue resection and risks injury to neurovascular structures that are often intimately associated with HO. As a result, there is considerable interest in understanding the cellular mechanism of HO and to develop strategies of addressing this condition before the formation of bone can occur.

HO is not limited to time-of-war injuries in the military population. Traumatic HO also occurs in the civilian population after polytraumatic injuries, especially when the polytrauma is accompanied by neurologic injury. The incidence of HO following severe blunt trauma is approximately 42.7% [3]. When traumatic brain injury (TBI) occurs in the absence of extremity trauma, HO formation in the extremities has been reported in approximately 25% of patients [4]. HO also occurs in approximately 1.2% of burn patients [5]. Thus the findings from the proposed study are also relevant to civilian trauma to better predict when and where HO is likely to occur. Furthermore, with a model that can be used to identify and analyze the action of factors that initiate HO, it may be possible to design targeted therapies designed to disrupt the osteogenic process and prevent ectopic bone formation.

4.1.1 Current Treatment Options for HO Are Inadequate to Prevent or even Mitigate the Disease

Current treatment of HO in traumatically injured patients has significant limitations and serious side effects. As such, treatment modalities cannot be initiated without careful consideration and should be limited to those patients who are at the highest risk of developing HO; however, it is almost impossible to predict which patients will develop HO. Early diagnostic tools such as serum alkaline phosphatase levels, bone scans and computerized tomography (CT) scans are often impractical and non-specific. Therefore, diagnosis typically is made late in the disease process by physical examination and routine radiographic evaluation. At this late stage, options include symptomatic treatment or surgical resection.

The two main prophylactic treatments involve systemic medical therapy or localized radiation therapy (XRT) and both modalities have significant drawbacks preventing their widespread use. The mainstay of medical therapy uses nonsteroidal anti-inflammatory drugs (NSAID) to attenuate the initial inflammatory response and prevent extracellular matrix (ECM) mineralization [3]. This method effectively reduces the incidence of HO formation, but gastrointestinal side effects and the negative impact on wound and fracture healing limit their usefulness in multi-trauma patients [1]. XRT is thought to inhibit proliferation and differentiation of cells that may contribute to the pathogenesis of HO. It is a localized method of controlling HO formation, but it requires the treating physician to predict the area of HO formation, may impede the wound healing process, and carries a risk of postradiation cancer. Therefore, XRT is most effective in atraumatic wounds without soft tissue compromise and in areas of predictable HO formation. Alternative methods of preventing HO formation include treatment with coumadin, calcitonin, bisphosphonates and free radical scavengers (i.e., *N*-acetylcystine and allopurinol) [6], but the effectiveness of these treatments has been anecdotal with few clinical reports to support their efficacy following orthopaedic trauma. The development of more effective prophylactic treatments for HO will require a more complete understanding of the cellular mechanism that contributes to this disease.

4.2 A Multi-step Approach to Investigating Traumatic HO Etiology

Fibrodysplasia ossificans progressiva (FOP) is a rare genetic disorder that is characterized by the formation of HO. FOP provides the best-characterized model of HO disease progression and demonstrates the complex nature of HO etiology. Similar to traumatic HO, patients suffering from FOP generate a pathological soft tissue environment that is inductive to bone formation. FOP has been linked to a mutation in the bone morphogenetic protein (BMP) receptor ACVR-1 (activin A receptor, type 1) [7]. The mutated ACVR-1 becomes constitutively activated in response to minor trauma, and this has been correlated with upregulation of BMP-4 [8] and inappropriate tissue-level regulation of BMP antagonists, including noggin and gremlin [9]. A fibroproliferative lesion forms over time and generates an environment where osteoinduction factors can be concentrated in the vicinity of progenitor cells that possess the potential to differentiate and form ectopic bone [10]. This example demonstrates how dysregulated cytokine signaling can result in pathological wound healing, although it is not clear to what extent any FOP-related mechanisms occur in muscle tissue following musculoskeletal injury.

FOP and traumatic HO are also similar in that these clinical conditions impose weighty limitations on the investigators who try to study the etiologies of these diseases. In the case of FOP, minor damage to the mesenchymal tissue leads to formation of ossified tissue, and it is therefore almost impossible to obtain soft tissue samples, either before or after the formation of HO. Causative factors related to the disease must be inferred from blood samples, and in some cases, skin biopsies. Given these limitations, FOP investigators have made substantial headway in predicting the disease mechanism by identifying genetic mutations, recapitulating these mutations in animal models and validating these observations against the available clinical data. In the case of traumatic HO, tissue availability is not a problem, as there is ample available tissue that can be collected when the injuries are cleaned and debrided prior to definitive closure of the wound. However, the investigators are limited by the timing of tissue collection, as it occurs during a prescribed window based on the clinical needs of the patient, usually around 7–21 days following the initial injury. A definitive diagnosis of traumatic HO can also lag substantially behind the tissue collection period by as much as 9–15 months, thereby making it impossible at the time of collection to distinguish between tissues from patients that develop HO from those who do not. The study of traumatic HO is also hindered by the high degree of variability between the available patients, such as patient age, mechanism of injury, severity of injury, and anatomical location of the injury.

Given these limitations, our laboratory has developed strategies to investigate the etiology of traumatic HO from multiple perspectives. We can broadly classify our research efforts into three primary areas:

1. Identify the cellular basis of HO formation
2. Characterize the biochemical promoters of HO
3. Investigate the role of matrix mechanics and topography in HO

These strategies are designed to take advantage of the core competencies of our team, which include clinical investigation, cell and molecular biology, and bioengineering. Mirroring the process that has been employed to study FOP, we can infer the etiology of traumatic HO by evaluating potentially pathological deviations that occur between traumatized and untraumatized muscle at the cell and tissue level, and then we can attempt to recapitulate the environments *in vitro* to assess their ability to induce an environment that can promote ectopic osteogenesis. Our eventual goal is to use our findings from these research areas to develop an animal model of traumatic HO that can recapitulate the cellular and molecular factors that initiate HO in our patient population.

4.3 Multipotent Progenitor Cells in Muscle Following Orthopaedic Trauma

Our initial priority at the beginning of this project was to identify the cell types that participated in HO. Working in collaboration with investigators at the Walter Reed Army Medical Center (WRAMC) and the National Institutes of Health (NIH), we identified a population of mesenchymal progenitor cells (MPCs) from within the war-traumatized muscle tissue on the basis of adherence selection. The morphology and cell surface epitope profile of MPCs are similar to those of bone marrow-derived mesenchymal stem cells (MSCs), which are the resident osteoprogenitor cells from the marrow stroma [11, 12]. We also verified that the traumatized muscle-derived MPCs are capable of giving rise to colony-forming-unit-fibroblasts (CFU-Fs), an indicator of a clonogenic, multipotent cell population. The MPCs can also undergo differentiation to become osteoblasts, adipocytes, or chondrocytes (Fig. 4.1).

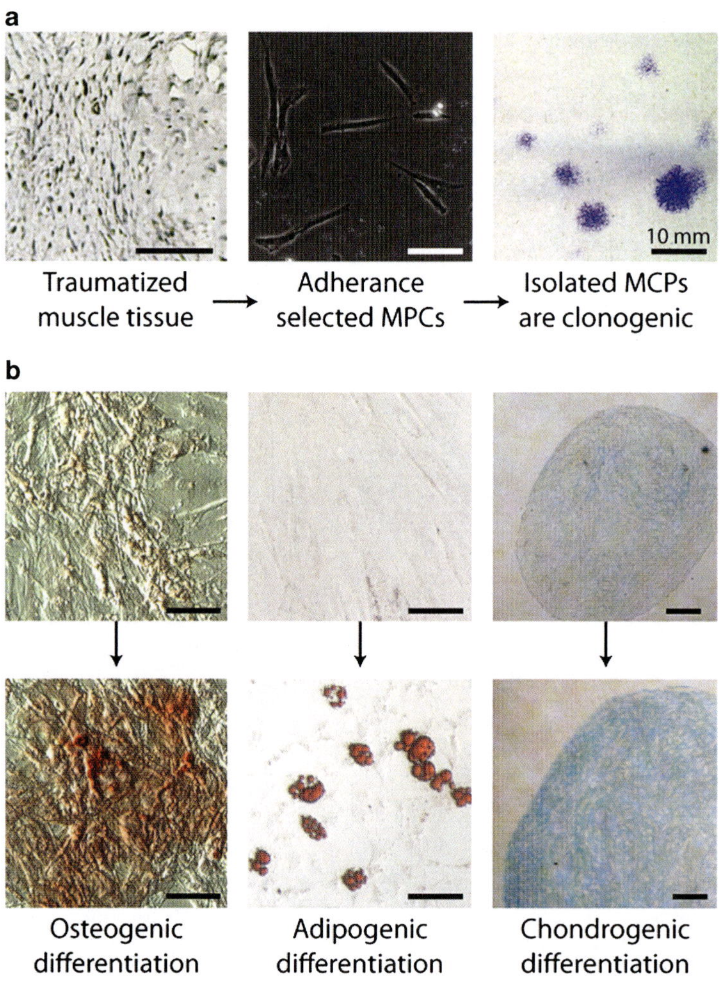

Fig. 4.1 (**a**) We have identified the MPCs as a population of CD105-positive cells (*brown*) in traumatized muscle tissue that can be isolated by adherence selection to tissue culture plastic. The fibroblastic cells are clonogenic, and some of the colonies are ALP-positive, which is an indicator of a multipotent cell population. (**b**) Monolayers of MPCs can be induced to undergo osteogenic differentiation (alizarin red indicates mineralize matrix) and adipogenesis (*oil red O* indicates lipid vacuoles). Pelleted MPCs can be induced to undergo chondrogenesis (*alcian blue* indicates sulfated glycosaminoglycans). Scale bars = 100 μm, except where noted

4.3.1 MPCs Are an Inherent Property of Muscle Tissue During Wound Healing

Based on the apparent homogeneity of MPCs harvested from different patients, we have concluded that the presence of MSC-like cell type is an inherent property of traumatized tissue [13]. The MPCs are immunohistochemically detected as a population of CD105-positive cells localized within the damaged muscle fibers. It has been hypothesized that MPCs are nonmuscle progenitor cells that participate in the process of muscle regeneration by facilitating the reparative function of myoblasts and myofibroblasts [14]. Although multipotent stem cells have previously been isolated from untraumatized muscle using immuno-selective techniques [15], the plastic-adherent MPCs isolated from traumatized muscle can be harvested in significantly higher numbers and may be the descendants of those stem cells that have been activated within their niche by injury-derived signals and have begun to proliferate within the tissue. It has also been hypothesized that pericytes, which occupy a perivascular niche *in vivo*, are the cells that exhibit an MSC phenotype *in vitro* [16]; thus the MPCs might also be activated pericytes that have entered the wound bed. These scenarios provide further justification that the migration of MPCs into the traumatized muscle is a part of the normal wound healing response.

4.3.2 MPCs from Traumatized Muscle Tissue Exhibit Regenerative Functions

We have recently initiated a new line of investigation to study the "trophic" behaviors of the MPCs. A new theory has emerged in the last 3–4 years on the mechanism by which bone marrow-derived MSCs render at least part of their regenerative benefit by secreting trophic factors, i.e., factors that promote the growth, survival, and differentiation of surrounding cells (reviewed in [17]). Our recent findings indicate that traumatized muscle-derived MPCs also exhibit several trophic behaviors, which could be applied in regenerative applications [18]. The timely attenuation of the local inflammatory response following injury is critical to enable the body to regenerate itself [16]. Although inflammation is an important part of the wound healing process, a traumatic injury may initiate an over-exuberant response in the wound bed, particularly given that the patient is also likely being treated with antibiotics to control the risk of infection. The endogenous wound healing/tissue regeneration mechanism is compromised in a highly inflammatory environment, and therefore functional healing is unlikely to occur. Instead, the body will respond by sealing off the wound with fibrotic scar tissue, which will act as a physical barrier that prevents functional tissue regeneration [19]. MPCs express genes for anti-inflammatory and immunosuppressive cytokines (e.g., *IL-6*, *IL-10*, and hepatocyte growth factor (*HGF*)) that can limit the amount of inflammation in the tissue. We have performed functional studies showing that the factors secreted by MPCs can attenuate the proliferation of T cells, an important cellular mediator of inflammation, using a mixed lymphocyte reaction. We have also determined that MPCs express high levels of the TGF-β3 gene, which may prevent the formation of fibrotic scar tissue.

MPCs also express factors that are specific to angiogenesis, the process of new blood vessel formation. They express genes associated with vascular maintenance, including vascular endothelial growth factor A, and we have verified the protein level expression of this gene in a Western blot assay. We have also verified that the factors secreted by the MPCs promote the proliferation of vascular endothelial cells. Interestingly, these factors do not lead to tubulogenesis of the endothelial cells into primitive vessels as do the factors secreted by bone marrow-derived MSCs. This difference in function between the two cell types suggests that while the MPCs promote an increase in the building blocks that are required to generate blood vessels, they are not the cell type responsible for signaling the final steps that are required for angiogenesis. Improving blood flow into the region of tissue regeneration should also limit the formation of scar tissue [16].

Given the recently discovered trophic functions of traumatized muscle-derived MPCs, we believe that this cell type will be effective for a variety of regenerative medicine applications. In all of the experiments we have performed, we compared the trophic effects of the MPCs to that of bone marrow-derived MSCs, and we found that many of the effects were similar and, in fact, more pronounced in the latter. However, the MPCs are present at substantially greater numbers at the time of harvest compared to bone marrow-derived MSCs. Thus, while on a per cell basis, the MPCs may elicit a lower response from the surrounding cells, but this will be made up for by delivering approximately 4,000 more MPCs at the time of surgery than would be possible for MSCs [20]. Given the immense tropic potential of the MPCs harvested from traumatized muscle, and their ready availability from otherwise surgical waste materials, their use as a cellular therapy to limit scar fibrosis would be an obvious application for these cells.

4.3.3 MPCs as Putative Osteoprogenitors in HO

Our laboratory has also rigorously characterized the osteogenic potential of these cells [21]. We have demonstrated that the traumatized muscle-derived MPCs are capable of forming ALP-positive colonies, which is an indicator of a cell population with osteogenic potential [13]. After osteogenic induction, the MPCs increased their ALP activity and began to generate a mineralized matrix, and the osteogenic gene expression profile of the MPCs was characteristic of early differentiation into osteoblasts. Taken together, our findings indicate that traumatized muscle-derived MPCs are the putative osteoprogenitor cells responsible for HO following traumatic injury. We believe that MPCs present in the wound have a primary function to facilitate the repair and regeneration of the injured tissues. However, under pathological conditions that are typical of high-energy trauma, the MPCs become dysregulated by the biochemical signaling in the local environment, which initiates their osteogenic programming. We aim to identify the pathological conditions that stimulate this cellular transformation.

4.4 Factors Expressed Post-trauma Influence Progenitor Cell Function

Wound healing is a dynamic and interactive process involving multiple cell populations, ECM and the soluble mediators such as growth factors and cytokines in order to replace damaged tissues with viable, functional tissue. This sequential process has three phases (inflammatory phase, proliferative phase, maturation and remodeling phase with scar tissue formation), involves a variety of biochemical and physical interactions that must occur in a prescribed sequence between diverse immunological and mesenchymal tissue systems [22, 23]. Within few hours of the initial inflammatory phase, neutrophils and macrophages move to the injured area and create a chemotaxic gradient to modulate the local inflammatory response and to direct the movement of reparative cells and fluid exudate [24]. During the proliferative phase, fibroblast and endothelial cells migrate into the damaged tissue from surrounding tissue to proliferate, repair the connective tissue, and construct vasculature to restore blood circulation. The macrophages continue to provide a source of growth factors necessary to stimulate fibroplasia and angiogenesis; the fibroblasts produce the new ECM necessary to support myoblast ingrowth; and endothelial cells generate blood vessels to carry oxygen and nutrients necessary and sustain cell metabolism [23]. Dysregulation of these processes can result in pathological cellular differentiation, which results in nonfunctional or inappropriate tissue production, as occurs during HO.

4.4.1 Evaluating the Gene Expression Profile of HO

Our laboratory has previously investigated the tissue-level gene expression of cytokine and growth factors associated with muscle regeneration [25]. In this study, we compared the cytokine gene expression profile in muscle tissue that was isolated from the wound margins of patients approximately 14 days following their initial injury to normal, uninjured muscle biopsies. Our findings indicated that muscle trauma resulted in a uniform upregulation of cytokines associated with fibrosis (i.e., *TGFB1*) and osteogenic induction (i.e., *BMP1*), and defined a novel cytokine gene expression profile to characterize severely injured muscle tissue. The upregulated expression of these cytokines was verified and localized in the traumatized tissue relative to the resident CD105-positive MPCs. Taken together, the results of this study describe the cytokine signaling profile of traumatized muscle tissue, which includes a set of cytokines that are variably upregulated in response to the trauma and should be investigated further to determine whether they have a role in the formation of traumatic HO.

Upon further inspection of our immunohistochemistry data, we observed that *TGFB1* was upregulated uniformly in all of the traumatized tissue samples. The *TFGB1* was localized in the regions surrounding fibrotic and fibroproliferative lesions, which has been assumed to be initiation sites of the HO, as occurs in FOP. By comparison, *TGFB3* was variably regulated in the traumatized tissue. We found that high expression levels of *TGFB3* correlated significantly with tissues that contained little histological evidence of fibroproliferative lesions, and the ratio of *TGFB1/TGFB3* was significantly upregulated in the tissue with fibrotic lesions (Fig. 4.2). These findings were consistent with the literature suggesting that the ratio of *TGFB1/TGFB3* is involved in regulating the balance between normal tissue regeneration and fibrosis [26, 27].

We have further investigated the tissue-level gene expression that is associated specifically with HO formation. By comparing the results from patients that developed HO to those without symptoms of HO, it was possible to identify factors associated with dysregulated wound healing. We observed that genes associated with both osteogenesis and odontogenesis (i.e., *BMP6* and *MSX1*) were expressed at higher levels in the traumatized tissues that developed HO (Fig. 4.3). We are cautious not to conclude that these factors initiate HO, as BMP-6 is upregulated in osteoprogenitors at early stages of osteogenic differentiation. Therefore, we are currently evaluating whether the expression of osteogenic genes is a dysregulated response to trauma that may initiate HO or an early indicator that the tissue repair process has gone awry.

4.4.2 Epigenetic Regulation of HO

TGFB3, or more specifically, the ratio of *TGFB1/TGFB3* appears to be a predictor of fibrosis in the tissue following trauma. We are currently

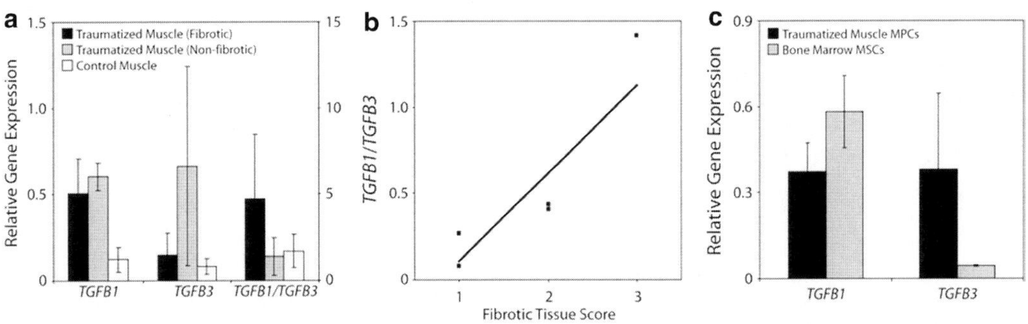

Fig. 4.2 TGF-beta gene expression profiles in traumatized muscle. (**a**) Relative gene expression (*left axis*) of *TGFB1* in traumatized muscle was not dependent on whether the tissue was fibrotic, but *TGFB3* was more highly expressed in nonfibrotic tissue. The ratio of *TGFB1/TGFB3* (*right axis*) was higher in the fibrotic tissue ($n>3$). (**b**) The severity of fibrosis (scored with 3 being most fibrotic) could be correlated to the ratio of *TGFB1/TGFB3* gene expression in traumatized muscle tissue ($n=7$). (**c**) Relative gene expression of *TGFB3* is substantially greater in MPCs than in bone marrow-derived MSCs ($n=3$)

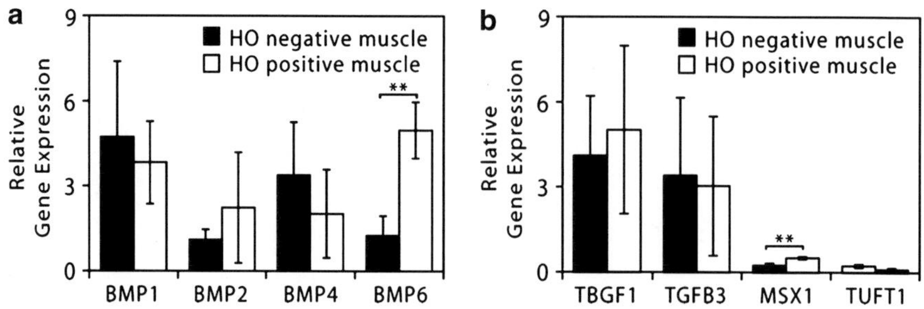

Fig. 4.3 Cytokine and osteogenesis gene expression profile of traumatized HO positive and negative muscle. (**a**) Relative gene expression from cytokine gene expression profile. (**b**) Relative gene expression from osteogenesis gene expression profile. $^{**}p<0.05$, Student's t-tests with $n=3$

applying whole genome techniques, including RNA-Seq, micro-RNA profiling and pathway analysis to determine whether the gene expression associated with HO may be initiated by more primitive mechanisms of cellular regulation that alter the epigenetic state of the MPCs and/or other cells involved in the regeneration of the traumatized muscle tissue. These analyses have allowed us to identify several novel pathways associated with HO, including the epithelial to mesenchymal transition pathway and the NF-κB signaling pathway, which will be useful in reconstructing the complex etiology of this disease.

4.5 Mechanical and Topographical Control of Cellular Function

Another key factor influencing stem/progenitor behavior is their microenvironment. Dr. Ray Schofield first postulated the idea that stem cells are regulated by a collection of cellular and acelluar components in specialized "niches" in 1978 [28], and this complex array of biochemical and physical cues are spatially and temporally coordinated in such a way as to control cell fate specification. Modern interpretation of the niche retains this view of the microenvironment directing cell proliferation, differentiation, migration and survival [29], and further derives the cell–matrix interaction into physical and soluble signals, working synergistically to steer cellular change [30].

The physical cues of the three-dimensional microenvironment are principally associated with the ECM. From a bioengineering perspective, the primary forms of physical cues in the ECM are topography and mechanical forces [31]: topography can be thought of as the spatial arrangement of molecular signals and proteins; mechanical forces are typically a combination of matrix elasticity and shear forces from flowing fluids. These physical cues serve to mediate cell attachment, cytoskeletal rearrangement and extracellular protein interactions, modulating the downstream cell signaling that determines cell behavior. The inherent complexity of the stem cell niche requires simplification of the physical and biochemical aspects for *in vitro* study [32], but does not lessen the value gained from utilizing biomaterials to dictate cellular response, both as a tool for guiding cell behavior and understanding mechanisms governing cell control. Since this cellular response is nonuniform across tissue types [33], to best understand a particular cell type and its corresponding niche, tissue engineers are best served to examine and replicate a specific microenvironment of interest [34].

In the case of musculoskeletal trauma and HO, MPCs are of keen interest, due to their wound proximity and role in healing (whether trophic or regenerative-based). MPCs are isolated in high concentrations from debrided muscle tissue, making this tissue architecture of high interest. Electron microscopic examination of surgically debrided muscle tissue reveals a prevalence of nanofibrous topography (Fig. 4.4a), likely a key physical

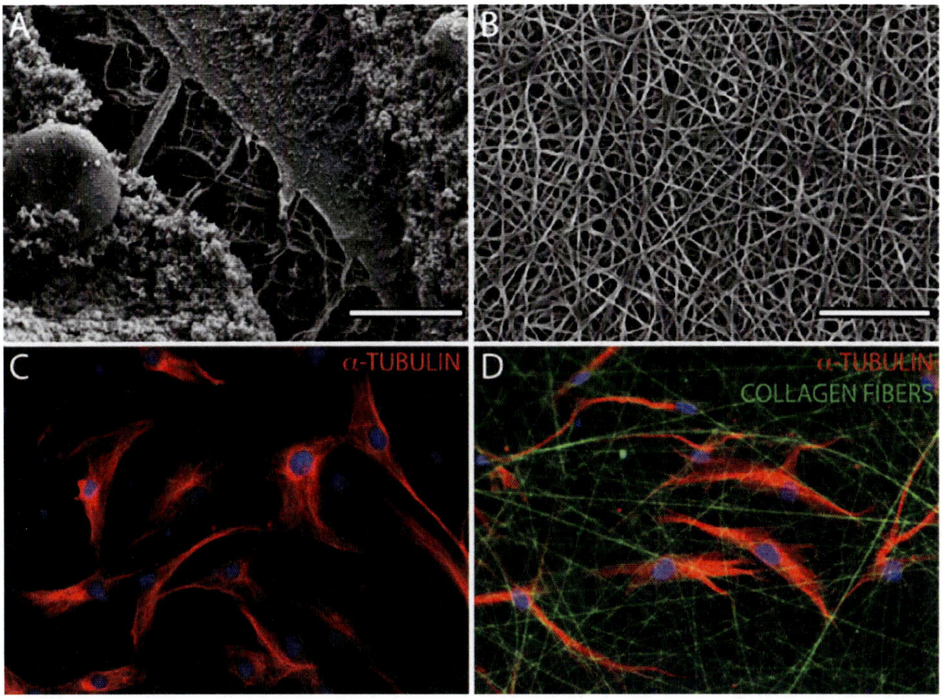

Fig. 4.4 (**a**) SEM micrograph of debrided muscle tissue, a source of MPCs; the bulk of the tissue is comprised of nanofiber extracellular matrix. (**b**) SEM micrograph of electrospun collagen nanofiber matrix, a biomimetic substrate used for in vitro MPC cultures. Scale bars = 5 μm. (**c**, **d**) Immunofluorescence images of MPCs cultured on traditional plastic and collagen fiber substrates; there are noticeable differences in cellular attachment and arrangement of alpha tubulin – known to play a key role in cell transport and movement

component of the MPC niche. Nanofiber structures have been observed across a range of native basement membranes, and are a popular topographical feature for experimentation [35], typically fabricated by an electrospinning process.

The impact of nanofiber topography on MSC differentiation pathways has been well examined, having been shown to support early multilineage differentiation of MSCs along adipogenic, chondrogenic, osteogenic [36], and even neuronal pathways [37]. Alterations in fiber alignment [38] and culture conditions [39], along with increasing complexity of scaffold designs [40], have better focused MSC differentiation returns, highlighting the long-term potential of nanofiber substrates for directing tissue growth. Despite the increasing attention paid to nanofiber substrates and their role in modulating MSC gross-phenotypic change, little emphasis has been placed on underlying gene regulation and cell signaling mechanisms, two key aspects to understanding MPC trophic behavior. The best experimental evidence of the impact physical cues have in MSC gene regulation and lineage specification has thus far been seen by studies correlating substrate stiffness to phenotypic response [41]. This provides a clue to examining nanofiber cultures, and underscores the importance of physical cues in mesenchymal development.

4.5.1 The Role of Tissue Topography During Wound Healing

Despite evidence that fibers can help guide MSC differentiation, it is our hypothesis that MSC-like cells isolated from *traumatized* tissue play a more supportive role, providing trophic guidance and protective support during tissue regeneration, and that local microtopography may play a crucial role in driving the trophic behavior. To more carefully investigate MPC changes induced by altering the structural microenvironment, we have recreated the fibrous architecture present in debrided muscle ECM (Fig. 4.4b) based on previous experimental conditions [42], and are in the preliminary stages of examining how this biomimetically-native topography regulates gene expression. Type I collagen was electrospun into a nanofiber mesh with an average diameter comparable to the native architecture seen in debrided muscle tissue (~100 nm average diameter), and stabilized by glutaraldehyde vapor-induced cross-linking. MPCs isolated from debrided muscle tissue were cultured on the electrospun matrix for 7 days, and evaluated using a combination of immunofluorescence and gene arrays. Immunofluorescence imaging reveals differences in early-passage MPC shape and internal structure when cultured on fibers vs. traditional polystyrene flasks – as seen by α-tubulin immunostaining – indicating the fiber structures affect key components involved with cell movement and transport (Fig. 4.4c, d). The magnitude and mechanisms by which this occurs are currently being explored. Early analysis using RT-PCR shows cells grown on the fiber matrix have an upregulation of factors associated with angiogenesis and cell homing, compared with a more osteogenic and chondrogenic response on traditional polystyrene plates. Expanded genetic analysis is currently underway, to better ascertain what specific factors and genes may be most significantly altered by the structural microenvironment.

4.6 Conclusions and Future Work

There are currently several animal models for HO that have been chosen to investigate a variety of clinical conditions. One commonly used model generates an ectopic bone lesion by forced manipulation of the muscle tissue interspaced by limb immobilization [43], which leads to myositis fibrosis and, in some cases, HO. HO can also be formed in an animal model by performing an achilles tenotomy to generate muscle damage [44]. There are also a variety of methods to generate HO by injecting the muscle tissue with demineralized bone matrix or calcium chloride [44, 45]. Finally, HO can be formed following a hip arthroplasty in an animal model. Each of these models provides a generalized model of HO in response to muscle damage, inflammation and foreign body response, and may be useful for investigating the etiology of HO in cases such as muscle overuse, chronic inflammation, and postoperative sequelae. However, these models cannot account for the high incidence of HO following combat-related trauma, which occurs at a substantially higher frequency than can be reproduced in these models [2]. Furthermore, these models do not investigate any specific cellular mechanisms, and therefore can only yield generalized treatment strategies (i.e., XRT and NSAID treatment), which are contraindicated in the patients with severe combat trauma who are at high risk of developing HO [1].

Given the currently available models, there is clear indication that a new model is needed to more effectively study the relationship between extremity combat trauma and HO. This model must recapitulate the key factors in the traumatized muscle environment that result in the formation of the fibrotic lesions, which become the initiation sites for HO as they undergo endochondral ossification [46]. Such a model would make it possible to study the cell and molecular events that are specific to HO formation following trauma and identify specific targets for therapeutic intervention to prevent the formation of ectopic bone without interfering with the other wound healing mechanisms. Traumatic orthopaedic injury causes substantial damage to the surrounding muscle tissues, which includes disruption and disorganization of the muscle fibers and infiltration of mononuclear cells into the site of injury, including myoblasts, MPCs and other immunological cell types. A valid animal model should recapitulate the cellular and molecular events that we have observed in the human traumatized tissue specimens in order to generate the disease state. Our laboratory is currently compiling the results of our *in vitro* experiments in order to develop a novel animal model to study traumatic HO.

Caution must be exercised in the development of such a model, as there are a number of mechanisms and agents known to cause ectopic bone formation. A valid model must incorporate precisely the same mechanisms as those occurring in traumatic HO; otherwise, clinical interventions developed with the animal model will unlikely have the predicted effects in human patients. We believe that our overall, multi-faceted approach to studying this disease will enable us to penetrate the complex interactions at the cell and tissue layers that lead to the formation of HO. Once an appropriate animal model can be developed to meet these requirements, it may be applied toward the identification and validation of novel targets and leads the way toward clinical trials for prophylactic or therapeutic interventions for HO.

Acknowledgements This work was supported acknowledged as part of the NIH Intramural Research Program. (Z01 AR41131 and 1ZIAAR041191), grants from the Department of Defense Military Amputee Research Program at WRAMC (PO5-A011), Comprehensive Neurosciences Program (CNP-2008-CR01) and Peer-Reviewed Orthopaedic Research Program (W81XWH-10-2-0084).

References

1. Potter BK et al (2007) Heterotopic ossification following traumatic and combat-related amputations. Prevalence, risk factors, and preliminary results of excision. J Bone Joint Surg Am 89(3):476–86
2. Potter BK et al (2006) Heterotopic ossification in the residual limbs of traumatic and combat-related amputees. J Am Acad Orthop Surg 14(10 Spec No.):S191–7
3. Pape HC et al (2004) Current concepts in the development of heterotopic ossification. J Bone Joint Surg Br 86(6):783–7
4. Stover SL, Hataway CJ, Zeiger HE (1975) Heterotopic ossification in spinal cord-injured patients. Arch Phys Med Rehabil 56(5):199–204
5. Peterson SL et al (1989) Postburn heterotopic ossification: insights for management decision making. J Trauma 29(3):365–9
6. Vanden Bossche L et al (2008) Free radical scavengers have a preventive effect on heterotopic bone formation following manipulation of immobilized rabbit legs. Eur J Phys Rehabil Med 44(4):423–8
7. Shore EM et al (2006) A recurrent mutation in the BMP type I receptor ACVR1 causes inherited and sporadic fibrodysplasia ossificans progressiva. Nat Genet 38(5):525–7
8. Shafritz AB et al (1996) Overexpression of an osteogenic morphogen in fibrodysplasia ossificans progressiva. N Engl J Med 335(8):555–61
9. Ahn J et al (2003) Paresis of a bone morphogenetic protein-antagonist response in a genetic disorder of heterotopic skeletogenesis. J Bone Joint Surg Am 85-A(4):667–74
10. Shore EM, Kaplan FS (2008) Insights from a rare genetic disorder of extra-skeletal bone formation, fibrodysplasia ossificans progressiva (FOP). Bone 43(3):427–33
11. Nesti LJ et al (2008) Differentiation potential of multipotent progenitor cells derived from war-traumatized muscle tissue. J Bone Joint Surg Am 90(11):2390–8
12. Chamberlain G et al (2007) Concise review: mesenchymal stem cells: their phenotype, differentiation capacity, immunological features, and potential for homing. Stem Cells 25(11):2739–2749
13. Owen ME, Cave J, Joyner CJ (1987) Clonal analysis in vitro of osteogenic differentiation of marrow CFU-F. J Cell Sci 87(5):731–738
14. Grefte S et al (2007) Skeletal muscle development and regeneration. Stem Cells Dev 16(5):857–68
15. Zheng B et al (2007) Prospective identification of myogenic endothelial cells in human skeletal muscle. Nat Biotechnol 25(9):1025–34
16. Lumelsky NL (2007) Commentary: engineering of tissue healing and regeneration. Tissue Eng 13(7):1393
17. Caplan AI, Dennis JE (2006) Mesenchymal stem cells as trophic mediators. J Cell Biochem 98(5):1076–84
18. Jackson WM et al (2010) Differentiation and regeneration potential of mesenchymal progenitor cells derived from traumatized muscle tissue. J Cell Mol Med 2010 Dec 3. doi: 10.1111/j.1582-4934.2010.01225.x. [Epub ahead of print]
19. Sivakumar P, Das A (2008) Fibrosis, chronic inflammation and new pathways for drug discovery. Inflamm Res 57(9):410–418
20. Jackson WM et al (2009) Mesenchymal progenitor cells derived from traumatized human muscle. J Tissue Eng Regen Med 3(2):129–38
21. Jackson WM et al (2009) Putative heterotopic ossification progenitor cells derived from traumatized muscle. J Orthop Res 27(12):1645–51
22. Velnar T, Bailey T, Smrkolj V (2009) The wound healing process: an overview of the cellular and molecular mechanisms. J Int Med Res 37(5):1528–42
23. Singer AJ, Clark RA (1999) Cutaneous wound healing. N Engl J Med 341(10):738–46
24. Quintero AJ et al (2009) Stem cells for the treatment of skeletal muscle injury. Clin Sports Med 28(1):1–11
25. Jackson WM et al (2011) Cytoke expression in muscle following traumatic injury. J Orthop Res 2011 Mar 30. doi: 10.1002/jor.21354. [Epub ahead of print]
26. Cowin AJ et al (2001) Expression of TGF-beta and its receptors in murine fetal and adult dermal wounds. Eur J Dermatol 11(5):424–31
27. Lu L et al (2005) The temporal effects of anti-TGF-beta1, 2, and 3 monoclonal antibody on wound healing and hypertrophic scar formation. J Am Coll Surg 201(3):391–7

28. Schofield R (1978) The relationship between the spleen colony-forming cell and the haemopoietic stem cell. Blood Cells 4(1–2):7–25
29. Jones DL, Wagers AJ (2008) No place like home: anatomy and function of the stem cell niche. Nat Rev Mol Cell Biol 9(1):11–21
30. Lutolf MP, Hubbell JA (2005) Synthetic biomaterials as instructive extracellular microenvironments for morphogenesis in tissue engineering. Nat Biotechnol 23(1):47–55
31. Discher DE, Mooney DJ, Zandstra PW (2009) Growth factors, matrices, and forces combine and control stem cells. Science 324(5935):1673–7
32. Place ES, Evans ND, Stevens MM (2009) Complexity in biomaterials for tissue engineering. Nat Mater 8(6):457–70
33. Guilak F et al (2009) Control of stem cell fate by physical interactions with the extracellular matrix. Cell Stem Cell 5(1):17–26
34. Lutton C, Goss B (2008) Caring about microenvironments. Nat Biotechnol 26(6):613–4
35. Lim SH, Mao HQ (2009) Electrospun scaffolds for stem cell engineering. Adv Drug Deliv Rev 61(12):1084–96
36. Li WJ et al (2005) Multilineage differentiation of human mesenchymal stem cells in a three-dimensional nanofibrous scaffold. Biomaterials 26(25):5158–66
37. Yim EK, Pang SW, Leong KW (2007) Synthetic nanostructures inducing differentiation of human mesenchymal stem cells into neuronal lineage. Exp Cell Res 313(9):1820–9
38. Baker BM, Mauck RL (2007) The effect of nanofiber alignment on the maturation of engineered meniscus constructs. Biomaterials 28(11):1967–77
39. Sefcik LS et al (2008) Collagen nanofibres are a biomimetic substrate for the serum-free osteogenic differentiation of human adipose stem cells. J Tissue Eng Regen Med 2(4):210–20
40. Nie H et al (2008) Three-dimensional fibrous PLGA/HAp composite scaffold for BMP-2 delivery. Biotechnol Bioeng 99(1):223–34
41. Engler AJ et al (2006) Matrix elasticity directs stem cell lineage specification. Cell 126(4):677–89
42. Shih YR et al (2006) Growth of mesenchymal stem cells on electrospun type I collagen nanofibers. Stem Cells 24(11):2391–7
43. Michelsson JE, Granroth G, Andersson LC (1980) Myositis ossificans following forcible manipulation of the leg. A rabbit model for the study of heterotopic bone formation. J Bone Joint Surg Am 62(5):811–5
44. Lin L et al (2006) Adenovirus-mediated transfer of siRNA against Runx2/Cbfa1 inhibits the formation of heterotopic ossification in animal model. Biochem Biophys Res Commun 349(2):564–72
45. O'Connor JP (1998) Animal models of heterotopic ossification. Clin Orthop Relat Res 346:71–80
46. Kaplan FS et al (2004) Heterotopic ossification. J Am Acad Orthop Surg 12(2):116–25

Comparative Proteomic Analysis of Mesenchymal Stem Cells Derived from Human Bone Marrow, Umbilical Cord and Placenta: Implication in the Migration

5

Guo Li, Xiao-Ai Zhang, Hua Wang, Xin Wang, Chun-Ling Meng, Chu-Yan Chan, David Tai Wai Yew, Kam Sze Tsang, Karen Li, Sau-na Tsai, Sai-Ming Ngai, Zhong Chao Han, Marie Chia-Mi Lin, Ming-Liang He, and Hsiang-Fu Kung

Abstract

Umbilical cord (UC) and placenta (P) have been suggested as alternatives to bone marrow (BM) as sources of mesenchymal stem cells (MSC) for cell therapy, with both UC- and P-MSC possess immunophenotypic and functional characteristics similar to BM-MSC. However, under defined conditions, the migration capacity of BM- and P-MSC was found to be 5.9- and 3.2-folds higher than that of UC-MSC, respectively. By the use of 2-DE and combined MS and MS/MS analysis, six differentially expressed proteins were identified among these MSC samples, with five of them known to be involved in cell migration as migration enhancing or inhibiting proteins. Interestingly, the expression levels of those proteins reflect perfectly the migration capacity of corresponding MSC, which is also proved by in vitro overexpression and silencing techniques. Our study indicates that a bunch of migration-related proteins are pivotal in governing the migration capacity of MSC.

Abbreviations

α-MEM	Minimum essential medium, alpha medium
BM	Bone marrow
CTSB	Cathepsin B
CTSD	Cathepsin D
GAPDH	Glyceraldehyde-3-phosphate dehydrogenase
GFP	Green fluorescence protein

H.-F. Kung (✉)
Stanley Ho Centre for Emerging Infectious Diseases, The Chinese University of Hong Kong, Hong Kong, People's Republic of China

State Key Laboratory in Oncology in South China, Sun Yat-Sen University Cancer Center, Guangzhou, People's Republic of China
e-mail: b110473@mailserv.cuhk.edu.hk

J.S. Rhim and R. Kremer (eds.), *Human Cell Transformation: Role of Stem Cells and the Microenvironment*, Advances in Experimental Medicine and Biology 720, DOI 10.1007/978-1-4614-0254-1_5, © Springer Science+Business Media, LLC 2011

GRP75	Stress 70 protein
MnSOD	Manganese superoxide dismutase
MSC	Mesenchymal stem cells
neo	Neomycin
P	Placenta
PAI-1	Plasminogen activator inhibitor-1
PE	Phycoerythin
PHB	Prohibitin
SiRNA	Small interfering ribonucleic acid
UC	Umbilical cord

5.1 Introduction

Mesenchymal stem cells (MSC) play important roles in tissue regeneration based on their abilities of self-renewal and multi-lineage differentiation. Bone marrow (BM) is the major source of MSC, which has been widely employed in experimental and clinical studies. However, the clinical application of BM-derived cells is limited for the relatively invasive procedure of sample collection, and a markedly reduction in cell number, proliferation and differentiation capacity with age [1]. Therefore, any alternative source of MSC other than BM would be invaluable and beneficial for future cell therapy. Placenta (P) and umbilical cord (UC) have been postulated as new sources of MSC for their relative ease of collection and more importantly, the much younger cellular age compared to the adult origins. It has been demonstrated that the UC- and P-MSC share similar features, including the expression of cell surface antigens and multi-lineage differentiation potential, to BM-MSC [2–4].

For common therapeutic purpose, systemic infusion of MSC is regarded as the most practical mode of administration. Apparently, the homing process of MSC is similar to that of leukocyte. Upon delivery into blood stream, the MSC keep close contacts with the endothelial cells. They engraft into the endothelium, and eventually pass and leave the endothelium for target tissues [5, 6]. The migration and homing capabilities are essential features of MSC for tissue regeneration process, but these characteristics are yet to be elucidated for both P- and UC-MSC. Using in vitro migration assay, we found that the migration capacity of UC-MSC was significantly lower than those of P- and BM-MSC, while BM-MSC have the highest migration capacity.

In an earlier study, the trans-endothelial migration of BM-MSC has been demonstrated [7]. Several other studies using in vitro migration assay have also been carried out and suggested that various molecules including adhesion molecules [8], chemokines [9, 10], growth factors [9], and proteases [10, 11] might be involved in the MSC migration mechanism. The WNT signaling was also suggested to be critically associated in regulating the migration capacity of BM-MSC [12]. Although considerable studies have been devoted to the effectors of MSC migration, no global analysis has been done to elucidate the complete underlying mechanisms.

Proteomic tools have been widely used in the studies of cancer metastasis, stem cell differentiation, and in deriving related molecular mechanisms [13, 14]. To identify candidate proteins responsible for the differential migration capacity, the protein expression profile among BM-, P- and UC-MSC was compared by 2-DE. With combined MS and MS/MS analysis, we identified six proteins including plasminogen activator inhibitor-1 (PAI-1), that are differentially expressed among three sources of MSC. The overexpression of PAI-1 in BM- and P-MSC, which produced little endogenous PAI-1, was noted to become sluggish in trafficking through Matrigel-coated membrane. On the other hand, the silencing of PAI-1 in UC-MSC, which produced much endogenous PAI-1, was found to significantly promote their migration capacity. The present study is the first report on the differential protein expression among BM-, P- and UC-MSCs, which possessed similar features and characteristics of stem cells yet different migration capacities. The finding provides us an in vitro platform of mechanistic investigation of cell migration and invasion. The discovery of molecular events associated with MSC migration will be pivotal to elucidate the roles of MSC in tissue repair and regeneration.

5.2 Materials and Methods

5.2.1 Isolation and Culture of Human MSC

Human umbilical cords and placentas were donated by two pregnant women with prior formal consents. The samples were washed thoroughly with PBS to remove contamination. Vessels with the Wharton's jelly were cautiously pulled away from the umbilical cords. Each dissected vessel was tied together at the ends, then stored in a 50 mL sterilized tube containing serum-free medium with 0.2 mg/mL collagenase for overnight digestion at 37°C. The vessels were then removed from the mixture, and the suspension was diluted with PBS to lower the viscosity. After centrifugation, supernatant was removed. The cell pellet was washed twice with PBS and re-suspended in culture medium.

For placenta sample collection, the whole tissue was cut into 1–2 mm pieces and then incubated in 0.2 mg/mL collagenase for 90 min. Digested tissue was squeezed through a 100 μm cell strainer to get rid of cell aggregates. Sieved cell suspension was washed twice with PBS and re-suspended in culture medium.

Bone marrow aspirations were obtained from two healthy donors, a 38-year-old male (BM1) and a 17-year-old female (BM2), with formal consents. Mononuclear cells were enriched by using gradient centrifugation in Ficoll-Hypaque with specific gravity of 1.077 g/mL. After washing with PBS twice, cells were re-suspended at a concentration of 1×10^6 cells/mL in culture medium.

The culture medium used in this study was minimum essential medium, alpha medium (α-MEM) supplemented with 10% FBS, 2 mM l-glutamine, 100 U/mL penicillin, and 100 μg/mL streptomycin. The cultures were maintained in a humidified atmosphere of 5% CO_2 at 37°C. All reagents were purchased from Invitrogen, USA. All samples were processed within 24 h after collection. Experiments were carried out according to the protocols approved by The Chinese University of Hong Kong – New Territories East Cluster Clinical Research Ethics Committee. On day 2, the complete medium was replaced and all the nonadherent cells were removed. Thereafter, medium was replaced twice a week and cells were subcultured or cryopreserved prior to confluence. MSC at the seventh to ninth passages were used in this study. Taking account of individual variation, two donors were employed for each MSC collection in our studies with the corresponding cells labeled as BM1-MSC, BM2-MSC, UC1-MSC, UC2-MSC, P1-MSC, and P2-MSC, respectively. The same passage of MSC was used for all experiments, while each experiment repeated thrice.

5.2.2 Cell Surface Markers Analysis

Cultures were washed twice with PBS and enzymatical segregation was carried out using 0.05% trypsin/EDTA (Invitrogen) for 5 min at 37°C. Cells were washed and adjusted to $<1 \times 10^7$/mL in PBS supplemented with 1% BSA (Sigma, USA). One hundred milliliter cell suspensions were stained with 20 μL FITC- and phycoerythrin (PE)-conjugated mAbs from Beckman Coulter, USA (unless specified otherwise) for 30 min at room temperature in the dark. They were CD3-PE, CD16-FITC, CD19-FITC, CD33-FITC, CD34-PE, CD38-FITC, CD45-FITC, 10 μL CD133-PE (Miltenyi Biotec GmbH, Germany), HLA-DR-FITC, 10 μL CD29-Rhodamin, CD44-FITC, CD90-PE, CD105-PE (Serotec, UK) and CD166-PE. Nonspecific background was evaluated by parallel staining with isotype-matched IgG1-FITC and IgG1-PE. Cells were processed by using the ImmunoPrep reagent kit in the Q-Prep Immunology Workstation (Beckman Coulter) and then analyzed with the Coulter Epic XL MCL flow cytometry (Beckman Coulter).

5.2.3 Osteogenic and Adipogenic Differentiation

The induction medium for adipogenesis was α-MEM supplemented with 10% FBS, 10^{-7} M dexamethasone, 50 μg/mL ascorbic acid-2 phosphate,

50 μM indomethacin, 10 μg/mL insulin, and 0.45 mM 3-isobutyl-1-methyl-xanthinel. The induction medium for osteogenesis was α-MEM supplemented with 10% FBS, 10^{-8} M dexamethasone, 50 μg/mL ascorbic acid-2 phosphate, and 10 mM glycerolphosphate. All chemicals were purchased from Sigma. Medium was changed twice a week. The MSC samples cultured in the FBS-enriched α-MEM with the corresponding concentration of vehicles were served as controls. After 2 weeks of adipogenic differentiation, cells were fixed with 4% paraformaldehyde and stained with 50% saturated oil red O. The MSC cells, which had put for osteogenic differentiation for 3 weeks, were fixed with 4% paraformaldehyde and stained with 2% Alizarin red S. Images were captured using a phase contrast microscope (Nikon, Japan).

5.2.4 Trans-Matrigel Migration Assay

The migration ability of MSC was evaluated by using 48-well Boyden Chamber and polycarbonate membrane with 10 μm pores (Neuro Probe, USA). Membrane of 25 × 80 mm was precoated with 500 μg Matrigel (BD Biosciences, USA) and allowed to dry at room temperature for 3 h. α-MEM containing 10% FBS (27 μL) was added to the lower compartment as a chemo-attractant. MSC (4×10^4), suspended in 50 μL α-MEM containing 0.5% BSA, were loaded onto the upper compartment of each chamber. Triplicate studies were performed for each group. After 5 h incubation in a humidified atmosphere of 5% CO_2 at 37ºC, nonmigrated cells on the upper surface were carefully scraped away. Cells which migrated through the chamber or adhered to the lower surface of the membrane were fixed with 4% paraformaldehyde and later stained with hematoxylin. Images were captured using a microscope at 100× magnification. For each chamber, migrated cells in three random selected fields were counted. The total number of cells migrated through the whole cavity area was estimated by averaging the number of migrated cells in selected fields. The extent of cell migration was presented as a percentage of total cell number seeded onto the upper compartment at the beginning of the experiment.

5.2.5 Protein Extraction

MSC cultures at 90% confluence were harvested. Cell pellets were re-suspended in lysis buffer (8 M urea, 2 M thiourea, 2% CHAPS, 1% NP-40, 2 mM TBP, 1× Protease Inhibitor Mix, 1× Nuclease Mix, 1 mM PMSF, and 2% IPG buffer) and put on ice for 45 min. They were then centrifuged at 14,000×*g* for 15 min at 4°C. The supernatant was collected and stored at −80°C. The protein concentration of samples was determined by Bradford assay (Bio-Rad, USA).

5.2.6 Two-Dimensional Gel Electrophoresis

IEF was carried out using IPGphor II apparatus (Amersham, USA). All IPG strips (13 cm, pH 3–10, NL) were used according to the manufacturer's instructions. Samples containing 150 μg protein were diluted to 250 μL in rehydration solution (8 M Urea, 2% CHAPS, 0.4% DTT, 0.5% IPG buffer, 0.002% bromophenol blue). The rehydration step was carried out at 30 V for 10 h. IEF was performed following a step-wise voltage incremental manner: 500 and 1,000 V for 1 h each, and 8,000 V afterwards until 64 kVh. After IEF, IPG strips were placed in an equilibration buffer (6 M urea, 2% SDS, 30% glycerol, 0.002% bromophenol blue, 50 mM Tris–HCl, pH 6.8) containing 1% DTT for 15 min with agitation. The strips were transferred to the equilibration solution containing 2.5% iodoacetamide and further agitated for 15 min before submission onto 12.5% uniform polyacrylamide gel slabs (150 × 158 × 1.5 mm). Second dimensional separation was performed in Tris–glycine buffer (25 mM Tris, 0.2 M glycine, 0.1% SDS) at a constant current setting of 15 mA/gel for 30 min and 30 mA/gel thereafter. SDS–PAGE was terminated when the bromophenol blue dye front migrated to the lower ends of the gels.

5.2.7 Gel Staining and Image Analysis

Gels were stained by a modified silver-staining method which was compatible with the MS analysis. The whole staining procedure was carried out at room temperature with gentle agitation. Briefly, gels were fixed in 50% methanol containing 12% acetic acid and 0.0185% formaldehyde overnight. Fixed gels were rinsed thrice in water followed by an equilibration step in 50% ethanol twice. Sensitization was performed by immersing the gels in 0.02% sodium thiosulfate for 10 min. After brief wash in water thrice, gels were incubated in 0.15% silver nitrate containing 0.0278% formaldehyde for an hour. Color development was carried out in 3% sodium carbonate containing 0.037% formaldehyde until protein spots could be clearly seen against the background. Staining was stopped by transferring the gels into 5% acetic acid for at least 10 min. Gel images were digitalized using a scanner (GS-800 calibrated densitometer, Bio-Rad) and the Quantity One software (Bio-Rad). All images were further analyzed using the software PDQuest (version 8.0, Bio-Rad) according to the user guide. Briefly, the spot detection parameters are set as the follows: Ruby speckles, 50; smoothing, medium 7×7; background removal, floating ball 61; removal vertical streaks, 93; removal horizontal streaks, 93; sensitivity, 16.51; size scale, 5; and min peak, 868. After manual editing, spots on different gels were matched and analyzed. Statistical analysis was performed using the independent-samples *t*-test.

5.2.8 Tryptic In-Gel Digestion

Spots of interest were manually excised from the gels. Gel chips were soaked in an equal volume mixture of 30 mM potassium ferricyanide and 100 mM sodium thiosulfate for 5 min. After washing in water twice, gel chips were equilibrated in 50 mM ammonium bicarbonate for 20 min and then in 25 mM ammonium bicarbonate and 50% acetonitrile, followed by soaking in 100% acetonitrile until gels turned opaque. Vacuum-dried gel chips were rehydrated with 10 μg/mL of Trypsin (Promega, USA) in 25 mM ammonium bicarbonate (pH 8.0) and digested for 16–18 h at 37°C. After digestion, the liquid containing the tryptic peptides was extracted and transferred into a new tube.

5.2.9 MALDI-TOF/TOF Mass Spectrometric Analysis

Mass spectrometric analysis was carried out using a MALDI-TOF/TOF tandem mass spectrometer ABI 4700 proteomics analyzer (Applied Biosystems, USA). For acquisition of mass spectra, samples (0.5 μL) were spotted onto a MALDI plate, followed by 0.5 μL matrix solution (4 mg/mL α-cyano-4-hydroxycinnamic acid in 35% acetonitrile and 1% TFA). Mass data acquisitions were piloted by 4000 Series Explorer™ Software v3.0 using batched-processing and automatic switching between MS and MS/MS modes. All MS survey scans were acquired over the mass range 800–3,500 *m/z* in the reflection positive-ion mode and accumulated from 2,000 laser shots with acceleration of 20 kV. The MS spectra were internally calibrated using porcine trypsin autolytic products (842.509, 1,045.564, 1,940.935, and 2,211.104 *m/z*) and resulted in mass errors of less than 30 ppm. Peptide precursor ions corresponding to contaminants including keratin and trypsin autolytic products were excluded in a mass tolerance of ±0.2 Da. The filtered precursor ions with a user-defined threshold (S/N ratio ≥50) were selected for the MS/MS scan. Fragmentation of precursor ions was performed using MS/MS 1 kV positive mode with CID on and with argon as the collision gas. MS/MS spectra were accumulated from 3,000 laser shots using default calibration with Glu-Fibrinopeptide B from 4700 Calibration Mixture (Applied Biosystems).

5.2.10 Database Search

In order to perform protein identification, the MS (peptide-mass-fingerprint approach) and MS/MS (DeNovo sequencing approach) data were loaded into the GPS Explorer™ software v3.5 (Applied Biosystems) and searched against NCBInr 5825255 sequences (released on January 10, 2008) by Mascot search engine v1.9.05 (Matrix Science, UK). The following search parameters were employed: monoisotopic peptide mass (MH^+); 800–3,500 Da; one missed cleavage per peptide; enzyme, trypsin; taxonomy, *Homo sapiens*; p*I*, 0–14; precursor-ion mass tolerance,

50 ppm; MS/MS fragment-ion mass tolerance, 0.1 Da; variable modifications, carbamidomethylation for cysteine and oxidation for methionine were allowed. Known contaminant ions corresponding to keratin and trypsin were excluded from the peak lists before database search. In this study, identified proteins were accepted to report when their protein score CI% and sequence coverage was more than 99% ($p<0.01$) and 15%, respectively, with at least four matched peaks. The top ten hits were reported for each protein search.

5.2.11 pLVTHM-Neo-GFP and pLVTHM-Neo-PAI-1 Construction

Replication deficient lentiviral vector pLVTHM and its package plasmids, pAX2 and pMD2G, were gifts from Dr. Didier Trono (Laboratory of Virology and Genetics, University of Geneva, Switzerland). IRES-neo (neomycin) fragment, between NheI and XbaI sites of pIRESneo3 vector (Clontech, USA), was subcloned into the SpeI site of pLVTHM. Namely, pLVTHM-neo-GFP (green fluorescence protein), the orientation of which was confirmed by EcoRI digestion. The whole codon sequence of *PAI-1* (NM_000602) was PCR amplified using Platinum® Taq DNA Polymerase High Fidelity (Invitrogen) and primers: forward 5′ GTGATATCAACTTCAGG ATGCAGATG 3′, reverse 5′ TTGCGGCCGCTC AGGGTTCCATCACTTG 3′. PCR product was cloned into pIRESneo3 between EcoRV and NotI sites. Then the PAI-1-IRES-neo fragment, between EcoRV and XbaI sites of IRESneo3-PAI-1, was subcloned into pLVTHM in replacement of the PmeI and SpeI fragment of GFP, i.e., pLVTHM-neo-PAI-1 construct. The sequence integrity was confirmed by sequencing.

5.2.12 Lentivirus Package and Transduction of MSC

Recombinant lentivirus was produced by transient transfection of 293T cells using calcium phosphate transfection method (http://tronolab.epfl.ch/page58122.html). The viral titer of pLVTHM-neo-GFP supernatant was evaluated through transduction of 293T cells by GFP expression. The titer of pLVTHM-neo-PAI-1 has not been tested for lacking GFP yet our team assume that it is similar with that of pLVTHM-neo-GFP. MSCs of the fifth passage were transduced at MOI 20 in the presence of 4 μg/mL protamine sulfate (Sigma). Upon selection with 400 μg/mL G418 (Sigma), the third passage of transduced cells were examined for migration study, and also used in the RT-PCR and Western blot analyses.

5.2.13 Transfection of UC-MSC with Small Interfering RNA

Two PAI-1-specific SiRNAs (SiRNA1 and 2) and one nontarget-directed negative control small interfering RNA (SiRNA) (NC) were designed online (http://www.dharmacon.com/DesignCenter/DesignCenterPage.aspx) and synthesized by Sigma (St. Louis, USA). The sequences of the sense strands are listed in the following: SiRNA1, sense: 5′ CCAGCUGACAACAGGAGGAGAAA CCCA3′, SiRNA2, sense: 5′ UGCAGAAAGUGAA GAUCGAGGUGAACG 3′, NC, Sense: 5′ GGUA UCUUGAUGUGCCACGUGAGCUUG 3′.

UC-MSC at its sixth to eighth passage was seeded into 12-well plates at 5×10^4 cells per well a day before transfection. One hundred pmole of SiRNA duplex were combined with 2 μL lipofectamine 2000 (Invitrogen) in a total volume of 200 μL and allowed to incubate at room temperature for 20 min. The transfection mixture was then applied to cells cultured in 400 μL culture medium without antibiotics and left to incubate in a humidified atmosphere containing 5% CO_2 at 37°C. Six hours later, the transfection mixture was removed and cells were maintained in culture medium without antibiotics for 48 h. Samples were collected and examined for the expression level of PAI-1 and their migration capacity afterwards.

5.2.14 Preparation of Conditioned Media

MSC (1×10^5) were seeded onto a six-well plate. Culture medium was removed when cell confluence reached 70–80%. The samples were washed twice with PBS and replenished by 1.2 mL serum-free α-MEM. In SiRNA transfection assay, serum-free medium (0.4 mL) was added 24 h after the removal of transfection mixture. Conditioned medium was removed 24 h later and samples were used to assay for PAI-1 expression.

5.2.15 Western Blotting

Equal amount of cell extracts or equal volume of conditioned media were separated by SDS–PAGE (12%) and transferred to PVDF membranes. Membranes were blocked with 5% skim milk for an hour and left to react with anti-human cathepain B (CTSB), anti-human PAI-1, or anti-human actin antibody (Santa Cruz, USA) overnight. Antibody binding was revealed by incubating with the corresponding HRP-linked IgG (Santa Cruz) and the ECL plus Western blotting detection system (GE Healthcare, UK). Chemiluminescence was detected by autoradiography using film (Fuji, Japan).

5.2.16 Real-Time PCR

Stable transduced MSC at 90% confluence were harvested. Their total RNA was extracted using TRIZOL reagent (Invitrogen) according to the manufacturer's instructions. Reverse transcription was carried out using SuperScript™ III Reverse Transcriptase (Invitrogen) and the cDNA fragments were amplified using GoTaq® DNA Polymerase (Promega). Having denatured at 94°C for 2 min, amplification was carried out by 29 cycles of 94°C for 30 s, 53°C for 40 s and 72°C for 1 min. The thermo-profile for amplification of glyceraldehyde-3-phosphate dehydrogenase (GAPDH) was 20 cycles of 94°C for 30 s, 56°C for 40 s and 72°C for 30 s. The elongation step was at 72°C for 7 min. The PAI-1 primers used were the same that we used in validating the PAI-1 constructs. Primers used for GAPDH were: forward 5′ TCCATGACAACTTTGGTA TCG 3′ and reverse 5′ TGTAGCCAAATTCG TTGTCA 3′. Thermocycling was performed with the gradient thermocycler (Takara, Japan).

5.2.17 Quantitative Real-Time PCR

Quantitative real-time PCR was carried out on an ABI 7500 real-time PCR system using SYBR® Green PCR Master Mix (Applied Biosystems, USA) with a thermocycler profile at 50°C for 2 min; 95°C for 10 min; followed by 40 cycles of 95°C for 15 s and 60°C for 1 min. DNA content was measured by real-time fluorimetric intensity of SYBR green I incorporation after completion of the primer extension step in each cycle. The primers used were: sense 5′ TGC TGG TGA ATG CCC TCT ACT 3′ and anti-sense 5′ CGG TCA TTC CCA GGT TCT CTA 3′. GAPDH was used as an internal control.

5.2.18 Statistical Analysis

Statistical analysis was performed by using the independent-samples *t*-test (SPSS Inc., USA). Results were expressed in mean ± SD. A *p* value of less than 0.05 was regarded as statistically significant.

5.3 Result

5.3.1 Immunophenotype and Differentiation Potential of MSC

As revealed by the flow cytometry, around 90% of the MSC samples expressed CD29, CD44, CD90, CD105, and CD166 markers. However, hemic markers such as CD45, HLA-DR, T-lineage

CD3, NK-lineage CD16, B-lineage CD19, myeloid CD33 and CD38 and stem/precursor CD34 and CD133 were not found. Upon adipogenic and osteogenic inductions, all MSC samples from the three different sources could differentiate into adipose and bone cells as demonstrated by the presence of fat globules and mineral deposition in Fig. 5.1.

5.3.2 In Vitro Migration Capacity of MSC

In this study, BM-MSC was noted to have the highest trafficking ability, followed by P-MSC and lastly the UC-MSC. It was revealed by using the migration assay on two individual samples (Fig. 5.2). The migration rate of BM-MSC was 5.9- and 1.8-folds higher than that of UC- and P-MSC, respectively, whilst P-MSC is 3.2-folds higher than that of UC-MSC.

5.3.3 Differentially Expressed Proteins Identified by Proteomic Analysis

Total number of protein spots in one gel was 265 ± 21.2, 300.5 ± 10.6 and 293.5 ± 14.8 for BM-MSC, P-MSC, and UC-MSC, respectively. Proteins were regarded as differentially expressed between BM-, P-, and UC-MSC when the magnitude of difference was greater than twofold or more with the result that was reproducible twice. As a result, ten proteins were found differentially expressed among the three MSC samples and they were subjected to further analysis by the

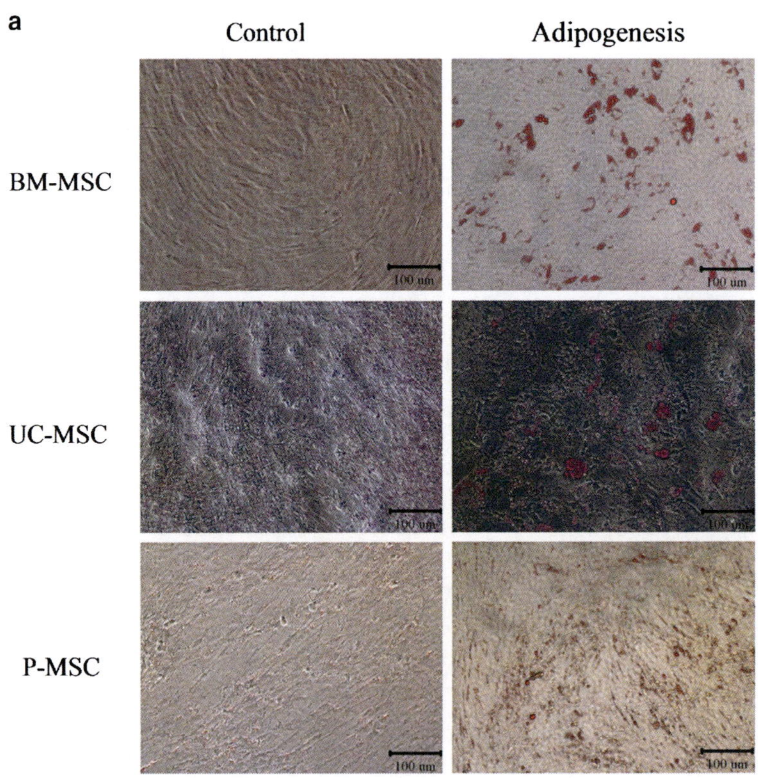

Fig. 5.1 In vitro differentiation of MSC cultures. (**a**) Adipogenesis of BM-, UC- and P-MSC. Cells were stained with Oil Red O after 2-week induction culture. The lipid globules of adipocytes were stained red whereas control cells displayed no positive staining. Scale bar, 100 μm. (**b**) Osteogenesis of BM-, UC-, and P-MSC. Cells were stained with Alizarin Red S after osteogenic induction for 3 weeks. Calcium deposits of osteocytes were stained in red. No calcium deposit formation was found in the control cells. Scale bar, 100 μm

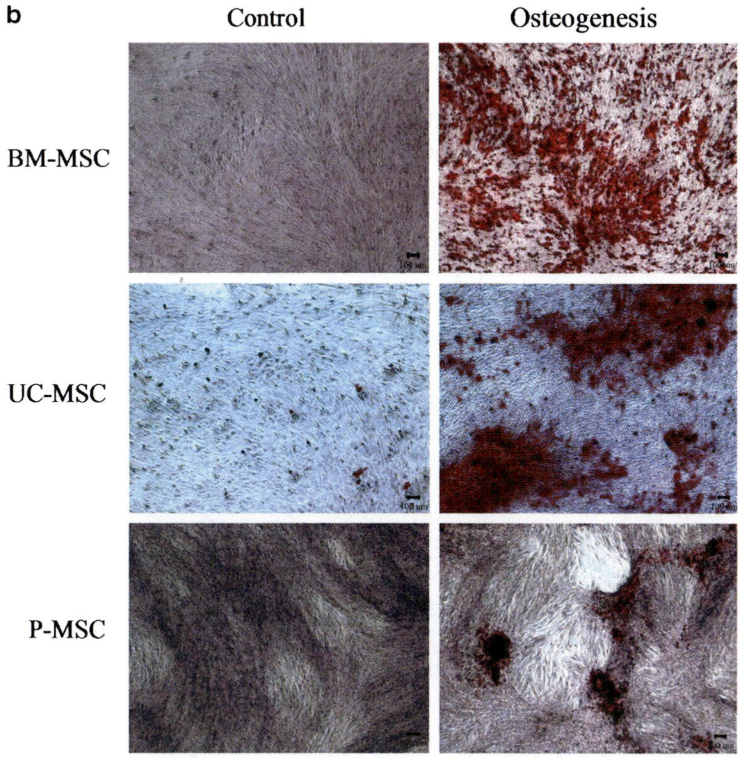

Fig. 5.1 (continued)

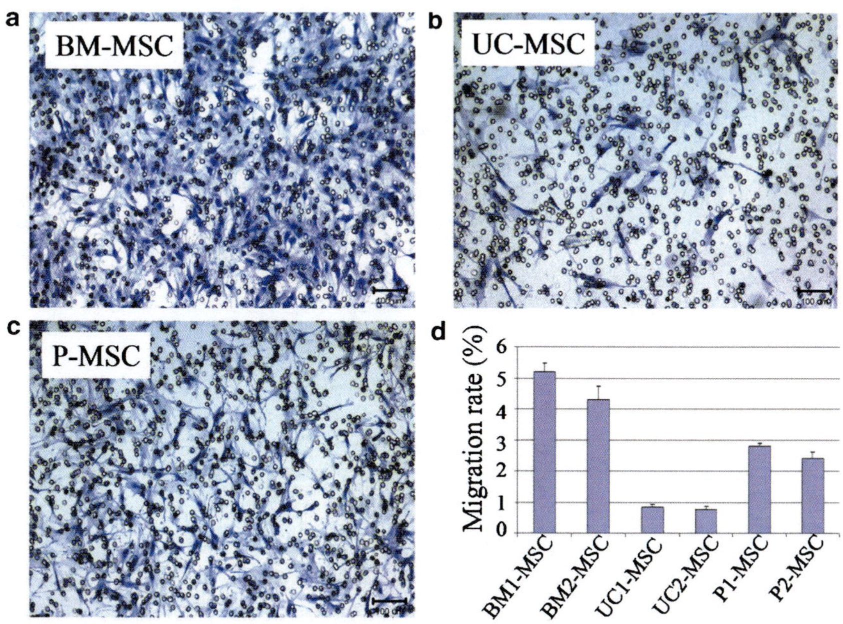

Fig. 5.2 In vitro migration of BM-, UC-, and P-MSC. Representative images of migration of (**a**) BM-MSC, (**b**) UC-MSC, and (**c**) P-MSC. The migrated cells spread on the membrane were stained blue with hematoxylin. Dots indicate the 10 μm pores through which the MSC migrated. Scale bar, 100 μm. (**d**) Quantitative analyses of migration capability of BM-, UC-, and P-MSC derived from two individuals ($n=6$; $p<0.01$)

Table 5.1 Differentially expressed proteins among BM-, UC-, and P-MSC

Spot no.	Expression fold changes: UC vs. BM	UC vs. P	P vs. BM	Protein name	NCBI no.	Theoretical MW/PI	Observed MW/PI	Seq. Cov. (%)	Matched peptides	Protein score (CI%)	Function
14	↑3.8*	↑1.6*	↑2.4*	MnSOD	gi|134665	24.71/8.35	20/8.12	35	9	201 (100%)	Migration inhibiting
29	↑2.1*	↑2.7*	↓1.3	GRP75	gi|21040386	73.73/5.97	58/5.84	26	14	141 (100%)	Cell proliferation
30	↑11.5**	↑4.8**	↑2.4*	PAI-1	gi|10835159	45.03/6.68	36/7.58	37	12	131 (100%)	Migration inhibiting
31	↓22.0*	↓22.7**	↑1.03	CTSB	gi|181192	37.78/5.88	25/5.43	36	9	215 (100%)	Migration enhancing
32	↓48.5**	↓49.2**	↑1.01	PHB	gi|4505773	29.79/5.57	24/5.59	47	10	243 (100%)	Migration enhancing
34	↓108.8**	↓38.2**	↓2.9**	CTSD	gi|4503143	44.52/6.1	24/5.95	30	10	171 (99.98%)	Migration enhancing

*$p < 0.05$, **$p < 0.01$ ($n = 2$)

↓ Lower expression; ↑ higher expression

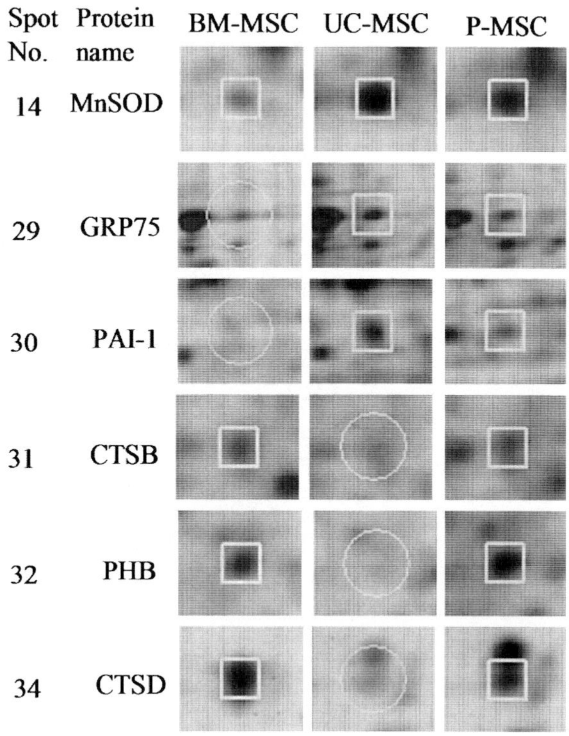

Fig. 5.3 Enlarged images of differentially expressed protein spots among three sources of MSC. Selected regions of 2-DE gels illustrate differentially expressed proteins among BM-, UC-, and P-MSC. Spots of interest with relative high intensity are shown in squares, vice versa are shown in *circles*

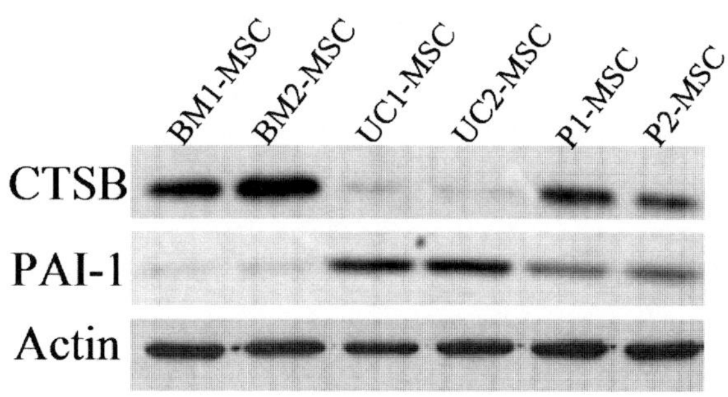

Fig. 5.4 Confirmation of differentially expression of CTSB and PAI-1. Western blot analysis showed that both CTSB and PAI-1 were differentially expressed among three sources of MSC. Actin was used as internal control

combined MS and MS/MS analysis. Six of the ten proteins were well recognized which include the manganese superoxide dismutase (MnSOD; spot 14), stress 70 protein (GRP75; spot 29), PAI-1 (spot 30), CTSB (spot 31), prohibitin (PHB; spot 32) and cathepsin D (CTSD; spot 34). They are shown as enlarged protein spots in Fig. 5.3.

Briefly, MnSOD, GRP75 and PAI-1 were expressed at a higher level in UC-MSC than that in P-MSC or BM-MSC. Lower expression level of CTSB, PHB, and CTSD was found in UC-MSC than the others. Both MnSOD and PAI-1 were found highly expressed in P-MSC than that in BM-MSC, but with CTSD expressed at a lower level. Table 5.1 also lists the details of the identified proteins, including their NCBI accession number, theoretical and observed MW/pI, sequence coverage, the number of matched peptides, protein score and functions.

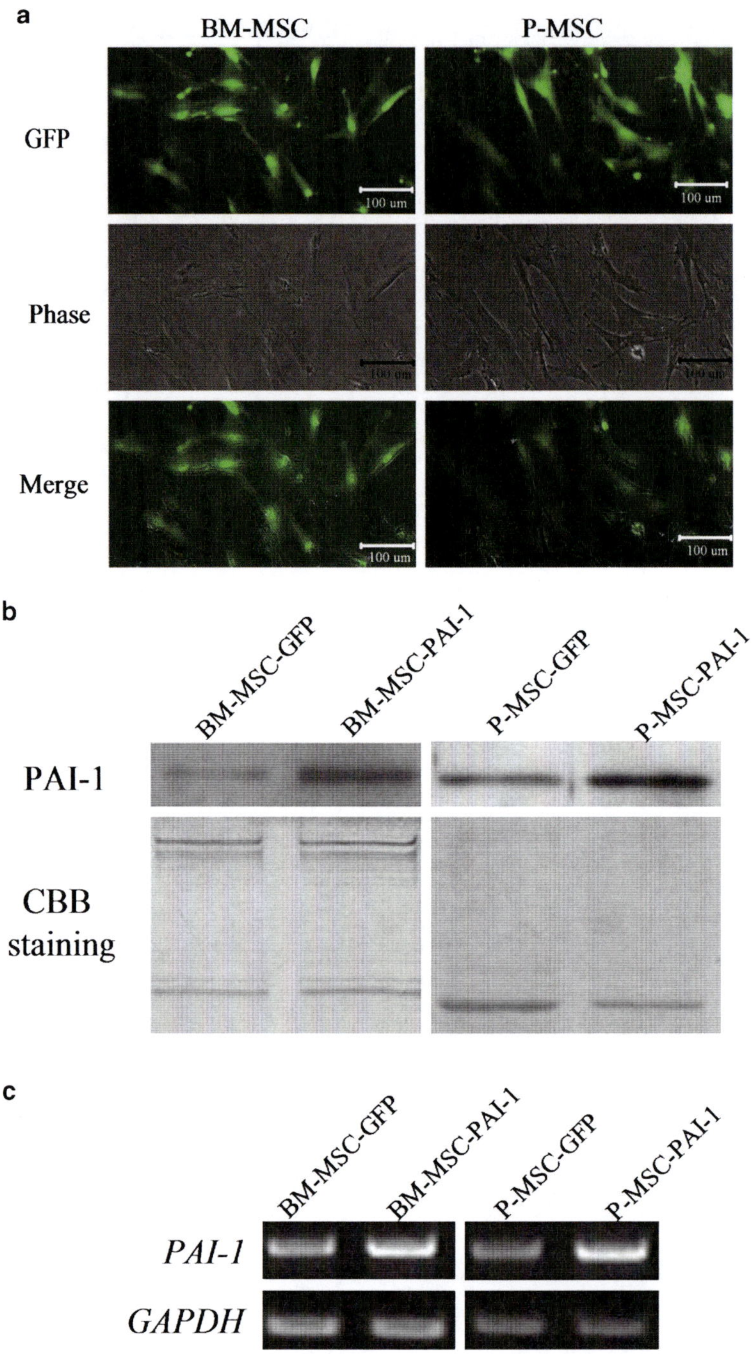

Fig. 5.5 Transduction of BM- and P-MSC. (**a**) Transduction efficiency of pLVTHM-neo-GFP. Nearly 100% transduction efficiency was achieved as shown by GFP expression. (**b**) PAI-1 secretion after transduction. Elevated PAI-1 secretion was found in transduced cells when compared with GFP control. Gel loading was stained with CBB as protein loading control. (**c**) mRNA expression of *PAI-1* after transduction. Higher *PAI-1* mRNA expression was found in transduced cells when compared with GFP control. *GAPDH* was served as internal control

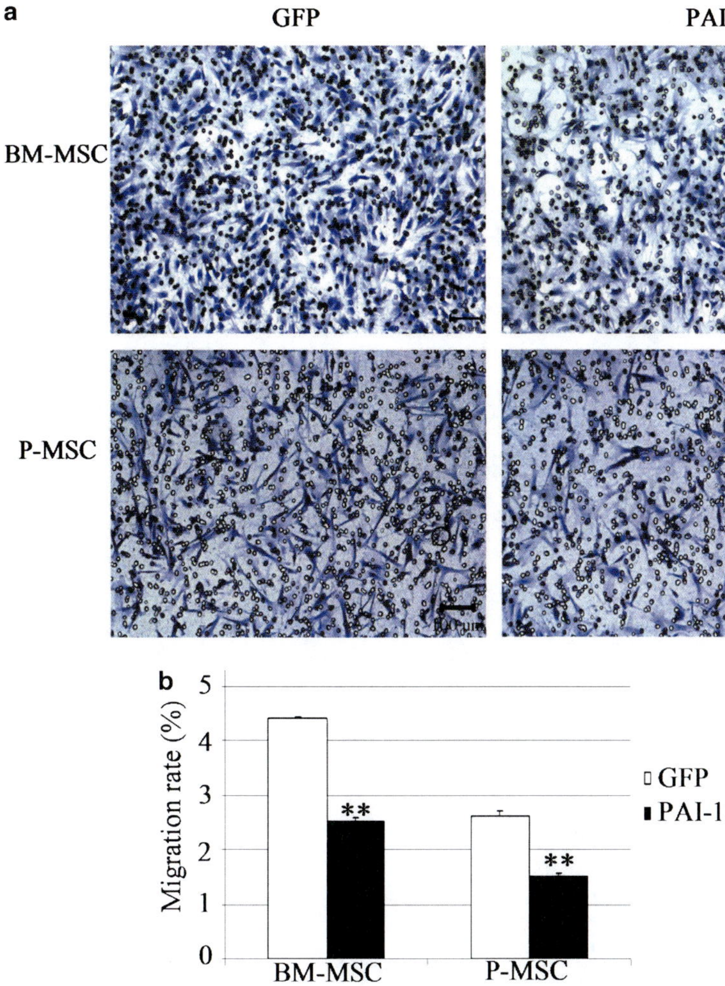

Fig. 5.6 The effect of PAI-1 overexpression on the migration ability of BM- and P-MSC. (**a**) Representative images of migration of transduced BM- and P-MSC. The migrated cells spread on the membrane were stained blue with hematoxylin. Scale bar, 100 μm. (**b**) Quantitative results of migration of transduced BM- and P-MSC ($n=3$; $^{**}p<0.01$)

5.3.4 Confirmation of the Differentially Expressed Proteins, CTSB and PAI-1

The expression level of CTSB and PAI-1 was further analyzed by western blotting on two individual samples of each MSC group (Fig. 5.4). Consistent with the 2-DE results, it demonstrated a relative abundance of PAI-1 but a lower expression level of CTSB in the UC-MSC. Besides, higher expression level of PAI-1 was found in the P-MSC than that in BM-MSC.

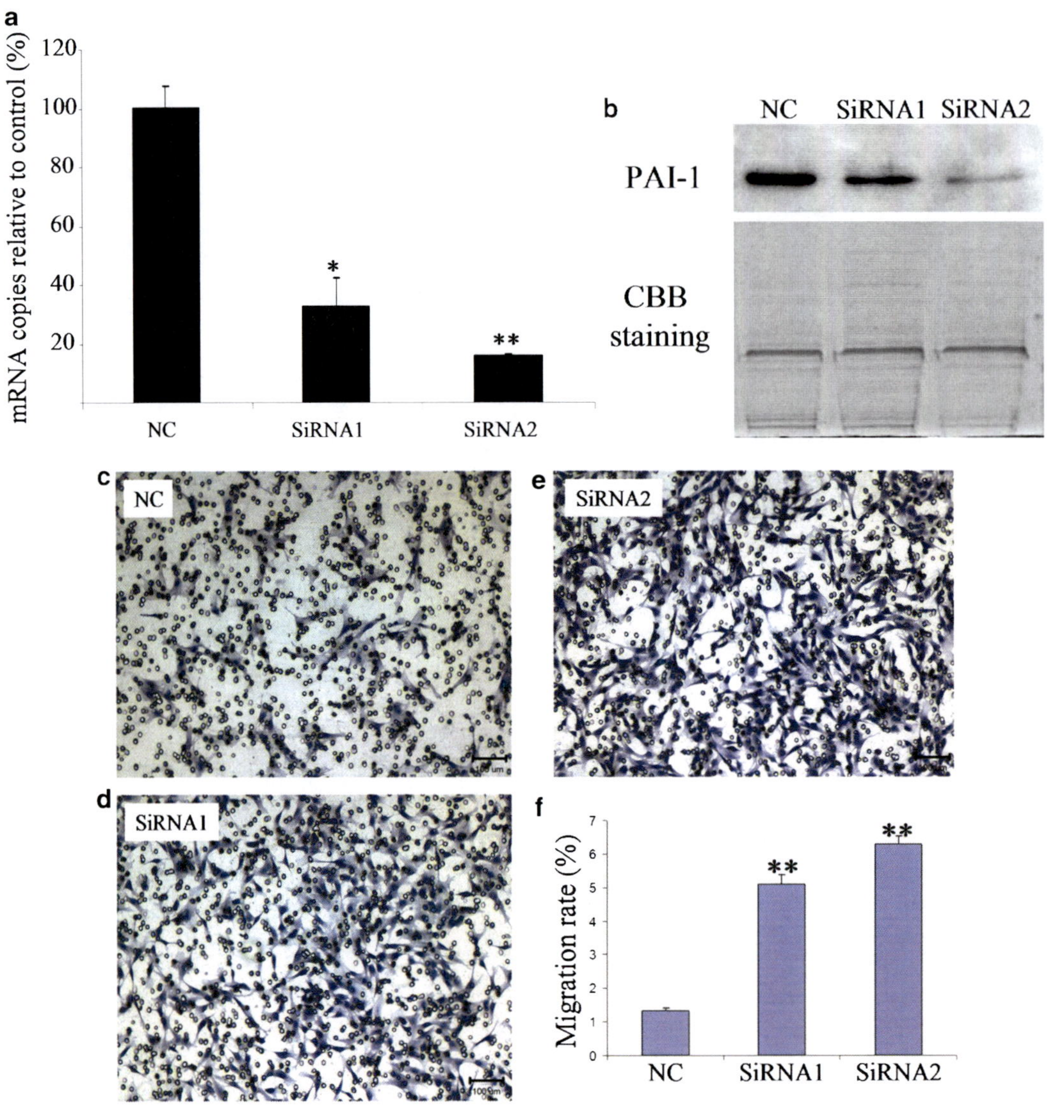

Fig. 5.7 The effect of PAI-1 knock-downs on migration ability of UC-MSC. (**a**) Quantitative real-time PCR detection of the mRNA expression of *PAI-1* after SiRNA transfection for 48 h. *GAPDH* was served as internal control ($n=3$; $*p<0.05$, $**p<0.01$). NC: control cells transfected with nontarget-directed SiRNA. (**b**) PAI-1 secretion after SiRNA transfection. CBB was served as loading control. (**c**–**e**) Representative images of SiRNA transfected UC-MSC migration. The migrated cells spread on the membrane were stained blue with hematoxylin. *Dots* are the 10 μm pores through which the MSC migrated. Scale bar, 100 μm. (**f**) Quantitative results of UC-MSC migration after SiRNA transfection ($n=3$; $**p<0.01$)

5.3.5 PAI-1 Overexpression Inhibited the Migration of P- and BM-MSC

To further investigate the role of the identified proteins in regulating MSC migration, PAI-1 was overexpressed in P- and BM-MSC, with GFP served as control. Using the lentivirus system and G418 selection, approximately 100% transduction efficiency was achieved as shown in Fig. 5.5a. Western blotting and RT-PCR demonstrated that the transduced cells secreted more PAI-1 when compared to the control (Fig. 5.5b, c). A significant decrease in the migration capacity was noted in the PAI-1-overexpressed P-MSC (i.e. 42.2%) and BM-MSC (42.9%) (Fig. 5.6).

5.3.6 Knock Down of PAI-1 Enhanced the Migration Capacity of UC-MSC

To further confirm the inhibitory activity of PAI-1 on MSC migration, the PAI-1 expression was knocked down in the UC-MSC sample. Compared with the control which was transfected with non-target-directed SiRNA, silencing efficiencies of 67 and 84% was achieved by using SiRNA1 and SiRNA2, respectively, as demonstrated by real-time PCR (Fig. 5.7a). The expression of PAI-1 was significantly reduced as seen in Fig. 5.7b. The UC-MSC samples with blocked PAI-1 expression by SiRNA1 or SiRNA2 showed a 3.9- and 4.8-fold increase of migration capacity, respectively (Fig. 5.7c–f).

5.4 Discussion

In this study, MSC was isolated from human BM, UC, and P Cells were enriched based upon their adherent property on plastic surface and propagating potential in culture medium containing 10% FBS. Concordant with previous reports [2–4], MSC derived from the three different sources expressed similar cell surface antigens as well as adipogenic and osteogenic potential. It has been reported that both P- and UC-MSC could preserve photoreceptor integrity and visual functions when injected into the subretinal space of the Royal College of Surgeons rat at their early stage of retinal degeneration [15]. Moreover, intra-cranial implantation of UC-MSC onto the striatum of hemiparkinsonian rats could induce an amelioration of apomorphine-induced rotations [16]. Taken together, P- and UC-MSC are potential alternatives to BM-MSC for cell therapy.

For effective in vivo systemic delivery, MSC is capable to migrate across the endothelial barrier lining of blood vessels in order to reach its target tissue. In earlier reports, BM-MSC has demonstrated its ability to engraft into a wide range of tissues after systemic administration [17–20]. The in vitro migration capacity of BM-MSC has also been well documented [8–12], yet similar details are scarcely available for both P- and UC-MSC. In this study, the migration capacity of the three MSC samples was found and a huge discrepancy was noted. Despite how similar their basic stem cell features are, BM possesses the highest migration capacity whilst the UC-MSC ranks the lowest. This finding provides us a platform to further investigate the underlying migration mechanisms of MSC. By the use of proteomic study, six proteins were found differentially expressed among the three MSC sources which might help to explain the discrepancy in their migration capacities.

In the present study, MnSOD was found highly expressed in UC-MSC, which indicate that it is unlikely a migration enhancing protein. As a matter of fact, MnSOD is known as a nuclear-encoded mitochondrial matrix enzyme that can scavenge toxic superoxide radicals [21]. Data suggested that the MnSOD is a candidate tumor suppressor gene, since abnormal MnSOD activity has been observed in transformed cells [22]. Artificial elevation on MnSOD expression has suppressed the metastasis of fibrosarcoma cells, which originally exhibited low endogenous levels of MnSOD [23].

GRP75 belongs to the heat shock protein 70 family, which may also acts as a chaperone. It encodes a heat-shock cognate protein, which plays a role in controlling cell proliferation. On the other hand, PHB is an evolutionarily conserved and ubiquitously expressed protein that was known to be involved in diverse cellular processes such as cell proliferation and metabolism [24]. In a previous study, HeLa cells lacking PHB expression were found defective in epidermal growth factor-induced in vitro migration. Moreover, PHB is required in the migration of epithelial cells through Ras signaling cascade [25]. The lower expression of PHB in the UC-MSC might be partly responsible for the inferior migration capability found in this type of MSC.

CTSB and CTSD are proteases while PAI-1 is protease inhibitor. Tumor cell invasiveness involves cell attachment, proteolysis of extracellular matrix components, and migration of cells through the disrupted matrix [26]. The proteolytic enzymes implicated in this process include metallo, serine, cysteine, and aspartic proteases [27–29]. CTSB is a lysosomal cysteine protease of the papain superfamily with endo- and exo-peptidase activities [30, 31]. Both tumor and stromal cell-derived CTSB can enhance tumor metastasis in transgenic mice [32]. Wickramasinghe et al. showed a direct role for CTSB in promoting oral cancer invasion [33]. CTSD, a lysosomal aspartyl protease, was found associated with cancer cell invasion and metastasis such as submucosal colorectal cancer and gastric cancer, etc. [34, 35]. By the use of antisence gene transfer technique, Glondu et al. demonstrated that the CTSD could facilitate the metastasis of breast cancer cells [36].

PAI-1 is an active serine protease inhibitor whose activity is stabilized by vitronectin binding. In general, the activation of serine protease is controlled by two physiological plasminogen activators: tissue-type plasminogen activator and urokinase plasminogen activator. Overexpression of PAI-1 reduces in vitro fibrosarcoma cell migration and in vivo metastasis by inhibiting urokinase plasminogen activator or by competing for an integrin binding site on vitronectin [37]. In our study, PAI-1 was minimally expressed in both P- and BM-MSC, which indicate that a lower level of PAI-1 might be associated with a higher migration capacity. Using lentivirus-based system, PAI-1 was overexpressed in both BM- and P-MSC. The migration capacity of PAI-1-transduced cells reduced significantly. The contribution of PAI-1 to cellular migration of MSC was further confirmed by SiRNA strategy. After the expression of PAI-1 was silenced in the UC-MSC, a significant increase on the migration capacity was noted. However, the underlying inhibitory mechanism of PAI-1 in MSC migration needs further elucidation.

It was hypothesized that the pathways and the mechanisms of cancer metastasis might be involved in MSC trafficking. Human MSC could utilize canonical Wnt signaling to mediate their migration behavior as tumor cells [12]. Besides, matrix metalloproteinase-2 and membrane-type matrix metalloproteinase-1 were shown to be associated with MSC migration [10, 11]. These findings are in concordance to our present observation that some tumor cell-related proteins might be functionally engaged in governing the trafficking of MSC. Hence, strategies targeting tumor invasion and metastasis, e.g., the use of cysteine protease inhibitors to reverse the deleterious effect of CTSB in fighting cancer [38], should be cautious, since the migration potential of endogenous MSC might be hindered which might then adversely affect the normal physiology of tissue or organ regeneration.

Taking together, five of the six identified proteins are known to be related to cell migration. Briefly, CTSB, CTSD and PHB can be regarded as migration-enhancing protein whilst MnSOD and PAI-1 are migration-inhibiting proteins. As revealed by the signal intensity derived from 2-DE and/or Western blotting, the preponderance was correlated to the MSC migration capability. Migration enhancing proteins were minimally expressed in UC-MSC. Conversely, UC-MSC expressed migration-inhibiting proteins profoundly. The same scenario was noted in P- and BM-MSC. Therefore, the present study provides evidence that some tumor metastasis-related proteins might play a pivotal role in MSC trafficking, yet further investigations would be necessary to derive the underlying molecular mechanisms. Moreover, it is the first report on the migration capability of UC- and P-MSCs, which might provide insight in testing the use of both MSCs as alternative sources in cell therapy.

Acknowledgements *Conflict of Interest Statement*: All authors declare no financial/commercial conflicts of interest.

The authors thank Dr. Didier Trono for kindly providing the lentiviral vector pLVTHM and its package plasmids. They also thank the donors of bone marrow, umbilical cord, and placenta. This work was supported by the Li Ka Shing Institute of Health Sciences Grant and RGC project: CUHK 7422_03M.

References

1. Panepucci RA, Siufi JL, Silva WA Jr, Proto-Siquiera R et al (2004) Comparison of gene expression of umbilical cord vein and bone marrow-derived mesenchymal stem cells. Stem Cells 22:1263–1278
2. Musina RA, Bekchanova ES, Belyavskii AV, Sukhikh GT (2006) Differentiation potential of mesenchymal stem cells of different origin. Bull Exp Biol Med 141:147–151

3. Miao Z, Jin J, Chen L, Zhu J et al (2006) Isolation of mesenchymal stem cells from human placenta: comparison with human bone marrow mesenchymal stem cells. Cell Biol Int 30:681–687
4. Lu LL, Liu YJ, Yang SG, Zhao QJ et al (2006) Isolation and characterization of human umbilical cord mesenchymal stem cells with hematopoiesis-supportive function and other potentials. Haematologica 91:1017–1026
5. Hordijk P (2003) Endothelial signaling in leukocyte transmigration. Cell Biochem Biophys 38:305–322
6. Nourshargh S, Marelli-Berg FM (2005) Transmigration through venular walls: a key regulator of leukocyte phenotype and function. Trends Immunol 26:157–165
7. Schmidt A, Ladage D, Steingen C, Brixius K et al (2006) Mesenchymal stem cells transmigrate over the endothelial barrier. Eur J Cell Biol 85:1179–1188
8. Zhu H, Mitsuhashi N, Klein A, Barsky LW et al (2006) The role of the hyaluronan receptor CD44 in mesenchymal stem cell migration in the extracellular matrix. Stem Cells 24:928–935
9. Lopez Ponte A, Marais E, Gallay N, Langonne A et al (2007) The in vitro migration capacity of human bone marrow mesenchymal stem cells: comparison of chemokine and growth factor chemotactic activities. Stem Cells 25:1737–1745
10. Son BR, Marquez-Curtis LA, Kucia M, Wysoczynski M et al (2006) Migration of bone marrow and cord blood mesenchymal stem cells in vitro is regulated by SDF-1-CXCR4 and HGF-c-met axes and involves matrix metalloproteinases. Stem Cells 24: 1254–1264
11. De Becker A, Van Hummelen P, Bakkus M, Vande Broek I et al (2007) Migration of culture-expanded human mesenchymal stem cells through bone marrow endothelium is regulated by matrix metalloproteinase-2 and tissue inhibitor of metalloproteinase-3. Haematologica 92:440–449
12. Neth P, Ciccarella M, Egea V, Hoelters J et al (2006) Wnt signaling regulates the invasion capacity of human mesenchymal stem cells. Stem Cells 24: 1892–1903
13. Ding SJ, Li Y, Tan YX, Jiang MR et al (2004) From proteomic analysis to clinical significance: overexpression of cytokeratin 19 correlates with hepatocellular carcinoma metastasis. Mol Cell Proteomics 3:73–81
14. Lee HK, Lee BH, Park SA, Kim CW (2006) The proteomic analysis of an adipocyte differentiated from human mesenchymal stem cells using two-dimensional gel electrophoresis. Proteomics 6:1223–1229
15. Lund RD, Wang S, Lu B, Girman S et al (2007) Cells isolated from umbilical cord tissue rescue photoreceptors and visual functions in a rodent model of retinal disease. Stem Cells 25:602–611
16. Weiss ML, Medicetty S, Bledsoe AR, Rachakatla RS et al (2006) Human umbilical cord matrix stem cells: preliminary characterization and effect of transplantation in a rodent model of Parkinson's disease. Stem Cells 24:781–792
17. Allers C, Sierralta WD, Neubauer S, Rivera F et al (2004) Dynamic of distribution of human bone marrow-derived mesenchymal stem cells after transplantation into adult unconditioned mice. Transplantation 78: 503–508
18. Bensidhoum M, Chapel A, Francois S, Demarquay C et al (2004) Homing of *in vitro* expanded Stro-1- or Stro-1+ human mesenchymal stem cells into the NOD/SCID mouse and their role in supporting human CD34 cell engraftment. Blood 103:3313–3319
19. Liechty KW, MacKenzie TC, Shaaban AF, Radu A et al (2000) Human mesenchymal stem cells engraft and demonstrate site-specific differentiation after in utero transplantation in sheep. Nat Med 6: 1282–1286
20. Gao J, Dennis JE, Muzic RF, Lundberg M, Caplan AI (2001) The dynamic in vivo distribution of bone marrow-derived mesenchymal stem cells after infusion. Cells Tissues Organs 169:12–20
21. Halliwell B, Gutteridge JMC (1986) Free radicals in biology and medicine. Oxford University Press, New York, pp 86–106
22. Oberley LW, Buettner GR (1979) Role of superoxide dismutase in cancer a review. Cancer Res 39: 1141–1149
23. Safford SE, Oberley TD, Urano M, St. Clair DK (1994) Suppression of fibrosarcoma metastasis by elevated expression of manganese superoxide dismutase. Cancer Res 54:4261–4265
24. McClung JK, Jupe ER, Liu XT, Dell'Orco RT (1995) Prohibitin: potential role in senescence, development, and tumor suppression. Exp Gerontol 30:99–124
25. Rajalingam K, Wunder C, Brinkmann V, Churin Y et al (2005) Prohibitin is required for Ras-induced Raf-MEK-ERK activation and epithelial cell migration. Nat Cell Biol 7:837–843
26. Aznavoorian S, Murphy AN, Stetler-Stevenson WG, Liotta LA (1993) Molecular aspects of tumor cell invasion and metastasis. Cancer 71:1368–1383
27. Del Rosso M, Fibbi G, Pucci M, D'Alessio S et al (2002) Multiple pathways of cell invasion are regulated by multiple families of serine proteases. Clin Exp Metastasis 19:193–207
28. Egeblad M, Werb Z (2002) New functions for the matrix metalloproteinases in cancer progression. Nat Rev Cancer 2:161–174
29. Turk V, Kos J, Turk B (2004) Cysteine cathepsins (proteases) – on the main stage of cancer? Cancer Cell 5:409–410
30. Podgorski I, Sloane BF (2003) Cathepsin B and its role(s) in cancer progression. Biochem Soc Symp 70:263–276
31. Yan S, Sloane BF (2003) Molecular regulation of human cathepsin B: implication in pathologies. Biol Chem 384:845–854

32. Vasiljeva O, Papazoglou A, Kruger A, Brodoefel H et al (2006) Tumor cell-derived and macrophage-derived cathepsin B promotes progression and lung metastasis of mammary cancer. Cancer Res 66: 5242–5250
33. Wickramasinghe NS, Nagaraj NS, Vigneswaran N, Zacharias W (2005) Cathepsin B promotes both motility and invasiveness of oral carcinoma cells. Arch Biochem Biophys 436:187–195
34. Oh-e H, Tanaka S, Kitadai Y, Shimamoto F et al (2001) Cathepsin D expression as a possible predictor of lymph node metastasis in submucosal colorectal cancer. Eur J Cancer 37:180–188
35. Ikeguchi M, Fukuda K, Oka S, Hisamitsu K et al (2001) Micro-lymph node metastasis and its correlation with cathepsin D expression in early gastric cancer. J Surg Oncol 77:188–194
36. Glondu M, Liaudet-Coopman E, Derocq D, Platet N et al (2002) Down-regulation of cathepsin-D expression by antisense gene transfer inhibits tumor growth and experimental lung metastasis of human breast cancer cells. Oncogene 21:5127–5134
37. Praus M, Collen D, Gerard RD (2002) Both u-PA inhibition and vitronectin binding by plasminogen activator inhibitor 1 regulate HT1080 fibrosarcoma cell metastasis. Int J Cancer 102:584–591
38. Lim IT, Meroueh SO, Lee M, Heeg MJ, Mobashery S (2004) Strategy in inhibition of cathepsin B, a target in tumor invasion and metastasis. J Am Chem Soc 126:10271–10277

Part II

Prostate Cancer/Stem Cells

6 Novel Human Prostate Epithelial Cell Culture Models for the Study of Carcinogenesis and of Normal Stem Cells and Cancer Stem Cells

Johng S. Rhim, Hongzhen Li, and Bungo Furusato

Abstract

Research into the mechanisms of prostate cancer progression has been limited by the lack of suitable in vitro systems. A hurdle in understanding the molecular genetic changes in prostate cancer has been the difficulty in establishing premalignant lesions and primary prostate tumors as in vitro cell cultures. Primary prostate epithelial cells grow for a finite life span and then senesce. Immortalization is defined by continuous growth of otherwise senescing cells and is believed to represent an early stage in tumor progression. To examine these early stages, we and others have developed in vitro models of prostate epithelial cell immortalization. Generation of primary human prostate epithelial (HPE) cells has been achieved using the serum-free condition. Retrovirus containing human telomerase reverse transcriptase (hTERT) was successfully used for the immortalization of primary HPE cells. Putative stem cell markers CD133 and CXCR4 were further identified in hTERT-immortalized primary nonmalignant and malignant tumor-derived HPE lines. In addition, an hTERT-immortlaized nonmalignant HPE cell were found to retain the properties of multipotent stem cells. These in vitro prostate cell culture models should be useful for the study of carcinogenesis and of normal and cancer stem cells.

Prostate cancer is the most common male cancer in the Western World and second leading cause of male cancer death in the United States [1]. The therapy most widely used against advanced disease is androgen ablation and, initially, it almost always produces objective clinical responses. However, most patients eventually relapse with ablation-resistant prostate cancer and develop metastatic disease; currently, there is no treatment that

J.S. Rhim (✉)
Department of Surgery, Uniformed Services University of the Health Sciences, Bethesda, MD 20814, USA
e-mail: jsrhim@verizon.net

J.S. Rhim and R. Kremer (eds.), *Human Cell Transformation: Role of Stem Cells and the Microenvironment*, Advances in Experimental Medicine and Biology 720,
DOI 10.1007/978-1-4614-0254-1_6,

will cure progressive hormone-refractory metastatic prostate cancer. The mechanisms of progression of prostate cancer have been extensively studied, yet are poorly understood. One of the concepts that has been evolved is that cancer arises from the neoplastic transformation of normal prostate epithelial stem cells or transit amplifying cells. Understanding normal stem cells and cancer stem cells (CSCs) may provide insight into the origin of and new therapeutics for prostate cancer. However, research in this field is limited by the lack of suitable in vitro systems.

6.1 Generation of Short-Term Cultures Derived from Benign and Prostate Cancer Patients

Studies of prostate cancer have been hampered by various factors including (a) restricted access to tissues, (b) slow in vivo growth, (c) difficulties in propagating tumor cells as well as normal cells in vitro, and (d) limited availability of prostate cancer cell lines or immortalized prostate epithelial cell lines for in vitro studies. In January 2000, we created a Prostate Cancer Cell Center in the newly established Center for Prostate Disease Research (CPDR) laboratory in the Department of Surgery, Uniformed Services University of the Health Sciences (USUHS). This center has successfully generated more than 100 primary prostate epithelial cells from primary tumors of prostate cancer patients as well as normal prostate tissue of the same patient using our established procedure (Table 6.1). We have for the first time found that a commercially readily available serum-free medium developed for human keratinocyte (K-SFM, Gibco, Grand Island, NY) is very useful in growing and maintaining primary HPE cells and for the cultivation of short-term cultures of primary HPE cells [2, 3].

6.2 Development and Characterization of Nonmalignant and Tumor-Derived Primary Tumor Prostate Epithelial Cells Immortalized by Introduction of Telomerase

Efforts spanning more than half a century, since the pioneering work of Burrows et al. [4], have produced only a few cell lines derived from human prostate epithelium. To date, only three readily and well-studied long-term human prostate cancer cell lines exist (DU-145, PC-3, and LNCaP). All were derived from metastatic lesions, thus leaving a void in reagents representing primary localized adenocarcinoma of the prostate. Nevertheless, their use has greatly contributed to current understanding of human prostate carcinogenesis and progression. Better understanding of the process of malignant transformation, the availability of recombinant DNA technology, and telomerase resulted in the successful establishment of novel primary nonmalignant and malignant tumor-derived HPE cell lines during the past decade. However, despite extensive work on the development of human prostate cancer cell lines, the proportion of

Table 6.1 Generating primary cells from biopsy specimen

Fresh radical prostatectomy specimen obtained by pathologist
Chop into small fragments (1–2 mm in size) with a sterile blade
Place pieces into type 1 collagen-coated dishes (Becton-Dickinson, MA) and mince thoroughly
Add growth medium with HEPES buffer solution with 5%FBS
Incubate at 37°C in 5% CO_2 for a week without changing medium until reaching semiconfluency
Aliquots of primary cultures: freeze in liquid nitrogen until cells were reestablished in secondary culture for further serial passages
For serial passages, routine trypsinization was used once a week in collagen-coated dishes (split ratio 1:2) containing growth medium without HEPES and serum
Growth medium: keratinocyte-SFM (#17005–42, GIBCO, MD) containing L-glutamine, human rEGF, BPE, low calcium concentration <0.1 nM

Table 6.2 Phenotypic characteristics of hTERT-immortalized human prostate epithelial cell lines

Cell line	Tissue derivation	Soft agar colony formation (%)	Androgen sensitivity	Tumor formation in SCID mice
957 E/hTERT	Malignant[a]	<0.001	No	0/5
RC-58T/hTERT/SA#4	Malignant	0.120	Yes	5/5
RC92a/hTERT	Malignant	0.155	No	2/5
RC-165N/hTERT	Benign[b]	<0.001	Yes	0/5
RC-170N/hTERT	Benign	<0.001	No	0/5

[a] Familial prostate cancer patient
[b] African American prostate cancer patient

patients that give rise to immortalized human prostate cancer cell lines is still disappointingly low. Since the inception of this cell center, we have successfully been able to establish for the first time a number of novel immortalized HPE cell lines derived from primary malignant prostate tumor as well as benign prostate tissues using telomerase, the gene that prevents senescence. Furthermore, we have succeeded in the establishment of HPE cell models for the study of prostate cancer in high risk populations, one focusing on African American prostate cancer and the other focusing on familial prostate cancer (Table 6.2). Telomerase is an enzyme responsible for replicating telomere and is composed of an RNA subunit containing an integral catalytic subunit, human telomerase reverse transcriptase (hTERT). Recent finding have implicated telomerase in the escape from cellular senescence. Transfection of hTERT into selected human cell type can itself induce immortalization. Interestingly, telomerase expression in human somatic cells does not induce changes associated with a transformed phenotype or an altered genetic phenotype.

A retrovirus vector expressing the human telomerase catalytic subunit (hTERT) [5] (a gift from Dr. V. Band, Nebraska Cancer Center, Omaha, NE) was used for immortalization. Expression of the malignant phenotype was examined for all cell lines by evaluating their proliferation rates, their ability to grow in low serum and to grow above the agar layer, to form colonies in soft agar and in vivo in SCID mice, and their sensitivity to androgen (Table 6.2). All the cells derived from primary tumor prostate epithelium exhibited the typical transformed morphology and had faster doubling times than nonmalignant prostate cell lines. The distinct phenotypic difference observed for the telomerase-immortalized cells was that two of three primary tumor-derived cell lines, but one of the nonmalignant tissue-derived cell lines, were able to form cell aggregates, grew above the agar layer, and form colonies in soft agar. Furthermore, these properties were correlated with the formation of tumors in SCID mice. Two of three telomerase-immortalized primary human prostate tumor-derived cell lines were able to form tumors in SCID mice whereas nonmalignant tissue-derived cell lines were nontumorigenic. RT-PCR assay results showed that all the cell lines expressed NKX3.1 and CK8. The PSA was not expressed in all the immortalized cell lines. The androgen receptor was also expressed in all the cell lines except in 957E/hTERT cell lines. In addition, production of androgen receptor was detected in all the cell lines except in 957E/hTERT cells by western immunoblot analysis. The growth of RC-58T/hTERT/SA#4 and RC-165N/hTERT cells exposed to DHT (1 nM) showed modest biphasic profiles which was similar to that of LNCaP cells with respect to DHT doses [6, 7]. No difference in growth was observed for 957E/hTERT, RC-92a/hTERT, and RC-170N/hTERT cells exposed to DHT compared to the control. All the cell lines expressed AMACR [8], a gene often expressed in prostate cancer by RT-PCR and western immunoblot analysis. The result obtained here has demonstrated that these telomerase-immortalized nonmalignant and primary prostate tumor-derived cell lines retained the original properties of the primary cells and express some of their prostate-specific markers. The androgen responsive properties of some cell lines should help answer questions related to androgen regulation of prostate cells. These novel in vitro models, generated from nonmalignant or primary prostate tumor epithelium, may offer unique models for the study of prostate carcinogenesis, and also provide the means for testing both chemopreventive and chemotherapeutic agents.

6.3 Identification of Putative Prostate Stem Cell Markers CD133 and CXCR4 in hTERT-Immortalized Primary Nonmalignant and Malignant Tumor-Derived HPE Cell Lines

Normal prostate epithelial stem cells have been demonstrated to exist in the basal component. The study proposed that androgen-independent stem cells give rise to two types of cells: stem cells and androgen-independent transit amplifying cells which can divide rapidly with limited proliferative capacity and differentiate into luminal cells through an intermediate phenotype. To date, several putative stem cell populations have been identified as prostate stem cells by means of clonal assay, identification of several cell surface markers, and side population (SP) analysis [9]. The first evidence for CSCs was shown in hematopoietic tumors, and this principle has been implicated in other tumors including prostate. Recent studies have postulated the existence of CSCs in the primary prostate cancer cells, prostate cancer cell lines, and animal models [9]. However, the hypothetical model for the hierarchical organization of CSCs remains unproven because of the lack of appropriate in vitro and in vivo models.

Primary cell cultures derived directly from tissues or tumors have a number of advantages because it is believed that primary cells reflect well the characteristics of the original tissues. However, primary cell culture does have difficulties because of the limited access, their finite lifespan, and the specific culture techniques. On the other hand, cell lines are widely used in many aspects of research as the most common in vitro culture model because they have a big advantage of being easy to handle for their infinite reproducible quantities. As described, most of the human prostate cancer cell lines have been established from metastatic lesions or from xenograft tumors. Despite extensive work on the development of many human prostate cancer cell lines, only a few patient sampled gave rise to immortalized HPE cell lines but none from primary HPE cells from prostate cancer patients.

Although it has been suggested that the study of CSCs should be performed using primary prostate cancer cells rather than prostate cancer cell lines, recent evidence using SP analysis and cell surface markers such as CD133 or clonal analysis shows that long-term cultured cell lines may retain a hierarchical proliferation or differentiation pattern as a potential of CSCs. We have shown that nonmalignant and malignant tumor-immortalized cell lines may contain a subpopulation of cells with stem cell properties [10, 11].

Several in vitro culture systems including primary cells, immortalized benign tissue-derived, and prostate tumor-derived cell lines have been reported as useful in the study of prostate normal stem cells and CSCs [10].

CSCs theory has now emerged as an innovative theory within the field of cancer biology, especially regarding solid tumors. It will be important to show the existence of normal stem cells and CSCs in prostate tissue. Although cell lines have some disadvantages or shortcomings compared to primary cells, in vitro culture system including primary cells, immortalized normal cell lines, and cancer cell lines may contain a heterogeneous and hierarchical subpopulation as described. Thus, the development of in vitro culture systems not only provides a useful novel tool for understanding key molecular pathways in prostate epithelial differentiation and prostate cancer progression, but also may have important implication for biological and pharmacological functions. Primary cells, immortalized normal and cancer cell lines as in vitro models for normal stem cells and CSCs in the prostate, may provide new insights into the mechanisms of prostate cancer development.

We demonstrated for the first time that hTERT-immortalized primary nonmalignant (RC-165N/hTERT) and malignant (RC-92a/hTERT) tumor-derived human prostate epithelial (HPE) cell lines retain stem cell properties with a CD133(+)/CD44(+)/alpha(2)beta(1)(+)/34betaE12(+)/CK18(+)/p63(−)/androgen receptor (AR)(−)/PSA(−) phenotype. Higher CD133 expression was detected in the hTERT-immortalized cells than in primary prostate cells. These immortalized cells exhibited “prostaspheres” in nonadherent culture systems and also maintained higher CD133 expression. The CD133(+) cells from these immortalized cell lines had high proliferative potential and were able to differentiate into AR(+) phenotype. In three-dimensional culture, the CD133(+) cells from RC-165N/hTERT cells produced branched structures, whereas the CD133(+) cells from RC-92a/hTERT cells produced large irregular spheroids with less branched structures. SDF-1 induced, but anti-CXCR4 antibody inhibited, migration of CD133(+) cells from RC-92a/hTERT cells, which coexpressed CXCR4. CXCR4/SDF-1 may sustain tumor chemotaxis in CSCs. Furthermore, immunostaining of clinical prostate specimens showed that CD133 expression was detected in a subpopulation of prostate cancer cells and corresponded to the loss of AR. Expression of CXCR4 was also detected in CD133(+) cancer cells [10].

6.4 Telomerase-Immortalized Nonmalignant HPE Cells Retain the Properties of Multipotent Stem Cells

There is now increasing evidence that tumor cells are organized as hierarchy originated from rare stem cells that are responsible for maintaining the tumor. A stem cell model for prostate organization and prostate cancer has been postulated [12, 13] and verified by using mouse model [14, 15], human embryonic stem cell [16], and adult prostate normal and cancer tissues [17]. The human prostate is composed of ducts and acini embedded in a stromal matrix of fibroblastic and myofibroblastic cells. The majority of cells in prostatic basal compartment of acini are transient amplifying cells (TA) that express higher level of cytokeratin 5/14 and lower level of cytokeratin 8/18. Only less than 1% of stem cells were identified by using $\alpha2\beta1$hi-integrin [18, 19] and CD133 markers [20] in the prostate tissues. However, stem cells are normally quiescent in normal prostate tissue and senescent in vitro primary culture. Therefore, the properties of stem cells are not known. It has been reported that other tissue stem cells express OCT-4 and BMI-1 [21–23]. OCT-4, a POU transcription factor, is essential for maintaining undifferentiated cell fate in embryonic stem cells, multipotent adult progenitor cells, and hematopoietic stem cells [21, 22]. BMI-1, a transcription factor B lymphoma Mo-MLV insertion region 1, is required for maintenance of adult self-renewing hematopoietic stem cells [22] and played a key role in the stem cell self-renewal pathways including Hedgehog, Notch, and Wnt signaling pathways [24, 25]. Recently putative prostate epithelial stem cells have been identified in primary and immortalized HPE cells [10, 26, 27]. The expression of telomerase is associated with stem cells. Telomerase is usually repressed in normal human somatic cells but is found only in the stem cell compartment of several adult tissues [28]. Low telomerase activity in normal prostate primary cells [29] and telomerase activation in 80% of cases of prostate cancer have been observed [30] suggesting that the presence of normal stem cells and the immortality conferred by telomerase play key roles in cancer development. Telomere shortening and telomerase activity seem to have a critical role in initiating carcinogenesis in CSCs [31]. We have reported the multipotent stem cells (subpopulation of RC170N/h/clone 7 cells) that resist senescence were developed by transduction of the cells with telomerase. The subpopulation of RC170 N/h/clone 7 cells was clonogenic growth in monolayer culture, sphere formation in suspension culture, and differentiated into progenitors of all cell lineages within prostate gland. RC170N/h/clone 7 cells differentiated into multiple tissues following the renal capsule and subcutaneous inoculation into NOD-SCID mice [11].

6.5 Potential Stem Cell Marker Expression in RC170N/h/Clone 7 Cells

The potential stem cell markers expressed in RC170N/h/clone 7 cells were detected by fluorescence immunostaining and RT-PCR (Table 6.3). Less than 5% of giant cells were strongly stained for the human embryonic stem cell marker Oct-4, Bmi-1, and integrin $\alpha 2\beta 1$hi in RC170N/h/clone 7 cells and RC170N/h parental cells. Subpopulation of cells was stained for integrin $\alpha 2\beta 1$hi [18, 19]. We verified expression of Oct-4 in primary culture cells, RC170N/h parental cells, and RC170N/h/clone 7 cells from the different passages by using RT-PCR method. The expression of Oct-4 mRNA was detected in primary culture cells, RC170N/h parental cells, and RC170N/h/clone 7 cells.

6.6 Proliferation of RC170N/h/Clone 7 Cells Under Different Culture Conditions

To observe the growth behaviors of RC170N/h/clone 7 cells, we carried out different culture process with different media. We compared the proliferate capacity of the parental cells and clonal cells in the KGM medium. The clonal cell line showed similar proliferate capacity compared to parental cells. Under suspension culture in KGM medium, the spheres were observed in RC170N/h parental and clone 7 cells. The sphere formation is the properties of stem cells in the suspension culture. We have also observed that the stem-like cells proliferated symmetrically and asymmetrically under suspension culture after culturing for 1 day. Similar results were observed in the RC170N/h parental cells and clone 7 cells in early passage (p18–p23), middle passage (p39–p45), and later passage (p55–p65). Similar soft agar colony formation efficiencies were observed with the parental and the RC170N/h/clone 7 cells.

6.7 Phenotypic Differentiation and Molecular Markers of Differentiation in the RC170N/h/Clone 7 Cells In Vitro

To evaluate differentiation abilities, we have cultured RC170N/h/clone 7 cells in the matrigel with different media. After culturing for 10 days with matrigel, the formation of branched ducts, spheroids were observed in both KGM and DMEM medium. The more differentiated morphology-like multidendrite, spindle cells in KGM medium as well as neurofilament-like net and ductal budding structure in DMEM were also observed. To identify prostate epithelial stem cells, the potential stem cell markers, CD133,

Table 6.3 The cell markers expressed in RC170 N/h/clone 7 cells

	RC170N/h/clone 7 cells	
Markers	KGM medium	DMEM (10% FBS + Ins.)
Stem cell markers		
Oct 4	Positive (<5%)	Positive (<5%)
Bmi-1	Positive	Positive
CD133	Positive	Positive
ABCG2	Positive	Positive
Integrin alpha 2	Positive	Positive
Integrin beta 1 (high)	Positive	Positive
Basal cell markers		
CK5/CK14	Positive (80–100%)	Positive
CD44	Positive	Positive
GSTPi	Positive	Positive
P63	Negative	Positive (<10%)
Luminal cell markers		
CK8	Positive	Positive
CK18	Positive	Positive
AR	Positive (PCR)	Positive (PCR and IHC)
PSCA	Positive	Positive
PSA	Negative	Positive
PAP	Negative	Positive
Neuroendocrine cell markers		
Chromogranin A	NT	Positive
Fibroblast cell markers		
Vimentin	NT	Positive
Myofibroblast		
Desmin	NT	Positive
Squamous cells or renal endothelial cells		
Involucrin	NT	Positive

The expression of markers was analyzed by flow cytometry, IHC, and RT-PCR. NT: not tested

intergrin α2β1, and CD44 were examined. FACS analysis and immunofluorescence staining showed that near to 80% of RC170N/h/clone 7 cells were CD133 positive and near to 100% of RC170N/h/clone 7 cells were intergrin α2β1 and CD44 positive. Less than 5% RC170N/h/clone 7 cells were stronger stained with intergrin α2 and β1 antibodies. To identify expression of differentiation markers in RC170N/h/clone 7 cells, we stained the cells with combination fluorescence staining in DMEM culture condition. The population of the cells that expressed prostate epithelial cell markers (Table 6.3); (1) a luminar cell lineage which expressed integrin β1/NKX3.1 [32], CK18/AR, prostate stem cell antigen (PSCA) [33], AR, E-cadherin, and PSA; (2) a basal cell lineage which expressed CK5, CD44, and p63; and (3) a neuroendocrine cell lineage which expressed chromogranin A. The luminal lineage cells expressed integrin α2β 1, CK5/14 (but not CK18), and they differentiated into cells that expressed CK18. Based on our observations, the intergrin α2β1 and CK18 positive cells gave rise to the intermediate cells CK18+/PSCA+ and terminally differentiated luminal cells AR+/PSA+. Since p63 is a marker for basal cells and an epithelial stem cell marker [34–36], we performed different combinations using P63 antibody staining. Double staining with p63 and AR antibodies showed that p63 negative cells can produce p63 positive cells, but cells positive for both p63 and AR were not found. In addition, we also found that cells derived from both ectoderm and mesoderm were presented in the RC170N/h/clone 7 cell population. The cells were cultured in DMEM + 10% FBS + Ins. medium, and we found chromogranin A-positive cells, which differentiate to neurons, were present in the cell population; involucrin positive cells, squamous cells, or renal endothelial cells were derived from endoderm or ectoderm during embryonic development. Connective tissue stem cells (mesenchymal cells) were also present in the cell population. The cells were positively stained for vimentin (fibroblast), integrin β-1 and AR, and desmin (myofibroblast). The terminal differentiation marker PAP also detected in the RC170N/h/clone 7 cells in DMEM cultured condition.

6.8 RC170N/h/Clone 7 Cells Differentiate into Multiple Tissues in Xenograft Tissues Following Subcutaneous and Subrenal Capsule Transplantation into NOD-SCID Mice

The graft tissues were harvested from subcutaneous inoculation of NOD-SCID mice after 6 months. The small tissues (0.2 × 0.3 × 0.2 cm) were observed at the inoculation site with RC170N/h/clone 7 (p58) cells and RC170N/h (p43) parental cell. The histological analysis results showed that multiple tissues were resided in the graft tissues. The molecular markers of EpCAM for human-specific epithelial cells, smooth muscle actin for muscles and 34βE12 as well as CK18 for epithelial cells were positive stained in the series of frozen or paraffin tissue sections by immunofluorescence analysis. The results showed that the xenograft tissues contains the tissues differentiated from multiple lineages. We have further stained the tissues with CD133, K903, CK18, p63, AR, PSA, PAP, smooth muscle actin, chromogranin A, and desmin. We have demonstrated that RC170N/h/clone 7 cells differentiate into CD133−/k903+/CK18+/p63+/AR−/PSA−/PAP− transit-amplifying cells in the graft tissues. Smooth muscle actin positive stained cells were also found in the tissues. Chromogranin A positive stained cells were not found. Irregular hierarchy of primary prostate acini-like morphology was also observed in the grafted tissues. The cells were strongly stained by K903 which identifies keratins 5 and 6, and by CK 18 also. In the acini-like tissues, P63 positive stained cells were dominant and p63 negative cells were also present. However, further differentiation markers AR, PAP, and PSA were not detected. Smooth muscle actin positive stained cells were observed around epithelial cells. Similar results were also obtained from grafted tissues following the subrenal capsule inoculation of RC170N/h/clone 7 into NOD-SCID mice.

6.9 Summary

We have described our efficient method for the processing of primary human biopsy samples and the generation of HPE cells in serum-free condition. We have recently generated five new immortalized HPE cell lines derived from both the benign and malignant tissues of prostate cancer patients with telomerase. Examination of these cell lines for their morphologies and proliferative capacities, to respond to androgen stimulation, to grow above the agar layer, and to form tumors in SCID mice, suggests that they may serve as valid, useful tools for the study of prostate carcinogenesis. In addition, we have identified putative stem cell markers CD133 and CXCR4 in hTERT-immortalized primary nonmalignant (RC-165N/hTERT) and malignant (RC-92a/hTERT) tumor-derived HPE cell lines and in prostate cancer specimen. Furthermore, we have found that RC170N/hTERT/clone 7 cells cloned from hTERT-immortalized nonmalignant HPE cells retain the properties of multipotent stem cells. These novel in vitro models may also offer useful tools for the study of the biological features and functional integration of normal and CSCs in prostate.

Acknowledgments This work was funded by grant from the US Army Medical Research and Material Command and also supported by a Department of Defence Prostate Cancer Research Program (PCO30694).

References

1. Jemal A, Siegel R, Ward E, Murray T, Xu J, Thun MJ (2007) Cancer statistics 2007. CA Cancer J Clin 57:43–66
2. Yasunaga Y, Nakamura K, Ewing CM, Isaacs WB, Hukku B, Rhim JS (2001) A novel human cell culture model for the study of familial prostate cancer. Cancer Res 61:5969–5973
3. Gu Y, Li H, Miki J, Kim KH, Furusato B, Sesterhenn IA et al (2006) Phenotypic characterization of telomerase-immortalized primary non-malignant and malignant tumor-derived human prostate epithelial cell lines. Exp Cell Res 312:831–8434
4. Rhim JS (2000) In vitro human cell models for the study of prostate cancer. Prostate Cancer Prostatic Dis 3:229–235
5. Ratsch SB, Gao Q, Srinivasan DE, Wazer DE, Band V (2001) Multiple genetic changes are required for efficient immortalization of different subtypes of normal human mammary epithelial cells. Radiat Res 155:143–150
6. Lee C, Sutkowski DM, Sensibar JA et al (1995) Regulation of proliferation and production of prostate-specific antigenin androgen-sensitive prostate cancer cells, LNCaP, by dihydrotestosterone. Endocrinology 136:796–803
7. de Launoit Y, Veilleux R, Dufour M et al (1991) Characteristics of the biphasic action of androgens and of the potent antiproliferative effects of the new pure antiestrogen EM-139 on cell cycle kinetic parameters in LNCaP human prostatic cancer cells. Cancer Res 51:5165–5170
8. Lou J, Zha S, Jage WR et al (2002) Alpha-methylacyl-CoA racemase; a new molecular marker for prostate cancer. Cancer Res 62:2220–2226
9. Miki J, Rhim JS (2008) Prostate cell culture as in vitro models for the study of normal stem cells and cancer stem cells. Prostate Cancer Prostatic Dis 11:32–39
10. Miki J, Furusato B, Li H et al (2007) Identification of putative stem cell markers, CD133 and CXCR4 in hTERT-immortalized primary nonmalignant and malignant tumor-derived human prostate epithelial cell lines and in prostate cancer specimens. Cancer Res 67:3153–3161
11. Li H, Zhou J, Miki J, Furusato B, Gu Y et al (2008) Telomerase-immortalized non-malignant human prostate epithelial cells retain the properties of multipotent stem cells. Exp Cell Res 314:92–102
12. Isaacs JT, Coffey DS (1989) Etiology and disease process of benign prostatic hyperplasia. Prostate Suppl 2:33–50
13. Bonkhoff H (1996) Role of the basal cells in premalignant changes of the human prostate; a stem cell concept for the development of prostate cancer. Eur Urol 30:201–205
14. Xin L, Ide H, Kim Y et al (2003) In vivo regeneration of murine prostate from dissociated cell populations of postnatal epithelia and urogenital sinus mesenchyme. Proc Natl Acad Sci USA 100(suppl 1): 11896–11903
15. Tsujimura A, Koikawa Y, Salm S et al (2002) Proximal location of mouse prostate epithelial stem cells: a model of prostate homeostasis. J Cell Biol 157:1257–1265
16. Taylor RA, Cowin PA, Cunha GR et al (2006) Formation of human prostate tissue from embryonic stem cells. Nat Methods 3:179–181
17. Schalken JA, van Leenders G (2003) Cellular and molecular biology of the prostate stem cell biology. Urology 62:11–20
18. Hudson DL, O'Hare M, Watt F, Masters JR (2000) Proliferative heterogeneity in the human prostate evidence from epithelial stem cells. Lab Invest 80:1243–1250
19. Collins AT, Habib FK, Maitland NJ, Neal DF (2001) Identification and isolation of human prostate epithelial stem cells based on alpha(2)beta(1)-integrin expression. J Cell Sci 114:3865–3872
20. Richardson GD, Robson CN, Lang SH et al (2004) CD133, a novel marker for human prostate epithelial stem cells. J Cell Sci 117:3539–3545
21. Nichols J, Zevnik B, Anastassiadis K et al (1998) Formation of pluripotent stem cells in the mammalian embryo depends on the POU transcription factor Oct4. Cell 95:379–391
22. Park IK, Qian D, Kiel M et al (2003) Bmi-1 is required for maintenance of adult self-renewing haematopoietic stem cells. Nature 423:302–305
23. Molofsky AV, Pardal R, Iwashita T et al (2003) Bmi-1 dependence distinguishes neural stem cell self-renewal from progenitor proliferation. Nature 425:962–967
24. Reya T, Morrison SJ, Clarke MF et al (2001) Stem cells, cancer, and cancer stem cells. Nature 414:105–111
25. Karhadkar SS, Bova GS, Abdallah N (2004) Hedgehog signalling in prostate regeneration, neoplasia and metastasis. Nature 431:707–712
26. Litvinov IV, Vander Griend DJ, Xu Y et al (2006) Low-calcium serum-free defined medium selects for growth of normal prostatic epithelial stem cells. Cancer Res 66:8598–8607
27. Gu G, Yuan J, Wills M et al (2007) Prostate cancer cells with stem cell characteristics reconstitute the original human tumor in vivo. Cancer Res 67:4807–4815
28. Harrington L (2004) Does the reservoir for self-renewal stem from the ends? Oncogene 23:7283–7289
29. Soda H, Raymond E, Sharma S (2000) Effects of androgens on telomerase activity in normal and malignant prostate cells in vitro. Prostate 43:161–168
30. Sommerfeld HJ, Meeker AK, Piatyszek MA et al (1996) Telomerase activity: a prevalent marker of malignant human prostate tissue. Cancer Res 56:218–222

31. Ju Z, Rudolph KL (2006) Telomeres and telomerase in cancer stem cells. Eur J Cancer 42:1197–1203
32. Bieberich CJ, Fujita K, He WW et al (1996) Prostate-specific and androgen-dependent expression of a novel homeobox gene. J Biol Chem 271:31779–31782
33. Reiter RE, Gu Z, Watabe T (1998) Prostate stem cell antigen: a cell surface marker overexpressed in prostate cancer. Proc Natl Acad Sci USA 95:1735–1740
34. Signoretti S, Waltregny D, Dilks J (2000) p63 is a prostate basal cell marker and is required for prostate development. Am J Pathol 157:1769–1775
35. Signoretti S, Pires MM, Lindauer M (2005) p63 regulates commitment to the prostate cell lineage. Proc Natl Acad Sci USA 102:11355–11360
36. Signoretti S, Loda M (2006) Defining cell lineages in the prostate epithelium. Cell Cycle 5:138–141

Prostate Tumor Cell Plasticity: A Consequence of the Microenvironment

7

Clayton Yates

Abstract

During each step of prostate cancer metastasis, cancer displays phenotypic plasticity that is associated with the expression of both epithelial and mesenchymal properties or an epithelial to mesenchymal transition. This phenotypic transition is typically in response to microenvironment signals and is the basis for basic cancer cell survival (e.g. motility and invasion versus proliferation). In this review we discuss the loss and gain of E-cadherin expression as a marker of tumor plasticity throughout the steps of metastasis, and particularly focus on dynamic tumor–stromal interaction that induce a cancer cell-associated mesenchymal to epithelial reverting transition in the bone and liver microenvironments.

7.1 Epithelial to Mesenchymal Transition

Histological evidence of distinct neoplastic cell types within a tumor mass were observed as early as 1978 [1]. There are two main cell types; epithelial and mesenchymal cells, albeit most tumor cells are derived from epithelial origins, however in 1987 the term epithelial to mesenchymal transition (EMT) was utilized. This was subsequently followed by Elizabeth Hay in 1995 [2], with a cellular characterization of transitioned cells, that is still currently utilized to identify phenotypic subtypes within the tumor mass. Several hallmarks to phenotypically characterize these transitioned cells, such as cellular morphogenesis, change in shape, and tissue organization have all been associated with EMT. However, loss of cell–cell connectivity appears to an essential step feature. Normal epithelial cells comprise a sheet of cells that adhere laterally to each other by cell-to-cell junctions. In addition, epithelial cells have apical–basolateral polarization that is maintained through organization of the actin cytoskeleton, which has intimate interactions with cell membrane adhesion molecules such as cadherins, tight junctions, and certain integrins. This allows the polarized cells to maintain cell–cell junctions as a lateral belt, preventing robust cell motility, while remaining within the epithelial layer.

Mesenchymal-like cells, on the other hand, are spindle-shaped cells that exhibit end-to-end polarity,

C. Yates (✉)
Department of Biology and Center for Cancer Research, Tuskegee University, Tuskegee, AL 36088, USA
e-mail: cyates@mytu.tuskegee.edu

J.S. Rhim and R. Kremer (eds.), *Human Cell Transformation: Role of Stem Cells and the Microenvironment*, Advances in Experimental Medicine and Biology 720,
DOI 10.1007/978-1-4614-0254-1_7, © Springer Science+Business Media, LLC 2011

and have fibroblast morphology. Mesenchymal cells do not form an organized cell layer, nor do they have the same apical–basolateral organization, polarization of cell surface molecules, and the actin cytoskeleton as epithelial cells. Cell–cell contacts with neighboring mesenchymal cells are possible, however limited to focal adhesion only. As such this provides the freedom to migrate and interact with the surround extracellular matrix (ECM). Cell migration results from dynamic remodeling of actin into filamentous filopodia, lamellipodia, stress fibers. These cell protrusions lead to dynamic interactions with ECM substrates, which are mainly intergrins. The onset of these cell extensions are a prerequisite for maintenance of cell motility in normal and cancer cells, whether they are initiated spontaneously or induced by chemokines and growth factors. Coincidently, the migration mechanisms that occur in normal, non-neoplastic cells, such as embryonic morphogenesis, wound healing and immune-cell trafficking are identical to neoplastic cells [3, 4].

7.1.1 Epithelial to Mesenchymal Transition in Prostate Cancer

EMT has been shown to be a necessary step in the dissemination of cancer cell from the primary tumor mass. During this process there have been documented changes in the phenotypic expression of the cancer cells including a reduction in the cell adhesiveness. In-depth analysis showed that reduced or aberrant expression of cytokeratin levels, and cell–cell contacts related proteins are observed over multiple cancer types including breast and prostate cancer. Adhesive complexes such as ZO-1, desmoplakin, and E-cadherin are typically loss, and serve as a prerequisite for dissemination. The clinical significance of E-cadherin loss has also been well documented. Decreased expression of cell adhesion molecule E-cadherin has been largely observed to be inversely correlated clinical characteristic including grade, local invasiveness, and biochemical failure after salvage radiotherapy. Furthermore, patients with biochemical failure after prostatectomy and aberrant E-cadherin expression are likely to have subclinical disseminated disease [5]. Thus, the mechanisms responsible for such changes in adhesion complexes are of great interest.

Majority of the reports focused on of *E-cadherin* gene (*CDH1*), suggest that hypermethylation of the *E-cadherin* promoter [6, 7], is the main mode of downregulation, however a combination of mutations in one allele with loss or inactivation (by DNA methylation) of the remaining allele [8, 9] has been observed. However, in many types of cancer including breast and prostate cancers, E-cadherin expression is lost without mutations in the gene [10], due to transcriptional repression of *E-cadherin.* Concomitant with the loss of E-cadherin, N-cadherin levels increases during prostate carcinomas. This increased expression of N-cadherin has also been observed in invasive prostate cancer cell lines, and is associated with androgen deprivation [11]. The decreases in E-cadherin expression and increases in N-cadherin expression have been shown to be correlated with increased metastatic ability [12, 13]. Up-regulation of N-cadherin, and cadherin-11, and OB (osteoblasts) cadherin are typically associated with high-grade E-cadherin negative tumors. Other EMT-related changes included transition from cubodial morphology to a spindle-shaped fibroblastic morphology, and genotypic changes including loss of cytokeratin, and increased vimentin, snail, collagen I, thrombospondin-I, and other mesenchymal genes. However, the most consistent marker of EMT has been E-cadherin.

The relevance of EMT-associated markers is supported by studies describing how expression is regulated. Many transcription factors such as the family of zinc finger proteins of the Slug/Snail family, EF1/ZEB1, SIP-1, and the basic helix-loop-helix E12/E47 factor that interact with E-box sequences in the proximal E-cadherin promoter region triggering repression. Of these transcriptional repressors, forced expression of SNAIL is sufficient to induce EMT in $ARCAP_E$ and LnCaP prostate cancer cell lines [14], while Slug acts to only regulate cell proliferation [15]. However, recent reports have suggested in PC-3 cells that SNAIL inhibition alters common EMT markers, but does not affect invasiveness [16].

Other transcription factors are implicated as EMT mediators as well. TWIST, a highly conserved bHLH transcription factor, is upregulated in 90% of prostate cancer tissues. RNAi interference of TWIST expression significantly increased sensitivity to the anticancer drug taxol-induced cell death [17]. Furthermore, in addition to EMT, TWIST may also promote prostate cancer to bone metastasis by modulating prostate cancer cell-mediated bone remodeling via regulating the expression of a secretory factor, DKK-1, and enhancing osteomimicry of prostate cancer cells [18]. Thus, multiple factors contribute EMT in prostate cancer cells. Although the complex mechanisms that regulate the expression of multiple factors simultaneously in prostate cancer one common observation is that targeting individual factor is sufficient to reverse step wise events associated with EMT, thus providing targets for the development of therapeutics.

Decreased cell–cell adhesion in many cancers may not only be the result of direct transcriptional regulation. Soluble factors such as epidermal growth factor (EGF), scatter factor/hepatocyte growth factor (SF/HGF), and members of the transforming growth factor, TGFβ1, and basic fibroblast growth factor (bFGF) families have been shown to promote EMT in several model systems. Most all of these have been shown to influence the downregulate of E-cadherin expression with subsequent increased cell proliferation, dedifferentiation, and induction of cell motility [19–21]. As cancer-associated EMT is reversible, the loss of cell–cell connections creates a situation where decreased E-cadherin levels concede the tight junctions and enable apically secreted soluble growth factors to establish an autocrine loop with the basolaterally sequestered receptors. [22]. This feed-forward mechanism supports the maintenance of the mesenchymal phenotype. Although decreased levels of E-cadherin mRNA occurs at the transcriptional level, E-cadherin stability is a direct result of phosphorylated catenins. Extensive investigations have revealed that increased phosphorylation of the preferential catenins, β-catenin and p120, destabilize the cadherin complex thus inducing scattering of cancer cell lines to a more invasive phenotype [23]. We have showed that DU-145 and PC-3 cells express aberrant p120ctn and ß-catenin, and this is reversible through blockage of EGFR signaling [24]. In addition to disrupting the cell–cell junctions and enabling a more migratory phenotype, EGF upregulates secretion of matrix metalloproteinases that degrade the ECM aiding in tumor dissemination. EGF upregulates matrilysin (MMP-7) that mediates extracellular cleavage of E-cadherin, thereby further disrupting cell–cell adhesion and switching of prostate cells from a lesser to a highly invasive phenotype [25]. Thus ADAM10, ADAM9 knockdown increased E-cadherin and integrins and modulates epithelial phenotype and functional characteristics of prostate cancer cells [26], further emphasizes the vast number of pathways regulating E-cadherin expression.

Accumulating evidence suggest that growth factor-induced EMT is the result of transcriptional reprogramming and chromatin remodeling. Of the soluble growth factors mentioned, TGFß-1 is the most noted, however for the focus of this review we will focus on tyrosine kinase growth regulation of EMT. IGF-I stimulation of $ARCaP_E$ cells upregulates ZEB1 expression in prostate cancer cells exhibiting a phenotype and increased cell migration. The authors also demonstrated that this is mediated through activation of MAPK/ERK pathway [27]. Similarly EGF, which is a robust stimulator of the MAPK pathway, resulted in activation of new EMT-related marker receptor activator of NF-κB ligand (RANKL), and enhances bone resorption and bone turnover, facilitating successful bone metastasis [14]. Findings from our laboratory, support these observations in DU-145 and PC-3, both of which undergo enhanced EMT upon EGF stimulation [14, 28, 29].

It is important to note that in addition to transcriptional repression, DNA methylation of key tumor suppressor and EMT-related genes has been observed. In the case of E-cadherin, available cell culture models DU-145 and PC-3 do not exhibit methylation of E-cadherin, however this is not observed clinically, as E-cadherin is methylated in 70% of late-stage prostate [30].

More recently microRNAs (miRNAs), small non-coding RNAs regulating gene expression, control large cohort of genes post-translationally. A number of miRNAs have been identified as either oncogenes or tumor suppressor genes. The importance EMT-related miRNAs was first discovered in breast cancer model where the mir-200 family was found to indirectly regulate EMT via targeted regulation of transcription factor 8 (ZEB1, DEF1, Nil-2-A). Further evidence of miRNA involvement in E-cadherin expression is increased expression of mir-9 [31] and mir-9-1 [32], which target E-cadherin [33]. Of the miRNA 200 family return of miR-200b levels in PC3 induced to overexpress PDGF-D cells led to reversal of the EMT phenotype, which was associated with the downregulation of ZEB1, ZEB2, and Snail2 expression. Moreover, this resulted in inhibited cell migration and invasion, with concomitant repression of cell adhesion to the culture surface and cell detachment [34].

7.2 Tumor–Stromal Interactions Influence Tumor Plasticity

Several reports have shown that inoculation of prostate cancer cells within the bone microenvironment induces reciprocal interactions that results in seemly phenotypic and genetic changes in the cancer cells. This is evidenced by the LNCaP cells, ARCaP, and PC-3 cells after injection into the bone, yielding cell line derivatives C4-2, ARCaP M, and PC-3 M cells, which exhibit more mesenchymal phenotypes and increased growth and invasiveness [35, 36]. This general concept was described by Paget in 1889 [37] who proposed that the seeding of metastatic cancer cells is dependent upon the host organ microenvironment (the "seed and soil" concept). The realization that the host microenvironment comprises a number of stromal cell types (fibroblasts, smooth muscle cells, endothelial cells, neural endocrine cells, inflammatory cells) and a host of growth factors (VEGF, IGF, FGF, EGF, and TGFβ) and extracellular matrices (laminin, fibronectin, collagen, and proteoglycans) and the fact that these molecules support cancer growth and progression emphasizes the possible directive role of the local host tumor microenvironment at both primary and metastatic sites that could affect the overall growth and malignant potential of the transformed cancer epithelial cells. Our findings of this is indeed the case, as EMT-related cellular behavior of invasive cancer cells is subject to regulation by the tumor microenvironment [29]. To demonstrate the dynamic influence we examined paired primary and liver metastasis prostate cancer patient samples and stained for E-cadherin. Figure 7.1 shows that E-cadherin is densely expressed within epithelial compartment of primary prostate tumors and liver metastases, with only the prostate cancer cells that have invaded the local stroma exhibiting complete lack of E-cadherin expression. With the prostate cancer liver metastasis, cell appear to be less differentiated than primary tumor, and morphologically similar to the hepatocytes, through heterotypic cell–cell interactions. That prostate cancer cells would appear morphogencially similar to liver cells is rooted in the concept of prostate cancer osteomimicry within bone microenvironment [38]. This is also supported by observation that breast cancer metastases to the liver seem to recreate hepatocyte cords with carcinoma cells [39]. We further observed that DU-145 and PC-3 cells reexpress E-cadherin through heterotypic interactions and display complete reversion of the mesenchymal phenotype, with decreased vitmentin and increased cytokeratin, when cocultured with primary rat hepatocytes [29]. We further observed a lack of total and active EGFR expression in cell lines and patient tumors.

Liver metastases express E-cadherin on tumor cells suggested that signals from the tumor microenvironment modulate E-cadherin expression. However, there was question of whether the transdifferenation was sustained during secondary tumor development. Therefore, we co-cultured E-cadherin positive RFP-MCF-7 cells with GFP primary rat hepatocytes and examined E-cadherin expression over a multiday period (Fig. 7.2). These co-cultures revealed that E-cadherin was stability expressed after 2 days, but was subsequently reverted at the leading edge after 8 days

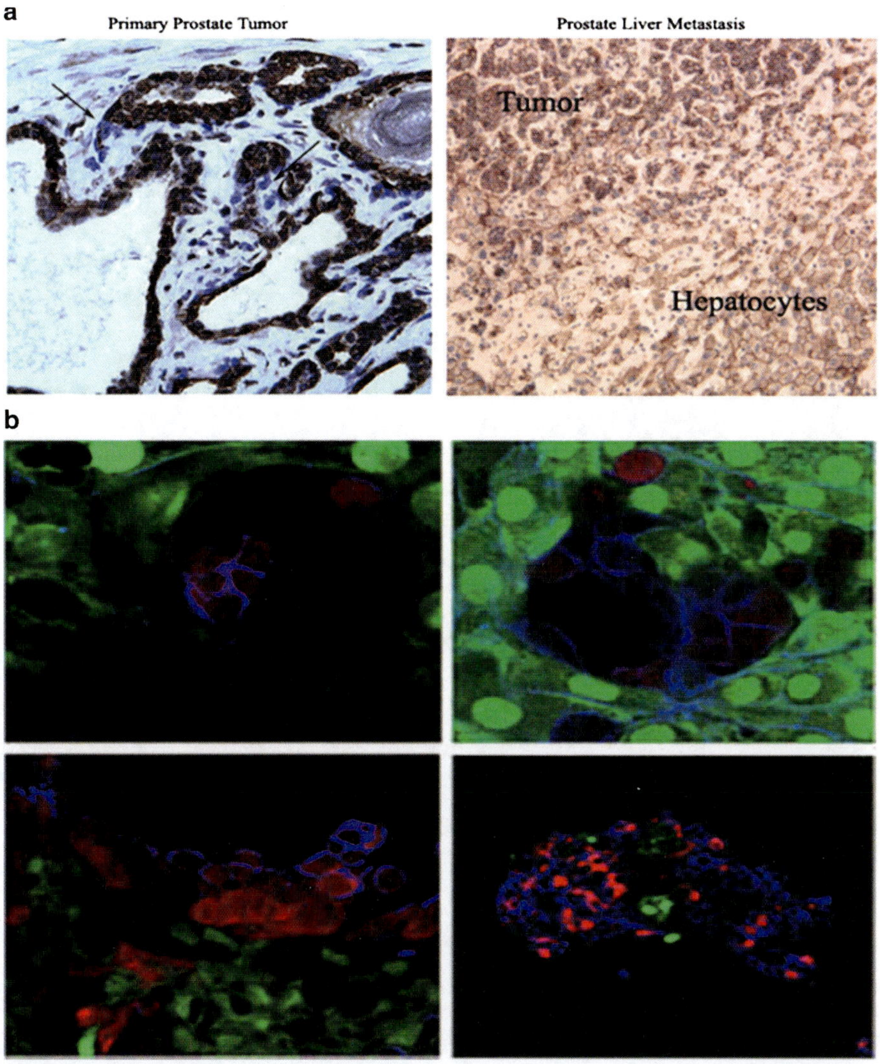

Fig. 7.1 Tumor cells exhibit phenotypic plasticity within the liver microenvironment. (**a**) Human primary prostate cancer (*left*) and metastases to liver (*right*) show expression of E-cadherin. Formalin-fixed, paraffin-embedded tissues were obtained from two well-defined prostate adenocarcinomas with liver metastasis, and stained with E-cadherin antibody. (**b**) Immunofluorescence of co-cultures shows subcellular location of E-cadherin re-expression. MCF-7 RFP (red) and GFP (*green*) primary rat hepatocytes were stained with human-specific anti-E-cadherin for a multiday period. *Top left* (Day 2), *top right* (Day 4), *bottom left* (Day 8), *bottom right* (Day 14). Cy5 secondary antibody (*blue*) was used for E-cadherin primary antibody

of co-culture. However, after long-term coculture (14 days) MCF-7 cells underwent three-dimensional organization. These findings are similar to our prostate cancer patient observations and provide the proof-of-principle that E-cadherin-associated EMT is the result of dynamic interactions of the tumor cell with its surrounding microenvironment.

Since we were able to observe stromal-induced reexpression of E-cadherin within the liver microenvironment, would suggest that a reepithelialization process is necessary for establishment

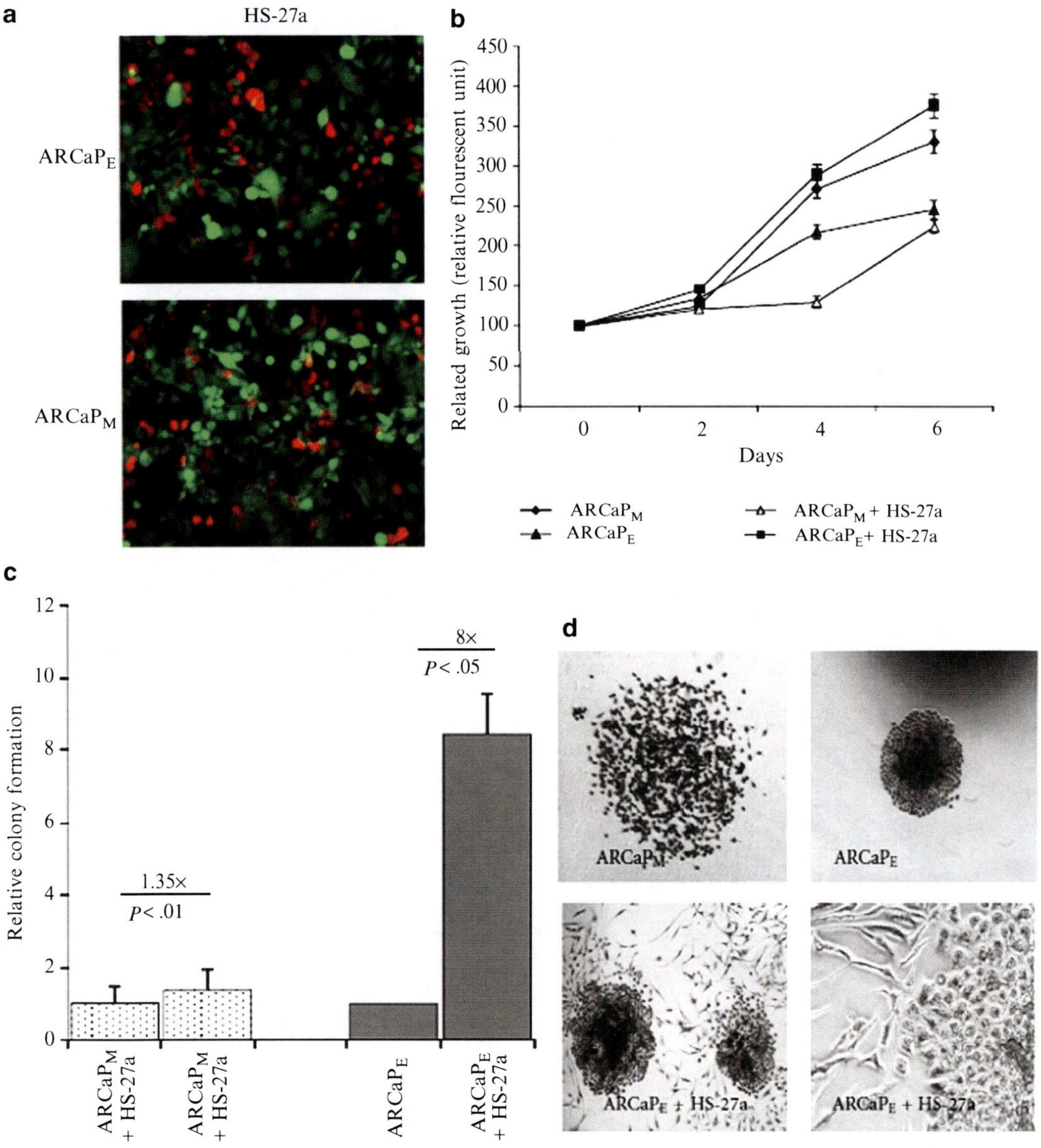

Fig. 7.2 ARCaP$_E$ cells show a growth and colony-forming capacity advantage in presence of HS-27a cells. (**a**) ARCaP$_M$ cells were cocultured in the presence of GFP-HS-27a cells over a 6-day period. Growth of RFP, ARCaP$_E$ or ARCaP$_M$ human prostate cancer cells was assessed by RFU (relative fluorescent units) in the presence cocultures over a 6-day period. Results are means ± SE of three independent experiments. *P, 0.05 (Student's t test) compared to cell number at day 1 ±SEM. (**b**) Clongenic colony-forming capacity of ARCaP$_E$ and ARCaP$_M$ prostate cancer cell after coculture ±SEM. ARCaP$_M$ data was normalized to ARCaP$_M$ control, and ARCaP$_E$ data was normalized to ARCaP$_E$ control. (**c**) Clonogenic colony forming capacity of ARCaP$_E$ and ARCaP$_M$ prostate cancer cell after coculture ± SEM. ARCaP$_M$ data were normalized to ARCaP$_M$ control, and ARCaP$_E$ data were normalized to ARCaP$_E$ control (Note HS-27a induced slightly (1.35×) the growth of ARCaP$_M$ cells but markedly (8×) the growth of ARCaP$_E$ cells). (**d**) ARCaP$_E$ or ARCaP$_M$ cells were cocultured with HS-27a cells. Shown are phase contrast images of colonies formed in the clonogenic assay

of secondary tumors. However, given the inherent differences in the stromal parenchyma of each organ, it is likely that multiple soluble factors can achieve similar effects.

For example, exogenous BMP-7 was able to induce E-cadherin of prostate tumors within the bone microenvironment, however failed to have any effect on tumors within the lymph nodes [40]. Furthermore, differential expression of a number of angiogenesis-associated genes and their proteins between prostate cancer metastasis to bone versus liver, and lymph nodes have been observed [41]. To determine if this is case with E-cadherin-associated EMT, we cocultured the ARCaP model with bone marrow stromal cells. Cocultured $ARCaP_M$ cells displayed a reversal of E-cadherin, and the more epithelial $ARCaP_E$ cells showing increased colony-forming capacity and growth advantage in presence of bone stromal cells [42]. Clinical evidence of E-cadherin expression in bone metastasis has been observed, and interesting is associated with a reversal of E-cadherin-specific methylation pattern [43].

7.3 Targeting Cell Adhesion for Therapeutic Intervention

Although we are just at the beginning of understanding the directive role of the stroma, more insight into how the stroma regulates tumor cells will lead to better therapies for late-stage metastatic disease. Multiple reports have suggested the benefits of targeting E-cadherin as a therapeutic approach. For example, E-cadherin neutralizing antibody (SHEP8-7) has been shown to sensitize multi-cellular spheroids to microtubule binding therapies in the taxane family in HT29 human colorectal adenocarcinoma cells [44]. A more recent observation that survival of androgen receptor-expressing differentiated prostate cells are dependent on E-cadherin and PI3K, but not on androgen, AR or MAPK [45]. Indeed this is the case because our findings suggest the blocking E-cadherin cell–cell interaction with E-cadherin neutralizing antibody, decreased both epithelial or mesenchymal-like prostate cells from reepithelialization and colonizing the bone microenvironment. The neutralizing antibody increases their sensitivity to radiation treatment [42]. Of further clinical benefit, recently a monoconal antibody to N-cadherin has been described as an effective treatment for prostate cancer limiting local invasion and metastasis both in vitro and in vivo [12]. Thus, blocking cellular adhesions appears to be a rationale strategy limiting prostate cancer metastasis.

7.4 Summary

In summary, we propose that the EMT required to "escape" from the primary tumor mass is transiently "reverted" during the initial stages of metastatic seeding, enabling the alien tumor cell to incorporate into the target tissue and derive survival signals thereof. Thus, tumor–stromal-interacts induce cellular plasticity gives rise to distinct populations of cancer cells within secondary site. This plasticity gives rise to distinct population, i.e. mesenchymal phenotype and its kinetic characteristics (motility/invasive), and the epithelial characteristics necessary for secondary tumor development. Our findings that epithelial cells are more successful in establishing secondary tumor suggest that dissemination from the primary tumor mass requires the mesenchymal phenotype, however a mesenchymal to epithelial transition is associated with initial metastatic seeding and subsequent formation of a cohesive tumor mass within the bone microenvironment (Fig. 7.3). Critical to our model of phenotypic mesenchymal-to-epithelial-reverting transitions is the underling signaling mechanisms that mediate this transition. Multiple-cell signaling pathways, most likely initiated by stromal-derived soluble factors, converge on an ever-expanding set of transcriptional and post-translation factors that epigenetically regulate specific proteins that ultimately serve as markers of the epithelial vs. mesenchymal phenotype. Understanding the events may offer new opportunity to target during and the reverting transition that appear to be essential to metastatic relapse.

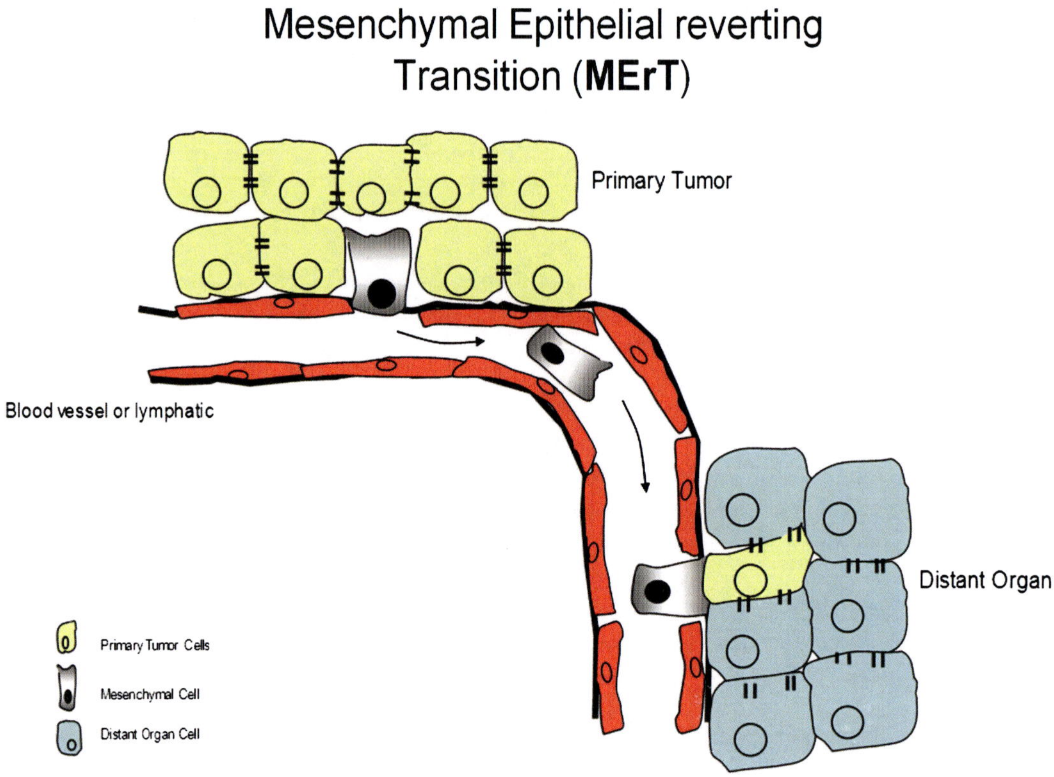

Fig. 7.3 Diagram of mesenchymal-to-epithelial reverting transition. Epithelial carcinomas cells and carcinoma cells undergo EMT, which involves a loss of adhesion and reinforcement of autocrine signaling that drive the cancer cell to escape the tumor mass and intravasate the blood or lymphatic vessels. At the secondary site mesenchymal cells extravasate the tissue parenchyma and phenotypic reversion occurs to form heterotypic interactions within foreign microenvironment, with subsequent development of micrometastases

Acknowledgments I would like to acknowledge all collaborators involved in the different phases of this work. This work was funded by grant from the Department of Defense Prostate Cancer Research Program (PC073977), and NIH/RCMI G12 RR03059-21A1.

References

1. Kahn LB, Uys CJ, Dale J, Rutherfoord S (1978) Carcinoma of the breast with metaplasia to chondrosarcoma: a light and electron microscopic study. Histopathology 2:93–106
2. Hay ED, Zuk A (1995) Transformations between epithelium and mesenchyme: normal, pathological, and experimentally induced. Am J Kidney Dis 26:678–690
3. Ehrlich JS, Hansen MD, Nelson WJ (2002) Spatio-temporal regulation of Rac1 localization and lamellipodia dynamics during epithelial cell-cell adhesion. Dev Cell 3:259–270
4. Akhtar N, Hudson KR, Hotchin NA (2000) Co-localization of Rac1 and E-cadherin in human epidermal keratinocytes. Cell Adhes Commun 7: 465–476
5. Ray ME, Mehra R, Sandler HM, Daignault S, Shah RB (2006) E-cadherin protein expression predicts prostate cancer salvage radiotherapy outcomes. J Urol 176:1409–1414, discussion 14
6. Hennig G, Behrens J, Truss M, Frisch S, Reichmann E, Birchmeier W (1995) Progression of carcinoma cells is associated with alterations in chromatin structure and factor binding at the E-cadherin promoter in vivo. Oncogene 11:475–484
7. Graff JR, Herman JG, Lapidus RG et al (1995) E-cadherin expression is silenced by DNA hypermethylation in human breast and prostate carcinomas. Cancer Res 55:5195–5199
8. Machado JC, Oliveira C, Carvalho R et al (2001) E-cadherin gene (CDH1) promoter methylation as the second hit in sporadic diffuse gastric carcinoma. Oncogene 20:1525–1528

9. Berx G, Becker KF, Hofler H, van Roy F (1998) Mutations of the human E-cadherin (CDH1) gene. Hum Mutat 12:226–237
10. Hirohashi S (1998) Inactivation of the E-cadherin-mediated cell adhesion system in human cancers. Am J Pathol 153:333–339
11. Jennbacken K, Tesan T, Wang W, Gustavsson H, Damber JE, Welen K (2010) N-cadherin increases after androgen deprivation and is associated with metastasis in prostate cancer. Endocr Relat Cancer 17:469–479
12. Tanaka H, Kono E, Tran CP et al (2010) Monoclonal antibody targeting of N-cadherin inhibits prostate cancer growth, metastasis and castration resistance. Nat Med 16:1414–1420
13. Tran NL, Nagle RB, Cress AE, Heimark RL (1999) N-cadherin expression in human prostate carcinoma cell lines. An epithelial-mesenchymal transformation mediating adhesion with stromal cells. Am J Pathol 155:787–798
14. Odero-Marah VA, Wang R, Chu G et al (2008) Receptor activator of NF-kappaB ligand (RANKL) expression is associated with epithelial to mesenchymal transition in human prostate cancer cells. Cell Res 18:858–870
15. Liu J, Uygur B, Zhang Z et al (2010) Slug inhibits proliferation of human prostate cancer cells via down-regulation of cyclin D1 expression. Prostate 70: 1768–1777
16. Emadi Baygi M, Soheili ZS, Schmitz I, Sameie S, Schulz WA (2010) Snail regulates cell survival and inhibits cellular senescence in human metastatic prostate cancer cell lines. Cell Biol Toxicol 26:553–567
17. Kwok WK, Ling MT, Lee TW et al (2005) Up-regulation of TWIST in prostate cancer and its implication as a therapeutic target. Cancer Res 65: 5153–5162
18. Yuen HF, Chan YP, Wong ML et al (2007) Upregulation of twist in oesophageal squamous cell carcinoma is associated with neoplastic transformation and distant metastasis. J Clin Pathol 60:510–514
19. Downing JR, Reynolds AB (1991) PDGF, CSF-1, and EGF induce tyrosine phosphorylation of p120, a pp 60src transformation-associated substrate. Oncogene 6:607–613
20. Hazan RB, Norton L (1998) The epidermal growth factor receptor modulates the interaction of E-cadherin with the actin cytoskeleton. J Biol Chem 273: 9078–9084
21. Andl CD, Mizushima T, Nakagawa H et al (2003) Epidermal growth factor receptor mediates increased cell proliferation, migration, and aggregation in esophageal keratinocytes in vitro and in vivo. J Biol Chem 278:1824–1830
22. Kassis J, Moellinger J, Lo H, Greenberg NM, Kim HG, Wells A (1999) A role for phospholipase C-gamma-mediated signaling in tumor cell invasion. Clin Cancer Res 5:2251–2260
23. Nakashiro K, Okamoto M, Hayashi Y, Oyasu R (2000) Hepatocyte growth factor secreted by prostate-derived stromal cells stimulates growth of androgen-independent human prostatic carcinoma cells. Am J Pathol 157: 795–803
24. Yates C, Wells A, Turner T (2005) Luteinising hormone-releasing hormone analogue reverses the cell adhesion profile of EGFR overexpressing DU-145 human prostate carcinoma subline. Br J Cancer 92:366–375
25. Davies G, Jiang WG, Mason MD (2001) HGF/SF modifies the interaction between its receptor c-Met, and the E-cadherin/catenin complex in prostate cancer cells. Int J Mol Med 7:385–388
26. Josson S, Anderson CS, Sung SY et al (2011) Inhibition of ADAM9 expression induces epithelial phenotypic alterations and sensitizes human prostate cancer cells to radiation and chemotherapy. Prostate 71(3):232–240
27. Graham TR, Zhau HE, Odero-Marah VA et al (2008) Insulin-like growth factor-I-dependent up-regulation of ZEB1 drives epithelial-to-mesenchymal transition in human prostate cancer cells. Cancer Res 68: 2479–2488
28. Yates C, Shepard CR, Papworth G et al (2007) Novel three-dimensional organotypic liver bioreactor to directly visualize early events in metastatic progression. Adv Cancer Res 97:225–246
29. Yates CC, Shepard CR, Stolz DB, Wells A (2007) Co-culturing human prostate carcinoma cells with hepatocytes leads to increased expression of E-cadherin. Br J Cancer 96:1246–1252
30. Li LC, Zhao H, Nakajima K et al (2001) Methylation of the E-cadherin gene promoter correlates with progression of prostate cancer. J Urol 166:705–709
31. Ma L, Teruya-Feldstein J, Weinberg RA (2007) Tumour invasion and metastasis initiated by microRNA-10b in breast cancer. Nature 449:682–688
32. Lehmann U, Hasemeier B, Christgen M et al (2008) Epigenetic inactivation of microRNA gene hsa-mir-9-1 in human breast cancer. J Pathol 214:17–24
33. Dalmay T, Edwards DR (2006) MicroRNAs and the hallmarks of cancer. Oncogene 25:6170–6175
34. Kong D, Li Y, Wang Z et al (2009) miR-200 regulates PDGF-D-mediated epithelial-mesenchymal transition, adhesion, and invasion of prostate cancer cells. Stem Cells 27:1712–1721
35. Wu TT, Sikes RA, Cui Q et al (1998) Establishing human prostate cancer cell xenografts in bone: induction of osteoblastic reaction by prostate-specific antigen-producing tumors in athymic and SCID/bg mice using LNCaP and lineage-derived metastatic sublines. Int J Cancer 77:887–894
36. Zhau HE, Li CL, Chung LW (2000) Establishment of human prostate carcinoma skeletal metastasis models. Cancer 88:2995–3001
37. Paget S (1889) The distribution of secondary growths in cancer of the breast. Cancer Metastasis Rev 8:98–101
38. Chung LW (2003) Prostate carcinoma bone-stroma interaction and its biologic and therapeutic implications. Cancer 97:772–778

39. Stessels F, Van den Eynden G, Van der Auwera I et al (2004) Breast adenocarcinoma liver metastases, in contrast to colorectal cancer liver metastases, display a non-angiogenic growth pattern that preserves the stroma and lacks hypoxia. Br J Cancer 90:1429–1436
40. Buijs JT, Rentsch CA, van der Horst G et al (2007) BMP7, a putative regulator of epithelial homeostasis in the human prostate, is a potent inhibitor of prostate cancer bone metastasis in vivo. Am J Pathol 171: 1047–1057
41. Morrissey C, True LD, Roudier MP et al (2008) Differential expression of angiogenesis associated genes in prostate cancer bone, liver and lymph node metastases. Clin Exp Metastasis 25:377–388
42. Josson S, Sharp S, Sung SY (2010) et al. Tumor-stromal interactions influence radiation sensitivity in epithelial – versus mesenchymal-like prostate cancer cells, J Oncol
43. Saha B, Kaur P, Tsao-Wei D et al (2008) Unmethylated E-cadherin gene expression is significantly associated with metastatic human prostate cancer cells in bone. Prostate 68:1681–1688
44. Green SK, Karlsson MC, Ravetch JV, Kerbel RS (2002) Disruption of cell-cell adhesion enhances antibody-dependent cellular cytotoxicity: implications for antibody-based therapeutics of cancer. Cancer Res 62:6891–6900
45. Lamb LE, Knudsen BS, Miranti CK (2010) E-cadherin-mediated survival of androgen-receptor-expressing secretory prostate epithelial cells derived from a stratified in vitro differentiation model. J Cell Sci 123:266–276

8 Role of Epigenetics in Cancer Initiation and Progression

Flora Chik, Moshe Szyf, and Shafaat A. Rabbani

Abstract

The epigenome which comprises DNA methylation, histone modifications, chromatin structures and non-coding RNAs controls gene expression patterns. In cancer cells, there are aberrant changes in the epigenome. The question in cancer epigenetics is that whether these changes are the cause of cell transformation, or rather the consequence of it. We will discuss the epigenetic phenomenon in cancer, as well as the recent interests in the epigenetic reprogramming events, and their implications in the cancer stem cell theory. We will also look at the progression of cancers as they become more aggressive, with focus on the role of epigenetics in tumor metastases exemplified with the urokinase plasminogen activator (uPA) system. Last but not least, with therapeutics intervention in mind, we will highlight the importance of balance in the design of epigenetic based anti-cancer therapeutic strategies.

8.1 Introduction

The epigenome consists of DNA methylation, various post-translational modifications of histone tails, nucleosome positioning and non-coding RNAs. The epigenetic marks are somatically heritable [1] but at the same time highly dynamic and are involved in embryonic development, parental imprinting and inactivation of the X chromosome [2]. The machinery regulating the epigenome comprises DNA methyltransferases (DNMTs), methyl-DNA binding proteins (MBDs), histone modifiers, proteins interacting with histone modifications and chromatin remodeling complexes. The epigenome is characterized by two seemingly contradictory features: stability and reversibility. DNA methylation that accompany cellular differentiation during embryogenesis are driven by innate processes that are highly organized, predictable and persistent. On the other hand, studies have shown that the state of DNA methylation could be modulated by environmental factors such as stress and diet [3–5]. Some of the environmental exposures in these studies occurred in utero but the effects sustained into adulthood. These studies demonstrate how early life experience could leave

F. Chik (✉)
Department of Pharmacology and Therapeutics, McGill University, 3655 Promenade Sir William Osler, Montreal, QC, Canada H3G 1Y6
e-mail: flora.chik@mail.mcgill.ca

J.S. Rhim and R. Kremer (eds.), *Human Cell Transformation: Role of Stem Cells and the Microenvironment*, Advances in Experimental Medicine and Biology 720,
DOI 10.1007/978-1-4614-0254-1_8, © Springer Science+Business Media, LLC 2011

a stable epigenetic mark in the genome and affect later life behavior and health [3, 4]. The dynamic nature of the epigenome is also evidenced when changes in DNA methylation were reported to occur within a few hours, between different phases of the cell cycle or in response to recruitment of transcription factors [6, 7].

One of the most widely studied components of the epigenome is DNA methylation. At first, it was believed that the DNA methylation patterns were set during embryogenesis by de novo DNMTs (DNMT3a and 3b) and were thereafter faithfully copied by the maintenance DNMT (DNMT1) after completion of cellular differentiation [8]. This view has changed when later studies discovered the cooperation between de novo and maintenance DNMTs occur in contexts other than during embryogenesis [9, 10]. The realization that the epigenome is dynamic has made it an interesting target for therapeutic intervention. Since then, the involvement of epigenetic changes has been studied in a variety of diseases such as cancers, diabetes, cardiovascular diseases and psychiatric disorders [11–15].

Cancer has long been regarded as a genetic disease, where inherited or spontaneous mutations lead to abnormal cellular functions. However, it is now well accepted that changes in epigenetic marks in the absence of changes in DNA sequence could be playing critical roles in cancer as well. In this review, we will focus on the possible epigenetic mechanisms that are involved in cellular transformation and cancer progression.

8.2 Diagnostic and Prognostic Significance of Epigenetics in Cancers

Most types of cancer cells display global DNA hypomethylation with regional gene hypermethylation compared to their normal counterparts [16]. The enzymes responsible for the epigenetic changes are also deregulated in many cancers. For example, members of the DNMT family, DNMT1, 3a and 3b have been shown to display significant changes in expression in various cancers including lung, liver, breast, gastric, leukemia and retinoblastoma [17–22]. In particular, DNMT1 overexpression is often associated with poor prognosis [17, 18]. Several studies have found positive correlations between DNMTs overexpression and hypermethylation of tumor suppressor genes [23, 24], although these studies did not demonstrate a causal relationship between DNMT overexpression and methylation of tumor suppressor genes. Nevertheless, methylation of tumor suppressor genes is involved in silencing of their expression and the methylation status of some tumor suppressor genes is found to be correlated with clinical outcomes. Examples of hypermethylated tumor suppressor genes are *p16*, *MGMT*, *RASSF1A* and *DAPK1* [25–27] and hypomethylated oncogenes are *uPA* and *MDR1* [28, 29]. These genes might serve as diagnostic and prognostic markers in different cancers.

Given the reversible potential of epigenetic marks, demethylating agents have become interesting candidates as anti-cancer drugs. 5-aza-2′-deoxycytidine (5-azaCdR) is the FDA-approved demethylating agent that is used to treat myelodysplastic syndrome (MDS) [30, 31] and is currently under clinical trials for treatment of solid tumors [32]. Although it was reported to have anti-cancer efficacy in phase III clinical trials with MDS patients [31], there are potential complications in the use of such a non-specific drug for treatment of cancer, especially solid tumors. This will be discussed in detail in the last section. The success in developing a cancer specific epigenetic drug hinges on our understanding of how different epigenetic factors are involved in the underpinning mechanisms of this disease.

8.3 DNA Methylation and Cancer Initiation: The Cause or the Consequence?

Cellular transformation refers to the process in which normal cells undergo changes in gene expression patterns and transform into immortal proliferating cells that do not differentiate into their mature post-mitotic state. Under certain conditions, transformed cells can be turned into malignant cancer cells. Given that the hallmark of cancer is global hypomethylation and regional hypermethylation at specific gene promoters, the question is whether this disruption of methylation patterns is a passive consequence of cellular transformation, or whether the aberrant changes in methylation are causing the transformation itself. The former view has been favored in the past, based on the belief that cancers are caused by accumulative mutations [33, 34]. Mutations cause rearrangements of large chromosomal regions, which confer the cells with growth advantage under selection pressure due to abnormal expression of oncogenes [33, 35]. The clonal expansion of the mutated cells leads to genomic instability and global demethylation, while the cell machinery progressively shuts down the anti-survival genes by hypermethylation. In this model, stochastic mutations cause genomic instability, which precedes methylation changes.

Due to the fact that most tumors display genomic instability, it is important to establish a model that could dissociate mutations from epigenetic changes to define the causal relationship between epigenetic aberrations and cancer. In a recent study, it was shown that aberrant changes in the epigenome could indeed lead to cancers that do not display genomic instability [36]. *Snf5* is a tumor suppressor gene and a core component of the chromatin remodelling complex SWI/SNF whose inactivation is detected in several types of tumors [37, 38], including the highly invasive malignant rhabdoid tumors (MRTs) [39]. Differing from most other tumors where the chromosomes are usually fragmented, MRTs often display intact genome. The authors generated *Snf5*-deficient primary mouse embryonic fibroblasts and showed that tumors derived from these cells were diploid and the cancer phenotype was correlated with the expression of the cell cycle protein cyclin D1, which was epigenetically upregulated by the SWI/SNF complexes [36]. An alternative approach to study the relationship between epigenetic changes and transformation is to study the epigenome of pre-cancerous cells. A series of studies on colon cancers found that global hypomethylation as well as regional gene promoter hypermethylation occur in pre-cancerous lesions or even benign colon polyps before they become malignant colon cancers [40–43]. Similar findings were observed in breast cancers, where normal tissues surrounding the tumors, up to a few centimeters apart, have been detected with aberrant DNA methylation patterns [44, 45]. These observations of methylation patterns change in pre-cancerous cells suggest that the loss in methylation can be an early event that precedes transformation, and it is not merely a consequence of genomic instability. In addition, it has also been shown that global hypomethylation leads to elevated mutation rates [46], suggesting that the epigenetic change may initiate downstream oncogenetic pathways. Studying these model systems may therefore aid our understanding of how epigenetic processes contribute to the process of oncogenic transformation.

8.4 The Driving Force

Hyperactivity of DNMT1 in cancers has a causal role in tumorigenesis [47, 48]. Antisense and siRNA against DNMT1 reversed tumorigenesis and inhibited tumor growth in mice [47, 48]. Nevertheless, the cause-and-effect relationship of DNMT1 overexpression and gene promoter hypermethylation has not been clearly demonstrated. Could other functions of DNMT1, in addition to DNA methylation, contribute to oncogenesis?

DNMT1 is a multifunctional protein with a C-terminal catalytic domain and an N-terminal regulatory domain where it is bound to a lot of known proteins. In addition to its methyltransferase activity, protein-protein interactions may play a role in the functioning of this maintenance methyltransferase. DNMT1 has been implicated in the DNA repair pathway in several studies. It was observed to localize at the sites of DNA double strand breaks (DSBs) together with other repair proteins soon after DNA damage was induced [49, 50]. One can think that the localization of DNMT1 to the damage sites is the cells' mechanism of recovering methylation loss due to

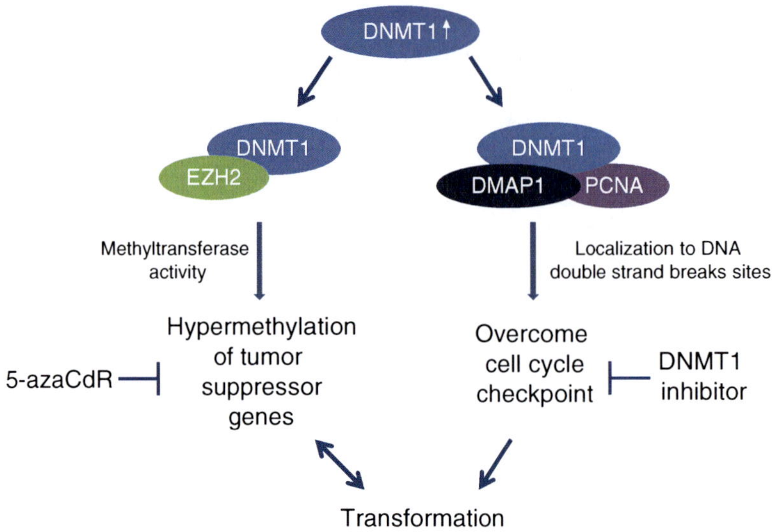

Fig. 8.1 The role of DNMT1 in transformation. DNMT1 is overexpressed in many cancers. Upregulation of DNMT1 may lead to gene promoter hypermethylation through targeting to EZH2-binding sites. Demethylating agent 5-aza-2′-deoxycytidine (5-azaCdR) exerts its anti-cancer effects by remethylating these suppressed genes. DNMT1 is also involved in the DNA damage repair pathway. DNMT1 is recruited to the DSBs sites by PCNA and DMAP1. It overcomes the cell cycle checkpoint by reducing the damage response. Specific DNMT1 inhibitor may block this effect and lead to a halt in the cell cycle in cancer cells

the damage repair process. But such an observation could have a more important implication in cancer. If DNMT1 overexpression allows the cells to continue replicate, despite the presence of damaged DNA, this may partly explain the ability of cancer cells to overcome cell cycle checkpoints and become terminally proliferating cells. In fact, overexpression of DNMT1 reduced the DNA damage markers, the phosphorylated form of histone variants H2A.X (γH2A.X) [51–53], and enabled the cells to overcome the replication barrier. Co-localized with DNMT1 and γH2A.X at the sites of DSBs are proliferating cell nuclear antigen (PCNA) and another important damage response protein Chk1, which both were later showed to be indispensible for the recruitment of DNMT1 to DSBs sites [50, 54]. DNMT1 also directly binds to 9-1-1 complex (RAD9. RAD1, HUS1) at DSBs sites which help augment the damage signals [50].

Another extensively studied binding partner of DNMT1, which is involved in the damage repair pathways is DNMT1-associated protein 1 (DMAP 1). DMAP1 was originally found to be co-localized with DNMT1 and PCNA in the replication fork [55]. Recent studies show that DMAP1 is essential for the repair system, and DMAP1-deficient cells display genomic instability and compromised proliferation [56, 57]. These cells are able to form tumors in p53-deficient mice [56]. It is of particular interest that when a full-length DMAP1, but not PCNA-binding site mutated one, is introduced to DMAP1-deficient cells, the levels of γH2A.X are reduced to normal levels. This suggested not only that PCNA and DMAP1 co-operated in the repair process, but that the ability to overcome damage checkpoints allowed the cells to proliferate without barrier, which eventually led to tumor formation. Given the importance DNMT1 in the damage response pathway, targeting DNMT1 in cancer might have additional benefits to the known outcome of re-activation of tumor suppressor genes. By reducing DNMT1 levels, we are taking advantage of the cells internal control machinery to stop cell cycle progression, which can result in an effective anti-cancer therapy (Fig. 8.1).

8.5 Epigenetic Reprogramming in Cancers

A relatively new and exciting topic in epigenetics is cell programming. In simple terms, epigenetic programming means the acquisition of novel epigenetic marks. It is a normal process during development, where most of the epigenome is erased and later reprogrammed in the embryos [58]. This early reprogramming event is believed to be important for the zygote to gain potency to be developed into a whole organism [58]. However, the idea of epigenetic reprogramming in fully differentiated cells was not widely accepted until recently, since such an event would require the cells to display plasticity. The term cellular plasticity refers to the ability of cells to respond to external factors and change their phenotypes accordingly. In fact, cellular plasticity in mature cells has been observed more than 20 years ago in a series of experiments showing fusion of human fibroblasts with mouse muscles cells resulted in human muscle proteins [59, 60]. This observation suggested that in the presence of suitable factors, differentiated cells were able to be epigenetically reprogrammed into other cell types. The underlying mechanism of such reprogramming in mature cells was not understood at the time, and the general scientific community believed that cells were only "reprogrammable" in the developmental stage. A breakthrough discovery has been made a few years ago where Takahashi and colleagues successfully generated induced pluripotent stem cells (iPSCs) from differentiated fibroblasts by introducing four transcription factors, *Oct4*, *Sox2*, *Klf4* and *cMyc* [61]. Such a discovery clearly demonstrated the presence of cellular plasticity in mature cell types.

In cancer epigenetics, the polycomb group (PcG) complexes have been widely studied and are believed to be one of the most important complex of proteins in the reprogramming event. PcG proteins were first discovered in *Drosophila* as protein complexes that silenced the *Hox* genes, a group of homeodomain-containing genes that regulate many developmental processes [62]. It was later found that the family of PcG proteins is also present in higher eukaryotes. PcG proteins display enzymatic activities and are associated with the repressive histone mark tri-methylation of histone 3 at position 27 (H3K27me3). This mark is responsible for gene silencing. In fact, studies have found that many lineage-specific genes in embryonic stem cells are bound by the PcG proteins and kept at a lowly expressed state by bivalent histone marks: the activating methylated lysine 4 in histone 3 (H3K4me) and the repressive H3K27me3 [63, 64]. These lineage-specific genes are poised to be expressed or repressed by losing the corresponding histone marks. They are not marked with DNA methylation in embryonic stem cells, in which would have completely silence them, but the genes are frequently methylated in adult cancer cells [65]. Recent evidence shows that EZH2, a PcG protein, binds directly to DNMT1 [66]. These findings may explain how DNMT1 is targeted to specific region and leads to regional hypermethylation in cancers (Fig. 8.1).

8.6 The Cancer Stem Cell Theory

"Stem cell" is the term for any cell that is pluripotent and is able to self-renew [67]. Given that cancer cells are abnormally proliferating cells, it is reasonable to propose that cancer cells and stem cells possess common features. The term cancer stem cells (CSCs) have gained significant attention in recent years. It was proposed that tumors are derived from a few tumor-initiating cells which display stem cell properties [68, 69]. There are two theories for the formation of CSCs (Fig. 8.2). The first hypothesizes that normal adult stem cells acquire accumulated mutations and become CSCs [70, 71]. Under normal conditions, stem cells divide asymmetrically into two parts, one daughter cell will undergo differentiation, while the other will possess self-renewal ability and further give rise to other stem cells. When mutations occur, these normal stem cells transform into CSCs. The many similarities between the two types of stem cells support the theory that CSCs are mutation-derived stem cell [72, 73].

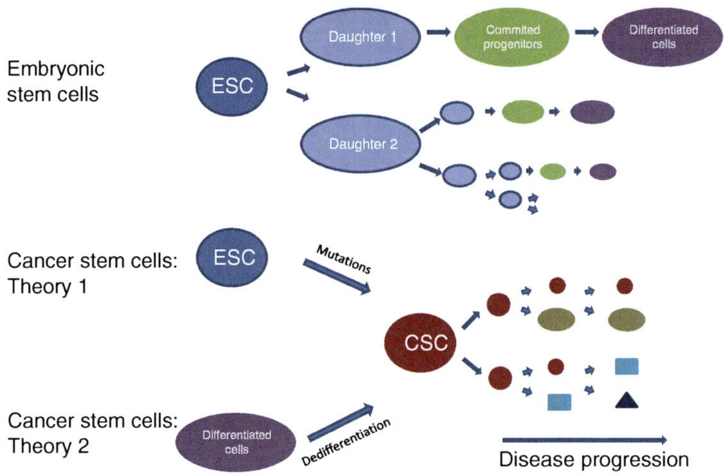

Fig. 8.2 The cancer stem cell theory. There are two theories in cancer stem cells (CSCs) formation. Theory 1 suggests mutations occur in normal embryonic stem cells, while theory 2 suggests dedifferentiation occur in mature cells and these cells become stem cells-like. While embryonic stem cells behave in defined principles, the phenotypes of CSCs may change throughout disease progression

The second theory suggests that the process of dedifferentiation, i.e. epigenetic reprogramming occurs in the more committed cell types, such as committed progenitors or mature cells. These cells acquire self-renewal properties and become stem cell-like which, at the very least, display self-renewal properties [74].

Notwithstanding whether one or both of the above theories is true, CSCs differ from normal stem cells. One of the major differences between the two cell types is that normal stem cells behave according to defined principles, in terms of replication, expression patterns and response to stimulus, while in CSCs these properties change throughout the progression of the disease (Fig. 8.2) [67]. If indeed CSCs are the origin of the whole tumor, this gives us hope that we could narrow down our targets in the search of anti-cancer drugs by focusing on the stem cells. In addition, CSCs are useful tools to study the pathways involved in self-renewal, the ultimate unwanted property we would like to eliminate in cancers. If we can target these specific pathways, anti-CSCs therapy can be very effective in avoiding relapses, which are believed to be brought about by CSCs that escape from the early therapy.

Another possible clinical use of the multipotent stem cells is to serve as delivery vehicles. Several studies suggest that the stem cells display tropism towards cancer cells [75, 76]. The mechanism of their behavior has only been recently discovered. A recent study shows that mesenchymal stem cells (MSCs) treated with histone deacetylase inhibitor (HDACi) specifically migrate towards tumor cells. One of the most important proteases involved in cell migration, the urokinase-type plasminogen activator (uPA), was found to be hyperacetylated and upregulated upon HDACi treatment. The ERK signaling pathway was also activated [75]. These findings suggest that HDACi induces MSCs-to-tumor cells tropism by inducing uPA expression. It also shows that the use of stem cells delivery system has tremendous potential in developing tumor-selective therapy.

8.7 Tumor Progression: From Neoplasticity to Metastases

If all tumors were localized, most cancers in non-vital organs such as breast and prostate would probably have much lower mortality rates. Unfortunately, tumor cells have various mechanisms by which to break away from their primary locations and support their growth in a distant site: local invasion, intravasation (entering the vascular or lymphatic system) and travel to distant organs, chemotaxis (guided movements by interactions with chemotactic molecules) and even angiogenesis (generating blood vessels) to support its own growth [77]. This colonization of tumor in a secondary site is termed metastasis, and is possible since cancer cells develop mechanisms to overcome normal numerous barriers. How do epigenetic processes play a role in this multi-step war? In this section, we will focus on the functions of some epigenetically regulated genes and their roles in the systemic changes that take place during the metastasis process.

8.8 The EMT: Epithelial to Mesenchymal Transition

Breast and prostate cancers are the most highly invasive cancers in women and men, respectively. They both have an epithelial origin and are hormone-dependent [78]. In the initial stage of the tumor, hormonal therapy is usually sufficient to inhibit tumor growth. However, as the disease progresses, they become hormone-insensitive and highly aggressive. The epithelial to mesenchymal transition (EMT), though not absolutely required, is considered an important step in tumor invasion and metastasis [77, 79, 80].

Epithelial cells contain a basal membrane on one end with adhesion proteins such as E-cadherins, ZO-1 (zona occludins) and cytokeratins [81]. This basal membrane is always attached to the epithelial wall, while the apical side interacts with the external environment. On the contrary, mesenchymal cells do not display polarity. Their cytoskeleton components change from E-cadherins and ZO-1 to vimentin and smooth muscle actin [81]. The normal functions of epithelial cells are, as indicated in their name, forming the lining of vessels or organs. On the other hand, mesenchymal cells are more mobile in their nature. The process of an epithelial cell losing its polarity and becoming mesenchymal-like is called EMT.

EMT can be induced by different pathways and exists in various forms. Nevertheless, there are some genes that are commonly affected in most types of EMTs. Interestingly, some of these genes contain CpG islands in their promoters, which provide them with the potential to be switched on or off by epigenetic mechanisms. One of most studied genes to be silenced in EMT is the adhesion molecule E-cadherin. Its silencing has been observed about 20 years ago in various aggressive cancers [82]. Genetic mutations of the E-cadherin allele are found in about 50% of the cancers. The epigenetic mechanism of E-cadherin silencing has later been proposed, where 5-azaCdR-treated cancer cells demethylated and re-expressed the gene [82]. It was also recently proposed that histone deacetylase (HDAC) is responsible for the suppression of E-cadherin, and that HDAC inhibitor was able to reactivate it [83, 84]. Another example of gene silencing in EMT is the secreted frizzled-related protein 5 (SFRP5). SFRP5 is an antagonist of the WNT signaling pathway, a pathway that is implicated in promoting cancer growth and invasion. Overexpression of SFRP5 in cancer cell lines decreased cell invasiveness [85]. Clinically, hypermethylation of SFRP5 was shown to correlate with overall survival in some cancers [85, 86]. On the other hand, there are mesenchymal-specific genes that are activated during EMT, which promote tumor progression. The typical EMT markers, vimentin and S100A4, contain large CpG islands and are found to be methylated in non-aggressive tumors but demethylated in more aggressive ones [87–89]. Another epigenetically regulated gene that is important in metastases is uPA, which will be discussed in the next section. In summary, abnormality in the epigenome may cause the expression changes of these EMT genes and contribute to tumor progression.

8.9 The uPA/uPAR System and Metastases

In order for the cancer cells to detach from their primary sites, they must interact with other molecules in the extracellular matrix (ECM). The ECM consists of different types of polysaccharides, the major one being collagen. In normal cell context, degradation of ECM proteins is essential for tissue remodeling in organogenesis, wound healing and inflammatory response [90]. As early as 30 years ago, it was suggested that

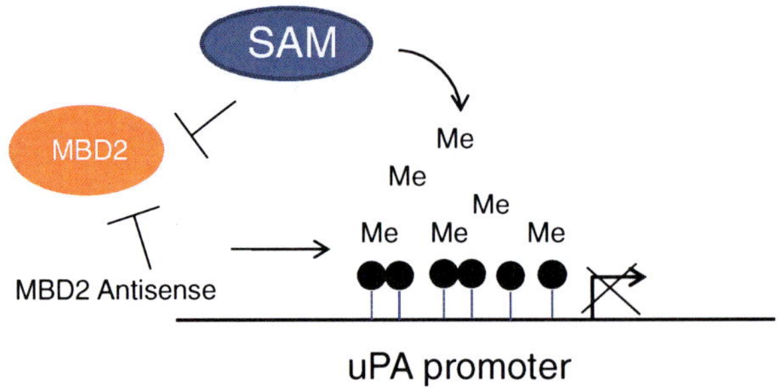

Fig. 8.3 SAM treatment and MBD2 antisense oligonucleotides reduce uPA expression. uPA is an important protease required for cancer metastasis. Demethylation in the promoter leads to its expression in invasive tumor. S-adenosyl-L-methionine (SAM), a methyl donor, is able to remethylate the uPA promoter and suppress uPA expression. SAM is shown to suppress MBD2, the putative DNA demethylase. Antisense oligonucleotides against MBD2 have the same effect on uPA suppression. This suggests the possible mechanism of SAM treatment on uPA suppression

metastatic potential of cancer cells correlated with their ability to degrade ECM proteins [91]. One of the most implicated proteases in promoting tumor invasion is the uPA. The proteolytic cascade starts when the inactive zymogen plasminogen is converted to its active form, plasmin. Plasmin is a broad spectrum protease that can catalyze proteolysis of a series of ECM proteins [92]. There are two enzymes that are responsible for this conversion, uPA and tPA (tissue-type plasminogen activator). Out of the various proteases present in the ECM, uPA stands out from the crowd by having a specific cell-surface receptor, uPAR. uPAR is a glycosyl-phosphatidylinositol (GPI)-receptor that has great mobility on cell surface due to the lack of transmembrane and cytoplasmic domains [90, 93]. Studies have shown that uPAR is localized in the leading edge of cell movement [94, 95], which supports the important role of uPA in cell invasion. Indeed, uPA expression is almost undetectable in most tissues while it is present in almost all solid tumors and is highly associated with poor prognostic invasive tumors [96–98].

Studies on the *uPA* gene have found that there is a large CpG island present in its promoter which is methylated in most tissues. The hypothesis that hypomethylation of the uPA promoter is responsible for its expression pattern was first proposed by Rabbani et al. [99]. They used methylation-specific PCR to compare methylation levels in the highly invasive, hormone-insensitive breast cancer cell line MDA-MB-231 and the non-invasive, hormone-sensitive breast cancer cell line MCF-7. The findings suggested that the more aggressive MDA-MB-231 cells had a fully demethylated uPA promoter while MCF-7 cells has highly methylated uPA promoter. They further proposed the use of the methyl donor, *S*-adenosyl-L-methionine (SAM), to re-methylate the uPA promoter and repress its expression. Indeed, later studies by them and other groups have shown that in addition to uPA suppression, SAM treatment was very effective in inhibiting invasion in vitro and metastasis in vivo [100–102].

The interesting questions arise, regarding the mechanism in which SAM methylates the uPA promoter. Does addition of SAM increases the methyl groups available and leads to DNMT1 hypermethylating the uPA promoter? Or does SAM suppresses an enhancer of uPA expression and therefore re-express the gene? SAM was previously shown to inhibit DNA demethylase activity [103]. It is worth to note that SAM highly suppresses MBD2, a putative DNA demethylase. MBD2, a family member of methylated-DNA binding protein, is deregulated in some tumors [104] and has been studied as a therapeutic target for cancers [105, 106]. This hypothesis is supported even more by the evidence that the antisense oligonucleotides against MBD2 has the same effect as SAM treatment, where it inhibited invasion in cell lines such as breast cancer MDA-MB-231 [100] and prostate cancer PC-3 [101]. MBD2 is an interesting candidate worth further investigation to understand the mechanism of SAM (Fig. 8.3).

8.10 The Balance of Methylation and Demethylation

As mentioned earlier, demethylating agents have become interesting candidates in the development of epigenetic anti-cancer drugs due to the reversibility of the epigenome. Re-expression of hypermethylated tumor suppressors has been intensely studied in the past 20 years and the demethylating agent 5-azaCdR is now in clinical use for MDS treatment [32]. 5-azaCdR is a cytidine analog. It is incorporated into the genome during DNA replication and physically trapped DNMTs at the site of incorporation by forming covalent bonds between the aza and the enzymatic pocket of DNMTs [107]. This leads to global loss of methylation as DNA replication continues without functional DNMTs. Due to the nature of this chemical reaction, 5-azaCdR is considered a non-specific drug that inhibited all family members of DNMTs. In that sense, any gene could be demethylated and re-expressed, including the pro-survival and pro-metastatic genes (Fig. 8.4). Indeed, 5-azaCdR treatment has been shown to increase cell invasion and metastases both in vitro and in vivo by upregulation of oncogenes such as uPA, MMP2, S100A4 and Ezrin [108–111]. Therefore, the use of 5-azaCdR in the treatment of early stage solid tumors might speed up the disease progression.

There are two solutions to this problem. The first is to develop a more specific DNMT1 inhibitor. As previously mentioned, DNMT1 antisense oligonucleotides and siRNA experiments show very promising results in tumor growth suppression [24, 52, 112]. In addition, a recent study also suggested that knockdown of DNMT1 significantly suppressed cancer growth without increasing cell invasion in breast cancer cell lines [109]. These observations may be due to the more specific targeting of the DNMT1 protein to its target genes by EZH2 recruitment, or due to its effects on the DNA damage repair system. Nevertheless,

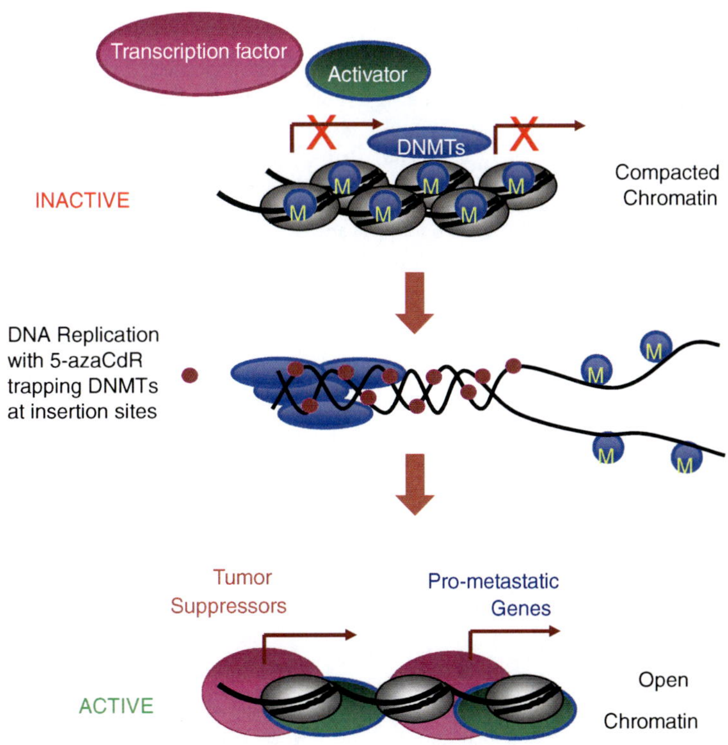

Fig. 8.4 5-azaCdR activates both tumor suppressors and pro-metastatic genes. 5-aza-2′-deoxycytidine (5-azaCdR) is a cytidine analog and it incorporates into the genome during DNA replication, trapping all DNA methyltransferases that have the catalytic domain. This leads to global hypomethylation and re-expression of both tumor suppressor genes and pro-metastatic genes

specificity is the key element in developing DNMT1 inhibitors. The second solution is to combine 5-azaCdR with other drugs that can inhibit its negative effect, of re-expressing pro-survival, pro-metastatic genes. As seen in the above-mentioned examples, uPA, vimentin and S100A4 are methylated in normal cell but get demethylated as cancer progresses. In order to maintain their methylated state, the use of a methylating agent such as SAM may be useful. In fact, a novel hypothesis of combining 5-azaCdR and SAM has recently been previously suggested by us [113–116]. Given that separately, either one of these drugs display very strong anti-tumor properties, combining the two may result in synergistic effects in tumor suppression while antagonizing the side effects. Combining the two already existing drugs seem to be an efficient way to develop a new therapy; however, the molecular mechanisms behind the treatment need to be carefully investigated in order to delicately balance the methylation and demethylation effects.

8.11 Concluding Remarks

Epigenetics is the most rapidly expanding field in biological science in recent years. The epigenome regulates gene expression pattern, which makes it a very important approach to study cell behavior. Cancer epigenetics is especially interesting because the reversible nature of the epigenome makes it a good target for therapeutic intervention. This review summarizes some of the processes in which the epigenetic machinery is involved in cancer initiation and progression. Studying the epigenome enables us to learn more about cancer as a disease. By understanding the molecular mechanisms behind the epigenetic machinery, we would hopefully be able to develop effective epigenetic therapies for cancer treatment.

References

1. Board E (2008) Moving ahead with an international human epigenome project. Nature 454:711–715
2. Reik W, Dean W, Walter J (2001) Epigenetic reprogramming in mammalian development. Science 293: 1089–1093
3. McGowan PO, Sasaki A, D'Alessio AC, Dymov S, Labonte B et al (2009) Epigenetic regulation of the glucocorticoid receptor in human brain associates with childhood abuse. Nat Neurosci 12:342–348
4. Wolff GL, Kodell RL, Moore SR, Cooney CA (1998) Maternal epigenetics and methyl supplements affect agouti gene expression in Avy/a mice. FASEB J 12:949–957
5. Jaenisch R, Bird A (2003) Epigenetic regulation of gene expression: how the genome integrates intrinsic and environmental signals. Nat Genet 33(Suppl): 245–254
6. Brown SE, Szyf M (2008) Dynamic epigenetic states of ribosomal RNA promoters during the cell cycle. Cell Cycle 7:382–390
7. Di Croce L, Raker VA, Corsaro M, Fazi F, Fanelli M et al (2002) Methyltransferase recruitment and DNA hypermethylation of target promoters by an oncogenic transcription factor. Science 295:1079–1082
8. Razin A, Riggs AD (1980) DNA methylation and gene function. Science 210:604–610
9. Kim G-D, Ni J, Kelesoglu N, Roberts RJ, Pradhan S (2002) Co-operation and communication between the human maintenance and de novo DNA (cytosine-5) methyltransferases. EMBO J 21:4183–4195
10. Szyf M, Schimmer BP, Seidman JG (1989) Nucleotide-sequence-specific de novo methylation in a somatic murine cell line. Proc Natl Acad Sci USA 86:6853–6857
11. Esteller M (2007) Cancer epigenomics: DNA methylomes and histone-modification maps. Nat Rev Genet 8:286–298
12. Pirola L, Balcerczyk A, Okabe J, El-Osta A (2010) Epigenetic phenomena linked to diabetic complications. Nat Rev Endocrinol 6(12):665–75
13. Ling C, Groop L (2009) Epigenetics: a molecular link between environmental factors and type 2 diabetes. Diabetes 58:2718–2725
14. Sharma P, Kumar J, Garg G, Kumar A, Patowary A et al (2008) Detection of altered global DNA methylation in coronary artery disease patients. DNA Cell Biol 27:357–365
15. Abel T, Zukin RS (2008) Epigenetic targets of HDAC inhibition in neurodegenerative and psychiatric disorders. Curr Opin Pharmacol 8:57–64
16. Feinberg AP, Vogelstein B (1983) Hypomethylation distinguishes genes of some human cancers from their normal counterparts. Nature 301:89–92
17. Ruo-Kai L, Han-Shui H, Jer-Wei C, Chih-Yi C, Jung-Ta C et al (2007) Alteration of DNA methyltransferases contributes to 5′CpG methylation and poor prognosis in lung cancer. Lung Cancer 55:205–213
18. Saito Y, Kanai Y, Nakagawa T, Sakamoto M, Saito H et al (2003) Increased protein expression of DNA methyltransferase (DNMT) 1 is significantly correlated with the malignant potential and poor prognosis of human hepatocellular carcinomas. Int J Cancer 105:527–532
19. Girault I, Tozlu S, Lidereau R, Bièche I (2003) Expression analysis of DNA methyltransferases 1,

3A, and 3B in sporadic breast carcinomas. Clin Cancer Res 9:4415–4422
20. Etoh T, Kanai Y, Ushijima S, Nakagawa T, Nakanishi Y et al (2004) Increased DNA methyltransferase 1 (DNMT1) protein expression correlates significantly with poorer tumor differentiation and frequent DNA hypermethylation of multiple CpG islands in gastric cancers. Am J Pathol 164: 689–699
21. S-I M, Chijiwa T, Okamura T, Akashi K, Fukumaki Y et al (2001) Expression of DNA methyltransferases DNMT1, 3A, and 3B in normal hematopoiesis and in acute and chronic myelogenous leukemia. Blood 97:1172–1179
22. Qu Y, Mu G, Wu Y, Dai X, Zhou F et al (2010) Overexpression of DNA methyltransferases 1, 3a, and 3b significantly correlates with retinoblastoma tumorigenesis. Am J Clin Pathol 134:826–834
23. Rhee I, Bachman KE, Park BH, Jair KW, Yen RW et al (2002) DNMT1 and DNMT3b cooperate to silence genes in human cancer cells. Nature 416: 552–556
24. Robert MF, Morin S, Beaulieu N, Gauthier F, Chute IC et al (2003) DNMT1 is required to maintain CpG methylation and aberrant gene silencing in human cancer cells. Nat Genet 33:61–65
25. Sato H, Oka T, Shinnou Y, Kondo T, Washio K et al (2010) Multi-step aberrant CpG island hypermethylation is associated with the progression of adult T-cell leukemia/lymphoma. Am J Pathol 176:402–415
26. Niklinska W, Naumnik W, Sulewska A, Kozłowski M, Pankiewicz W et al (2009) Prognostic significance of DAPK and RASSF1A promoter hypermethylation in non-small cell lung cancer (NSCLC). Folia Histochem Cytobiol 47:275–280
27. Braggio E, Maiolino A, Gouveia M, Magalhães R, Souto Filho J et al (2010) Methylation status of nine tumor suppressor genes in multiple myeloma. Int J Hematol 91:87–96
28. Pakneshan P, Têtu B, Rabbani SA (2004) Demethylation of urokinase promoter as a prognostic marker in patients with breast carcinoma. Clin Cancer Res 10:3035–3041
29. Sharma G, Mirza S, Parshad R, Srivastava A, Datta Gupta S et al (2010) CpG hypomethylation of MDR1 gene in tumor and serum of invasive ductal breast carcinoma patients. Clin Biochem 43:373–379
30. Muller CI, Ruter B, Koeffler HP, Lubbert M (2006) DNA hypermethylation of myeloid cells, a novel therapeutic target in MDS and AML. Curr Pharm Biotechnol 7:315–321
31. Kantarjian H, Issa JP, Rosenfeld CS, Bennett JM, Albitar M et al (2006) Decitabine improves patient outcomes in myelodysplastic syndromes: results of a phase III randomized study. Cancer 106:1794–1803
32. Anon (2003) Decitabine: 2'-deoxy-5-azacytidine, Aza dC, DAC, dezocitidine, NSC 127716. Drugs R D 4:352–358
33. Nowell PC (1976) The clonal evolution of tumor cell populations. Science 194:23–28
34. Loeb LA (1991) Mutator phenotype may be required for multistage carcinogenesis. Cancer Res 51: 3075–3079
35. Lengauer C, Kinzler KW, Vogelstein B (1997) DNA methylation and genetic instability in colorectal cancer cells. Proc Natl Acad Sci USA 94:2545–2550
36. McKenna ES, Sansam CG, Cho YJ, Greulich H, Evans JA et al (2008) Loss of the epigenetic tumor suppressor snf5 leads to cancer without genomic instability. Mol Cell Biol 28:6223–6233
37. Kohashi K, Oda Y, Yamamoto H, Tamiya S, Oshiro Y et al (2008) SMARCB1/INI1 protein expression in round cell soft tissue sarcomas associated with chromosomal translocations involving EWS: a special reference to SMARCB1/INI1 negative variant extraskeletal myxoid chondrosarcoma. Am J Surg Pathol 32:1168–1174
38. Trobaugh-Lotrario AD, Tomlinson GE, Finegold MJ, Gore L, Feusner JH (2009) Small cell undifferentiated variant of hepatoblastoma: adverse clinical and molecular features similar to rhabdoid tumors. Pediatr Blood Cancer 52:328–334
39. Biegel JA, Zhou JY, Rorke LB, Stenstrom C, Wainwright LM et al (1999) Germ-line and acquired mutations of INI1 in atypical teratoid and rhabdoid tumors. Cancer Res 59:74–79
40. Feinberg AP, Gehrke CW, Kuo KC, Ehrlich M (1988) Reduced genomic 5-methylcytosine content in human colonic neoplasia. Cancer Res 48:1159–1161
41. Finch PW, He X, Kelley MJ, Uren A, Schaudies RP et al (1997) Purification and molecular cloning of a secreted, Frizzled-related antagonist of Wnt action. Proc Natl Acad Sci USA 94:6770–6775
42. Suzuki H, Watkins DN, Jair KW, Schuebel KE, Markowitz SD et al (2004) Epigenetic inactivation of SFRP genes allows constitutive WNT signaling in colorectal cancer. Nat Genet 36:417–422
43. Zhang W, Glockner SC, Guo M, Machida EO, Wang DH et al (2008) Epigenetic inactivation of the canonical Wnt antagonist SRY-box containing gene 17 in colorectal cancer. Cancer Res 68:2764–2772
44. Yan PS, Venkataramu C, Ibrahim A, Liu JC, Shen RZ et al (2006) Mapping geographic zones of cancer risk with epigenetic biomarkers in normal breast tissue. Clin Cancer Res 12:6626–6636
45. Cheng AS, Culhane AC, Chan MW, Venkataramu CR, Ehrich M et al (2008) Epithelial progeny of estrogen-exposed breast progenitor cells display a cancer-like methylome. Cancer Res 68:1786–1796
46. Chen RZ, Pettersson U, Beard C, Jackson-Grusby L, Jaenisch R (1998) DNA hypomethylation leads to elevated mutation rates. Nature 395:89–93
47. MacLeod AR, Szyf M (1995) Expression of antisense to DNA methyltransferase mRNA induces DNA demethylation and inhibits tumorigenesis. J Biol Chem 270:8037–8043
48. Ramchandani S, MacLeod AR, Pinard M, von Hofe E, Szyf M (1997) Inhibition of tumorigenesis by a cytosine-DNA, methyltransferase, antisense oligodeoxynucleotide. Proc Natl Acad Sci USA 94: 684–689

49. Mortusewicz O, Schermelleh L, Walter J, Cardoso MC, Leonhardt H (2005) Recruitment of DNA methyltransferase I to DNA repair sites. Proc Natl Acad Sci USA 102:8905–8909
50. Ha K, Lee GE, Palii SS, Brown KD, Takeda Y et al (2011) Rapid and transient recruitment of DNMT1 to DNA double-strand breaks is mediated by its interaction with multiple components of the DNA damage response machinery. Hum Mol Genet 20(1): 126–40
51. Milutinovic S, Zhuang Q, Niveleau A, Szyf M (2003) Epigenomic stress response. Knockdown of DNA methyltransferase 1 triggers an intra-S-phase arrest of DNA replication and induction of stress response genes. J Biol Chem 278:14985–14995
52. Unterberger A, Andrews SD, Weaver IC, Szyf M (2006) DNA methyltransferase 1 knockdown activates a replication stress checkpoint. Mol Cell Biol 26:7575–7586
53. Chen T, Hevi S, Gay F, Tsujimoto N, He T et al (2007) Complete inactivation of DNMT1 leads to mitotic catastrophe in human cancer cells. Nat Genet 39:391–396
54. Palii SS, Van Emburgh BO, Sankpal UT, Brown KD, Robertson KD (2008) DNA methylation inhibitor 5-Aza-2′-deoxycytidine induces reversible genome-wide DNA damage that is distinctly influenced by DNA methyltransferases 1 and 3B. Mol Cell Biol 28:752–771
55. Rountree MR, Bachman KE, Baylin SB (2000) DNMT1 binds HDAC2 and a new co-repressor, DMAP1, to form a complex at replication foci. Nat Genet 25:269–277
56. Negishi M, Chiba T, Saraya A, Miyagi S, Iwama A (2009) Dmap1 plays an essential role in the maintenance of genome integrity through the DNA repair process. Genes Cells 14:1347–1357
57. Koizumi T, Negishi M, Nakamura S, Oguro H, Satoh K et al (2010) Depletion of Dnmt1-associated protein 1 triggers DNA damage and compromises the proliferative capacity of hematopoietic stem cells. Int J Hematol 91:611–619
58. Hochberg Z, Feil R, Constancia M, Fraga M, Junien C et al (2010) Child health, developmental plasticity, and epigenetic programming. Endocr Rev 32(2): 159–224
59. Chiu CP, Blau HM (1984) Reprogramming cell differentiation in the absence of DNA synthesis. Cell 37:879–887
60. Chiu CP, Blau HM (1985) 5-Azacytidine permits gene activation in a previously noninducible cell type. Cell 40:417–424
61. Takahashi K, Yamanaka S (2006) Induction of pluripotent stem cells from mouse embryonic and adult fibroblast cultures by defined factors. Cell 126:663–676
62. McGinnis W, Garber RL, Wirz J, Kuroiwa A, Gehring WJ (1984) A homologous protein-coding sequence in drosophila homeotic genes and its conservation in other metazoans. Cell 37:403–408
63. Bracken AP, Dietrich N, Pasini D, Hansen KH, Helin K (2006) Genome-wide mapping of Polycomb target genes unravels their roles in cell fate transitions. Genes Dev 20:1123–1136
64. Bernstein BE, Mikkelsen TS, Xie X, Kamal M, Huebert DJ et al (2006) A bivalent chromatin structure marks key developmental genes in embryonic stem cells. Cell 125:315–326
65. Ohm JE, McGarvey KM, Yu X, Cheng L, Schuebel KE et al (2007) A stem cell-like chromatin pattern may predispose tumor suppressor genes to DNA hypermethylation and heritable silencing. Nat Genet 39:237–242
66. Vire E, Brenner C, Deplus R, Blanchon L, Fraga M et al (2006) The polycomb group protein EZH2 directly controls DNA methylation. Nature 439: 871–874
67. Jordan CT (2009) Cancer stem cells: controversial or just misunderstood? Cell Stem Cell 4:203–205
68. Hong D, Gupta R, Ancliff P, Atzberger A, Brown J et al (2008) Initiating and cancer-propagating cells in TEL-AML1-associated childhood leukemia. Science 319:336–339
69. Mullighan CG, Phillips LA, Su X, Ma J, Miller CB et al (2008) Genomic analysis of the clonal origins of relapsed acute lymphoblastic leukemia. Science 322:1377–1380
70. Widschwendter M, Fiegl H, Egle D, Mueller-Holzner E, Spizzo G et al (2007) Epigenetic stem cell signature in cancer. Nat Genet 39:157–158
71. Passegue E, Jamieson CH, Ailles LE, Weissman IL (2003) Normal and leukemic hematopoiesis: are leukemias a stem cell disorder or a reacquisition of stem cell characteristics? Proc Natl Acad Sci USA 100(Suppl 1):11842–11849
72. Pardal R, Clarke MF, Morrison SJ (2003) Applying the principles of stem-cell biology to cancer. Nat Rev Cancer 3:895–902
73. Jordan CT, Guzman ML, Noble M (2006) Cancer stem cells. N Engl J Med 355:1253–1261
74. Krivtsov AV, Twomey D, Feng Z, Stubbs MC, Wang Y et al (2006) Transformation from committed progenitor to leukaemia stem cell initiated by MLL–AF9. Nature 442:818–822
75. Pulukuri SMK, Gorantla B, Dasari VR, Gondi CS, Rao JS (2010) Epigenetic upregulation of urokinase plasminogen activator promotes the tropism of mesenchymal stem cells for tumor cells. Mol Cancer Res 8:1074–1083
76. Zhao D, Najbauer J, Garcia E, Metz MZ, Gutova M et al (2008) Neural stem cell tropism to glioma: critical role of tumor hypoxia. Mol Cancer Res 6: 1819–1829
77. Thompson EW, Newgreen DF, Tarin D (2005) Carcinoma invasion and metastasis: a role for epithelial-mesenchymal transition? Cancer Res 65:5991–5995, discussion 5995
78. Risbridger GP, Davis ID, Birrell SN, Tilley WD (2010) Breast and prostate cancer: more similar than different. Nat Rev Cancer 10:205–212

79. Gupta GP, Massague J (2006) Cancer metastasis: building a framework. Cell 127:679–695
80. Thiery JP (2002) Epithelial-mesenchymal transitions in tumour progression. Nat Rev Cancer 2:442–454
81. Sarrio D, Rodriguez-Pinilla SM, Hardisson D, Cano A, Moreno-Bueno G et al (2008) Epithelial-mesenchymal transition in breast cancer relates to the basal-like phenotype. Cancer Res 68:989–997
82. Graff JR, Herman JG, Lapidus RG, Chopra H, Xu R et al (1995) E-cadherin expression is silenced by dna hypermethylation in human breast and prostate carcinomas. Cancer Res 55:5195–5199
83. von Burstin J, Eser S, Paul MC, Seidler B, Brandl M et al (2009) E-cadherin regulates metastasis of pancreatic cancer in vivo and is suppressed by a SNAIL/HDAC1/HDAC2 repressor complex. Gastroenterology 137:361–371
84. Lei W, Zhang K, Pan X, Hu Y, Wang D et al (2010) Histone deacetylase 1 is required for transforming growth factor-[beta]1-induced epithelial-mesenchymal transition. Int J Biochem Cell Biol 42:1489–1497
85. Su H-Y, Lai H-C, Lin Y-W, Liu C-Y, Chen C-K et al (2010) Epigenetic silencing of SFRP5 is related to malignant phenotype and chemoresistance of ovarian cancer through Wnt signaling pathway. Int J Cancer 127:555–567
86. Kawakami K, Yamamura S, Hirata H, Ueno K, Saini S et al (2010) Secreted frizzled-related protein-5 (sFRP-5) is epigenetically downregulated and functions as a tumor suppressor in kidney cancer. Int J Cancer 128(3):541–50
87. Xue C, Plieth D, Venkov C, Xu C, Neilson EG (2003) The gatekeeper effect of epithelial-mesenchymal transition regulates the frequency of breast cancer metastasis. Cancer Res 63:3386–3394
88. Nagaraja GM, Othman M, Fox BP, Alsaber R, Pellegrino CM et al (2006) Gene expression signatures and biomarkers of noninvasive and invasive breast cancer cells: comprehensive profiles by representational difference analysis, microarrays and proteomics. Oncogene 25:2328–2338
89. Ma X, Yang Y, Wang Y, An G, Lv G (2010) Small interfering RNA-directed knockdown of S100A4 decreases proliferation and invasiveness of osteosarcoma cells. Cancer Lett 299:171–181
90. Parfyonova YV, Plekhanova OS, Tkachuk VA (2002) Plasminogen activators in vascular remodeling and angiogenesis. Biochemistry (Mosc) 67:119–134
91. Liotta LA, Tryggvason K, Garbisa S, Hart I, Foltz CM et al (1980) Metastatic potential correlates with enzymatic degradation of basement membrane collagen. Nature 284:67–68
92. Mazar A, Henkin J, Goldfarb R (1999) The urokinase plasminogen activator system in cancer: implications for tumor angiogenesis and metastasis. Angiogenesis 3:15–32
93. Møller LB, Pöllänen J, Rønne E, Pedersen N, Blasi F (1993) N-linked glycosylation of the ligand-binding domain of the human urokinase receptor contributes to the affinity for its ligand. J Biol Chem 268: 11152–11159
94. Estreicher A, Muhlhauser J, Carpentier JL, Orci L, Vassalli JD (1990) The receptor for urokinase type plasminogen activator polarizes expression of the protease to the leading edge of migrating monocytes and promotes degradation of enzyme inhibitor complexes. J Cell Biol 111:783–792
95. Limongi P, Resnati M, Hernandez-Marrero L, Cremona O, Blasi F et al (1995) Biosynthesis and apical localization of the urokinase receptor in polarized MDCK epithelial cells. FEBS Lett 369:207–211
96. Quax PH, van Leeuwen RT, Verspaget HW, Verheijen JH (1990) Protein and messenger RNA levels of plasminogen activators and inhibitors analyzed in 22 human tumor cell lines. Cancer Res 50:1488–1494
97. Rabbani SA, Mazar AP (2001) The role of the plasminogen activation system in angiogenesis and metastasis. Surg Oncol Clin N Am 10:393–415, x
98. Rabbani SA, Xing RH (1998) Role of urokinase (uPA) and its receptor (uPAR) in invasion and metastasis of hormone-dependent malignancies. Int J Oncol 12:911–920
99. Guo Y, Pakneshan P, Gladu J, Slack A, Szyf M et al (2002) Regulation of DNA methylation in human breast cancer. Effect on the urokinase-type plasminogen activator gene production and tumor invasion. J Biol Chem 277:41571–41579
100. Pakneshan P, Szyf M, Farias-Eisner R, Rabbani SA (2004) Reversal of the hypomethylation status of urokinase (uPA) promoter blocks breast cancer growth and metastasis. J Biol Chem 279:31735–31744
101. Shukeir N, Pakneshan P, Chen G, Szyf M, Rabbani SA (2006) Alteration of the methylation status of tumor-promoting genes decreases prostate cancer cell invasiveness and tumorigenesis in vitro and in vivo. Cancer Res 66:9202–9210
102. Zhao Y, Li JS, Guo MZ, Feng BS, Zhang JP (2010) Inhibitory effect of S-adenosylmethionine on the growth of human gastric cancer cells in vivo and in vitro. Chin J Cancer 29:752–760
103. Detich N, Hamm S, Just G, Knox JD, Szyf M (2003) The methyl donor s-adenosylmethionine inhibits active demethylation of dna: a candidate novel mechanism for the pharmacological effects of S-adenosylmethionine. J Biol Chem 278: 20812–20820
104. Xing J, Stewart DJ, Gu J, Lu C, Spitz MR et al (2008) Expression of methylation-related genes is associated with overall survival in patients with non-small cell lung cancer. Br J Cancer 98:1716–1722
105. Campbell PM, Bovenzi V, Szyf M (2004) Methylated DNA-binding protein 2 antisense inhibitors suppress tumourigenesis of human cancer cell lines in vitro and in vivo. Carcinogenesis 25:499–507
106. Ivanov MA, Lamrihi B, Szyf M, Scherman D, Bigey P (2003) Enhanced antitumor activity of a combination of MBD2-antisense electrotransfer gene therapy and bleomycin electrochemotherapy. J Gene Med 5:893–899
107. Champion C, Guianvarc'h D, Senamaud-Beaufort C, Jurkowska RZ, Jeltsch A et al (2010) Mechanistic

insights on the inhibition of c5 DNA methyltransferases by zebularine. PLoS One 5:e12388
108. Ateeq B, Unterberger A, Szyf M, Rabbani SA (2008) Pharmacological inhibition of DNA methylation induces proinvasive and prometastatic genes in vitro and in vivo. Neoplasia 10:266–278
109. Chik F, Szyf M (2010) Effects of specific DNMT-gene depletion on cancer cell transformation and breast cancer cell invasion; towards selective DNMT inhibitors. Carcinogenesis 32(2):224–32
110. Frost P, Kerbel RS, Hunt B, Man S, Pathak S (1987) Selection of metastatic variants with identifiable karyotypic changes from a nonmetastatic murine tumor after treatment with 2′-deoxy-5-azacytidine or hydroxyurea: implications for the mechanisms of tumor progression. Cancer Res 47:2690–2695
111. Yu Y, Zeng P, Xiong J, Liu Z, Berger SL et al (2010) Epigenetic drugs can stimulate metastasis through enhanced expression of the pro-metastatic ezrin gene. PLoS One 5:e12710
112. Jung Y, Park J, Kim TY, Park JH, Jong HS et al (2007) Potential advantages of DNA methyltransferase 1 (DNMT1)-targeted inhibition for cancer therapy. J Mol Med 85:1137–1148
113. Szyf M, Pakneshan P, Rabbani SA (2004) DNA demethylation and cancer: therapeutic implications. Cancer Lett 211:133–143
114. Szyf M, Pakneshan P, Rabbani SA (2004) DNA methylation and breast cancer. Biochem Pharmacol 68:1187–1197
115. Szyf M (2009) Epigenetics, DNA methylation, and chromatin modifying drugs. Annu Rev Pharmacol Toxicol 49:243–263
116. Szyf M (2005) DNA methylation and demethylation as targets for anticancer therapy. Biochemistry (Mosc) 70:533–549

Cancer Stem Cells, Models of Study and Implications of Therapy Resistance Mechanisms

9

Fiona M. Frame and Norman J. Maitland

Abstract

There is now compelling evidence for tumour initiating or cancer stem cells (CSCs) in human cancers. The current evidence of this CSC hypothesis, the CSC phenotype and methods of identification, culture and in vitro modelling will be presented, with an emphasis on prostate cancer. Inherent in the CSC hypothesis is their dual role, as a tumour-initiating cell, and as a source of treatment-resistant cells; the mechanisms behind therapeutic resistance will be discussed. Such resistance is a consequence of the unique CSC phenotype, which differs from the differentiated progeny, which make up the bulk of a tumour. It seems that to target the whole tumour, employing traditional therapies to target bulk populations alongside targeted CSC-specific drugs, provides the best hope of lasting treatment or even cure.

9.1 The Cancer Stem Cell Hypothesis

The presence of a specialised tumour-initiating cell has been proposed for two centuries, stimulating experimental studies that resulted in interesting predications and insights with the technologies available at the time [1–4]. Huntly and Gilliland have presented an excellent timeline of cancer stem cell (CSC) research that starts in 1855 [5]. For example, Furth and Kahn described transmitting leukaemia between mice using a single cell in 1937 [3], which has now become a much-debated gold standard test for defining a CSC. The CSC hypothesis can be summed up by the statement that "not all cancer cells are equal" [6–9]. The hypothesis was confined to obscurity for a time, mainly due to the difficulties encountered in purification of the CSCs from complex and heterogenous tumour material, but also due to other "harder" lines of enquiry about the nature of the cancer cell. These included hugely significant discoveries in cancer research; cancer-associated viruses, cancer genetics, immune response, microenvironment, hormones and environmental factors, multistep mutations, oncogenes, tumour suppressors, genetic instability and epigenetics [10]. Incremental

F.M. Frame (✉)
Department of Biology, YCR Cancer Research Unit, University of York, Heslington, North Yorkshire YO10 5DD, UK
e-mail: fiona.frame@york.ac.uk

J.S. Rhim and R. Kremer (eds.), *Human Cell Transformation: Role of Stem Cells and the Microenvironment*, Advances in Experimental Medicine and Biology 720, DOI 10.1007/978-1-4614-0254-1_9, © Springer Science+Business Media, LLC 2011

progress on CSCs was made, but it has only been gathering significant momentum for the last 15 years. Following publication of Lapidot et al. in 1994 [11] describing a tumour-initiating cell in acute myeloid leukaemia (AML), and further findings in the haematopoietic system, where the ability to analyse single cells is greatest, discoveries in solid tumours were not far behind. So far, CSCs have been defined for AML [12], multiple myeloma [13], B-ALL [14, 15], T-ALL [16], colon [17], breast [18], prostate [19], glioblastoma [20], medulloblastoma [20, 21], bladder [22], melanoma [23, 24], head and neck [25, 26], liver [27], kidney [28], pancreas [29], ovarian [30], lung [31] and endometrial [32] tumours. The discovery of markers for CSCs in different tumour types and the ability to isolate stem cells from both cancer tissues and cell cultures has fuelled the current surge of research in this area.

9.1.1 Cancer Stem Cells: Identification and Isolation

CSCs were originally isolated on the basis of normal adult stem cell properties; self-renewal, differentiation and homeostasis, that is maintaining a balance between cell proliferation and cell death [33, 34]. It is perhaps pertinent at this juncture to discuss the term "cancer stem cell", since it is often a topic of debate. Alternative terms such as "tumour-initiating cells" and "clonogens", as used by radiation biologists, may be more accurate. The consensus view is that CSCs have the capacity to give rise to a tumour and indeed can be thought of as the driving force behind tumour growth. All cell types within a tumour mass should therefore be derived from a CSC and cellular heterogeneity in cancer may be the result of an aberrant hierarchy of cell differentiation and development observed in normal tissues. Importantly, like normal stem cells, CSCs should also have the ability to proliferate and self-renew and thus maintain themselves, as well as giving rise to daughter cells. But CSCs have the additional properties of tumour initiation and invasion and conversely have a disrupted homeostasis [8, 35, 36]. It is nevertheless important to remember that, despite the name, a CSC does not necessarily arise only from a normal adult stem cell but could arise from a more differentiated cell, as long as it displays the defined characteristics described above [37, 38]. This concept is illustrated by the existence of induced pluripotent stem (iPS) cells, where a differentiated somatic cell, e.g. a keratinocyte, can assume stem cell characteristics with the transduction of a defined set of transcription factors [39–42].

Like stem cells in a normal tissue, it has been assumed that CSCs would be a rare population of cells, but this has been questioned and indeed seems to vary between and even within different tumour types [43]. In melanoma studies with NOD/SCID mice tumour-initiating cell content was estimated at 0.001–0.1%, whereas when using an interleukin-2 receptor gamma chain null mouse, that is more immunocompromised and therefore more permissive for tumour initiation/growth, the percentage of tumour-initiating cells rose to 25% [44]. This finding first cast doubt on the rarity of CSCs. However, recent xenograft studies with pancreatic, non-small cell lung and head and neck cancers have shown that even under highly permissive xenograft conditions, the number of tumour-initiating cells was still low (1 in 2,500) [45]. These studies do highlight the variations between tumour types. In addition, a correlation between CD133 expression and the grade of tumour was observed by Singh et al. Lower-grade astrocytomas had a smaller fraction of $CD133^{+}$ cells than higher-grade medulloblastomas [20]. As well as biological testing, the CSC hypothesis has been mathematically modelled and the results indicate that the kinetics of treatment, regression and relapse, fit with the hypothesis [46]. The hope is that more sophisticated computer modelling combined with biological findings could contribute to data on drug discovery and drug response prior to in vivo testing.

9.1.2 Cancer Stem Cell Markers

There is no single CSC marker, and CSCs have been defined using a variety of markers that vary between tumour types. These include CD133,

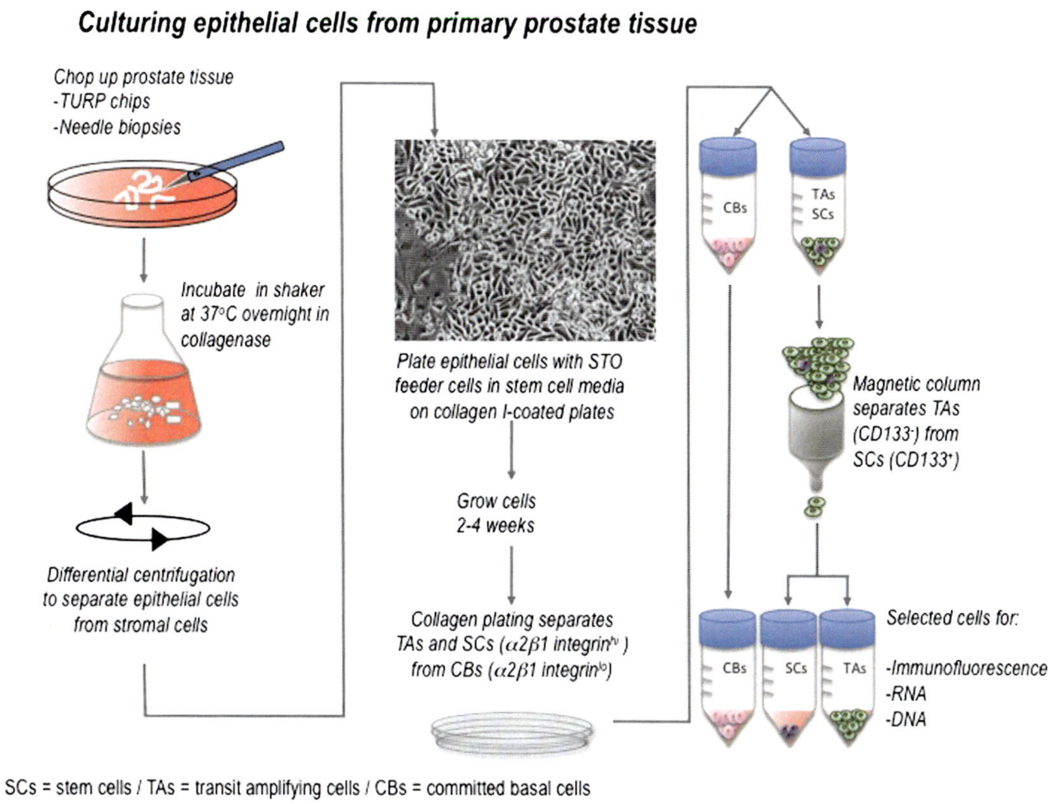

Fig. 9.1 How to culture epithelial cells from primary prostate tissue (*TURP* transurethral resection of the prostate)

CD44, CD24, CD90, CK5, CK20, CD47, ALDH1, ESA, CD166, β1 integrin, α6 integrin, CK8, β-catenin, BMI1, NANOG, SOX2, Musashi, OCT4, CD105, CD271, ABCB5, MyoD88, Lin28, CD117, α2 integrin, CD4, CD34, CD7, CD19 and CD20. Several recent reviews have listed comprehensive lists of markers for each tumour type [43, 47–49]. It is acknowledged that there is no perfect marker for all tumour types, and for isolation the use of several independent markers is required, such that a pure population of cells is obtained rather than just an enrichment [50]. To isolate prostate (cancer) stem cells in our laboratory we culture primary prostate epithelial cells, then select for α2βintegrin high and low expression by using rapid adherence to collagen I plates. The cells that do not adhere to collagen I are harvested as the α2β1integrin^{lo} cells (committed basal cells). Next, the α2β1integrin^{hi} fraction is collected and further selected using antibody beads and MACS columns to separate into α2β1integrin^{hi}/CD133^{-} (transit-amplifying cells) and α2β1integrin^{hi}/CD133^{+} (stem cells) populations (illustrated in Fig. 9.1). Culturing of primary prostate epithelial cells is described further below.

In terms of a therapeutic strategy, it is desirable (but not an absolute requirement in a nonessential organ like the prostate) to eliminate the CSCs without destroying healthy normal adult stem cells [51]. There are characteristics that are unique to adult stem cells but may not be relevant to CSCs, and vice versa, but there are also commonalities [52]. Therefore, it is important to examine each population in isolation. Several microarray studies have highlighted the importance of studying selected populations of cells from tumours [53, 54]. In prostate, for example, the proportion of normal stem cells is 1–3% and the proportion of CSCs is 0.1%, therefore any nuances of expression in these populations would be masked by the bulk populations if the tumour was analysed as a whole [53]. The microarray expression analyse have shown significant differences between stem cells and more differentiated cells, which strengthens the idea that stem cells are indeed different.

9.1.3 Criticisms of the Cancer Stem Cell Hypothesis

One criticism of the CSC hypothesis is that it focuses on the tumour-initiating capability of the stem cell, while ignoring the influence of the microenvironment [55]. This may be the case in some instances, although there are many studies that highlight the significance of microenvironment and stem cell niche, whilst also highlighting the difficulty of developing a true in vitro model. Another criticism is that most researchers are either proponents of the stochastic model or the hierarchical model of cancer and appear not to be open to both. In the stochastic model, all cells have the same potential to become tumorigenic, but it is the occurrence of random (mutational) events that induce one cell to be altered and not another and in this way heterogeneity of the tumour mass occurs. The work on melanoma, using severely immune-compromised mice [44] as described above, suggests that a very high proportion of cells are tumorigenic, in support of the stochastic model. However, there have been lineage-tracking studies in other tumours that support the CSC hypothesis. The likelihood is that all tumours have variable degrees of the features of each model, but not to the exclusion of the other. We must be open to all available evidences. A recent review by Gupta et al. describes the current evidence for the existence of CSCs [56]. To track the stem cell content and its ability to differentiate, we have carried out lineage-tracking studies in our lab using lentiviral transduction of both prostate cell lines and primary prostate epithelial cell cultures [57]. This strategy had been identified by ourselves in prostate and by others in leukaemias [58] as a means of monitoring stem cell differentiation in real time. Lentiviruses have the advantage that they infect all cells, including quiescent cells, and will integrate the lineage tracking expression cassettes into the genome, such that they will be passed to daughter cells. In our study, cell lines were used to confirm expression of differentiation-specific promoters within lentivirus vectors, prior to infection of primary prostate epithelial cells. We successfully infected the $CD133^{+}/\alpha2\beta1integrin^{hi}$ stem cell fraction of primary prostate epithelial cells. In addition, we showed that expression of the PSAPb promoter was activated following differentiation of basal cells into a luminal phenotype (culture conditions are described below). The strategy was successful, however in different systems the choice of promoters, the virus titers and efficiency of infection of primary cells would need to be optimised.

9.1.4 Prostate Cancer Stem Cells: Basal vs. Luminal Cells

The CSC hypothesis and indeed the specific markers to identify the CSCs are more established and more accepted in some tumour types than others. In prostate and breast cancer there remains a controversy about whether the stem cell is a basal cell or a luminal cell but here the discussion will focus on prostate cancer. The prostate is a slow-growing gland that does not have the turnover of, for example, the intestine or skin. Prostate cancer is principally a disease of ageing. Therefore, for a mutation to occur and be selected for and then be passed on to daughter cells such that a tumour is formed, it would be a long process in the prostate. It is likely that the differentiated cells that stop growing or indeed die, are less likely to be the initiators of tumours. A long-lived stem cell is more likely.

It was originally thought that the initiating cell of a prostate cancer would be a luminal cell because the majority of cells in the tumour are of a luminal phenotype. Indeed, the luminal:basal ratio changes from ~60:40 luminal:basal in normal prostate to ~99:1 luminal:basal in prostate cancer. In mouse prostate, Wang et al. suggested that a luminal cell is a cell of origin of prostate cancer by identifying CARN (castration-resistant Nkx3-1-expressing) cells [59]. These cells are defined as luminal because they do not express basal marker p63 but do express cytokeratin 18 and androgen receptor (AR). CARN cells are candidates for a stem cell because they are present in androgen-deprived, regressed prostates and form tumours when they are PTEN null.

However, there is equally compelling evidence that in fact it is a basal cell that is the origin of prostate cancer in mouse. Mouse prostate stem cells defined with $Lin^{-}Sca\text{-}1^{+}CD133^{+}CD44^{+}CD117^{+}$ markers, a basal cell, have been shown to form a prostate gland from a single cell [60]. These studies either mean that in mouse there are two stem cells, one basal and one luminal, or that further examination of both models is needed to find one definitive stem cell.

In normal human prostate, Robinson et al. [61] showed that luminal epithelial primary prostate cells arise from basal cells in culture. In 2004, CD133 was identified as a marker for normal human prostate stem cells from primary prostate cultures [62]. From this work it was then shown that CD133 could be used as a marker for CSCs using primary prostate epithelia from human tumours, along with $\alpha 2\beta 1$integrin [19]. This was concluded on the basis of their ability to self-renew, differentiate, proliferate and form secondary colonies. The primary cells were able to invade in matrigel assays at an even higher rate than the most invasive prostate cancer cell lines. All these experiments were carried out using primary epithelial cells cultured from patient tumours with no genetic manipulation. The one piece of evidence that is missing from this study is the demonstration of tumour initiation in vivo although work in this area is beginning to emerge (see below). Since then, other publications have supported the notion of a basal CSC in prostate [63–69]. Two elegant in vivo studies used a very different strategy to come to the same conclusion that the cell of origin of prostate cancer is a basal cell [64, 66]. Lawson et al. used mouse basal cells into which they engineered increased ERG expression and activation of the PI3K signalling pathway, resulting in tumours with a luminal phenotype. Similarly, Goldstein et al. used primary human benign prostate cells and lentiviruses to introduce activated AKT, ERG transactivator and AR in both basal and luminal cells, and assayed for tumorigenicity at each stage by injecting ≥5,000 cells into immunocompromised mice subcutaneously. Only the basal cells gave rise to a tumour, which was histologically similar to human prostate cancer, i.e. it contained a reduced basal cell content and had differentiated into a largely luminal mass. Currently, therefore the evidence is in favour of a basal phenotype for both normal and CSCs in the prostate [70].

9.1.5 In Vivo Evidence for Prostate Cancer Stem Cells

A precise identity for the CSC in prostate has not reached a consensus, primarily due to lack of in vivo results, to match those available from other tissue and tumour types. While xenografting of the human prostate into immunocompromised murine hosts is a complex procedure with, at best, moderate success rate [71], encouraging recent data from our laboratory indicates that tumour frequency is higher when using the $CD133^{+}$ stem cell fraction [70]. Primary tumour material, directly from patient samples is injected subcutaneously into $Rag2^{-/-}gammaC^{-/-}$ mice, xenograft tumours are harvested and lineage-depleted then total cells or fractionated cells are serially regrafted. $CD133^{+}$, $CD44^{+}$ and $CD24^{+}$ fractions were examined at cell numbers ranging from 10^2 to 10^5 cells per graft. Tumour latency ranged from 30 to 160 days depending on cell type and number of cells used and cancers developed in all fractions except the $CD24^{high}$ fraction, which typically is selected for luminal cells. The xenograft tumours generally had an intermediate phenotype including expression of $CD44^{+}$ and focal expression of AMACR, AR, PSA and p63. In vivo tumour-initiating assays with primary prostate epithelial cells has been challenging and even with an optimum in vivo model, the efficiency of tumour-take was still low (15%) although it is higher from the highest Gleason-grade tumours (Gleason 9) and highest from hormone-refractory cancers. This is most likely due to prostate tumours being slow-growing, and the requirement to adapt to the in vivo mouse model environment.

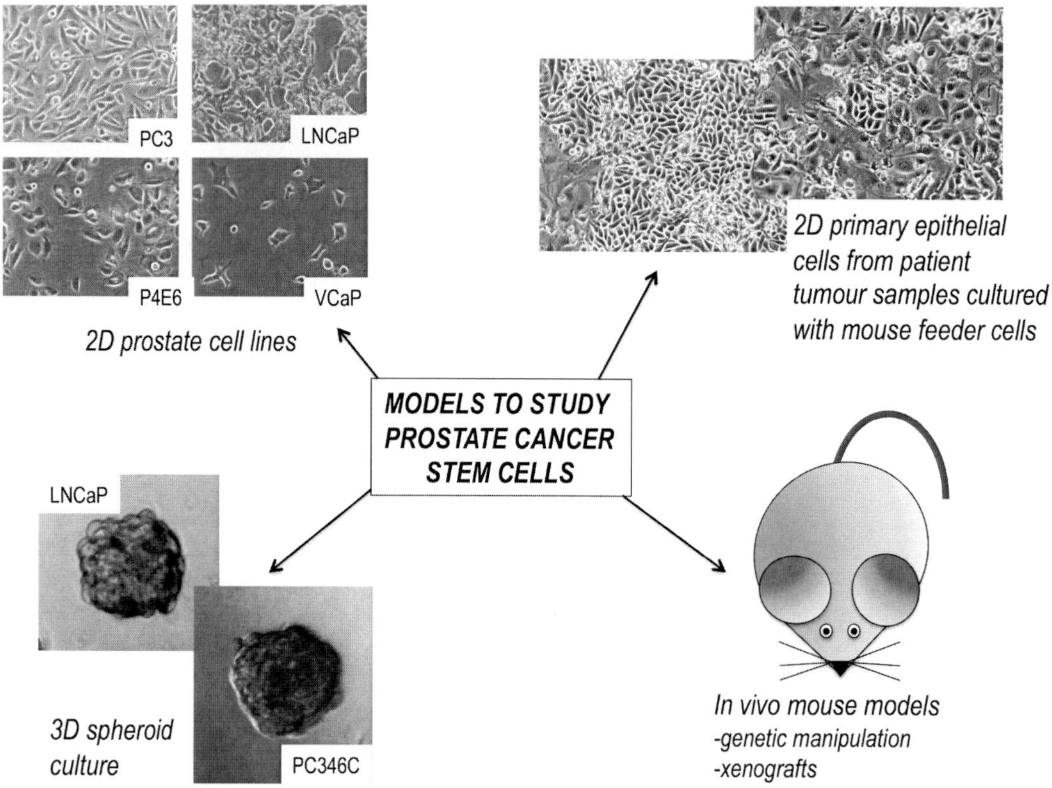

Fig. 9.2 Models available to study prostate cancer stem cells (CSCs) (spheroid images taken by Regina Leadley)

9.1.6 Culture Conditions to Study Cancer Stem Cells

There are a variety of models in use to study CSCs (see Fig. 9.2). Cells cultured from primary patient material without excessive manipulation are preferable, however this is not a possibility for all labs, primarily due to restricted availability of patient material. For this reason many laboratories have exploited cell lines that show some characteristics of CSCs [72–74]. Such models can be good for hypothesis building and testing, however they should be used with caution due to the genetic changes the cells have endured to induce immortalisation, and results should be followed up in primary samples. For prostate there is a broad panel of cell lines available [75], for example, androgen-independent PC3 cancer cells, androgen-responsive LNCaP cancer cells, PNT1a and PNT2C2 normal cells (transformed by large T antigen). LNCaPs and PC3 cells are derived from metastases, and so to provide a cell line that was derived from a primary localised prostate adenocarcinoma, Miki et al. derived two cell lines RC-92a/hTERT (cancer) and RC-165N/hTERT (normal) from primary prostate epithelial cells using a retroviral vector encoding hTERT to immortalise the cells [76]. These cell lines have a population of $CD133^{+}$ cells that retain stem cell properties and as a pair should be very useful tools to study normal vs. CSCs and cell differentiation hierarchy.

Stem cells from primary prostate epithelia can be maintained in culture, however it is to be acknowledged that long-term continuous culture actively selects for cells that have adapted to the growth conditions, plasticware and media components. In the case of prostate epithelium, the media used to maintain the stem cell population is called stem cell medium (SCM) and includes a base of KSFM (keratinocyte serum-free medium) with EGF (epidermal growth factor) and BPE (bovine pituitary extract) with LIF (leukaemia inhibitory factor), SCF (stem cell factor), GM-CSF (granulocyte-macrophage colony stimulating factor) and cholera toxin supplements [19]. The cells are maintained in SCM as a primitive (basal) phenotype, but can be induced to differentiate into a luminal phenotype by placing them in 50:50 DMEM (Dulbecco's Modified Eagles Medium):HAMsF12 with 10% foetal calf serum, DHT (dihydrotestosterone) and an insert of primary prostate stroma. This induces a double layer of cells with the top layer expressing luminal markers including AR and PSA. This manipulation in culture provides further evidence for the induction of an underlying basal cell phenotype to a luminal cell phenotype in prostate cancers [57, 77].

Somewhere between monolayer culture and in vivo models lie 3D in vitro culture models [78]. These models provide the means to investigate the responses of CSCs to environmental factors including cell–cell communication and physical constraints as well as additional features such as addition of stromal cells. Mammospheres, colospheres, neurospheres and for prostate, protaspheres are well developed in culture and sphere-forming assays are integral to some investigations as a test of stemness. Inhibition of stem cells can result in prevention of the formation of spheres. In terms of drug testing they are a more acceptable and affordable alternative for initial drug testing than in vivo models. A clear example of how the 3D model can be manipulated due to external influences comes from prostate [79, 80]. Without serum or stroma, prostate epithelial cells formed solid spheres, but with the addition of serum a lumen and double epithelial layer are formed, and with the further addition of stroma the epithelial cells are polarised, with respect to integrin expression and ultrastructural features. Markers of prostate epithelial cell hierarchy in the model with serum and stroma were observed, confirming a complete replication of the in vivo cellular differentiation. However, $CD133^+$ prostate epithelial cells in spheres appear to be more quiescent, whereas they are stimulated to divide in monolayer culture [81].

9.2 How are Cancer Stem Cells More Resistant to Therapy?

There are a number of mechanisms of therapy resistance that have been harnessed by CSCs as described below and illustrated in Fig. 9.3.

9.2.1 Considerations of Microenvironment and Niche

There is no doubt that environment plays a crucial role in whether or not a tumour will be "successful". It is highly likely that there are cells that over-proliferate or continue to exist, rather than undergo apoptosis and form the start of what could be a tumour but fail for some reason, e.g. lack of blood supply, clearance by the host immune system, lack of adaptation to environment or physical constraints. For a ball of uncontrolled cells to become a primary tumour and to further develop into a potentially fatal secondary tumour, several factors contribute and "niche" is one of those [81–83]. It has however, been very difficult to experimentally model the CSC niche.

Given the importance of microenvironment, a two-pronged approach to prostate cancer treatment, i.e. targeting both the cancer cells and the surrounding stroma may well be required. Here however, the ability of the cancer cells and stem cells to adapt to their environment, will render them something of a moving target [84]. This adaptation is highlighted most clearly by the

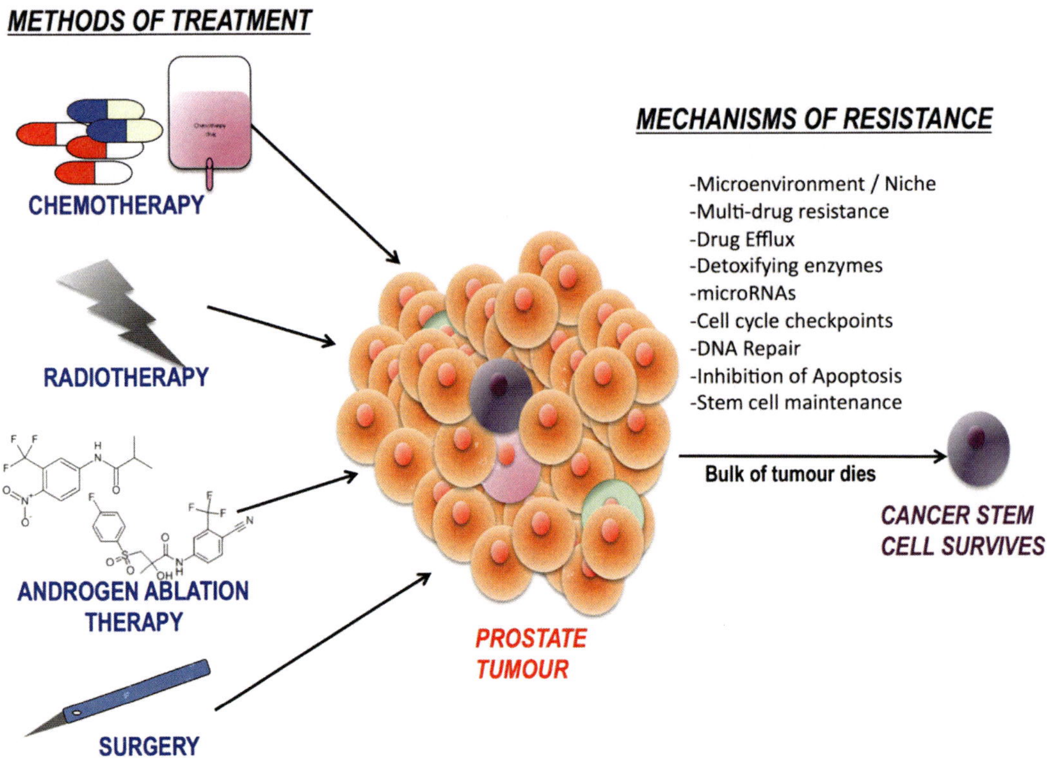

Fig. 9.3 Mechanisms of therapy resistance resulting in survival of cancer stem cells following treatment

epithelial-to-mesenchymal transition (EMT) that is necessary for invasion and metastasis [85–87]. However, there are other adaptations that occur; for example, adapting to inflammation, oxidative stress and hypoxic conditions, co-evolution of cancer cells and fibroblasts resulting in cancer-associated-fibroblasts (CAFs) and adaptation of prostate cancer cells to grow in bone (osteomimicry) [84]. More careful consideration of this plasticity will be necessary when deciding how best to target CSCs.

9.2.2 Multi-Drug Resistance Due to Drug Efflux and Detoxifying Enzymes

The ABC transporters are a family of >10 drug efflux proteins that are expressed in normal and CSCs and include ABCG2 and ABCB1 [88, 89]. They are a mechanism of resistance of CSCs to chemotherapy [90, 91] and their ability to pump out dyes as well as drugs means that they are used as a marker for stem cells [92]. ALDH1 (alcohol dehydrogenase) is able to detoxify drugs and has also been used as a marker of CSCs [93] and a measure of poor prognosis in the clinical setting, for example, in breast cancer [94]. The use of ABC transporter inhibitors has been suggested as an approach to usurp the CSCs [95, 96]. Unfortunately, finding inhibitors with limited toxicity that result in uptake of the drug with no loss of efficacy has proved challenging. New strategies to inhibit the ABC transporters including nanoparticles, antibodies, siRNA, antisense oligonucleotides and transcriptional regulation are being developed [97].

9.2.3 DNA Damage Response: Cell Cycle Checkpoints and DNA Repair

Traditional cancer treatments, such as chemotherapy and radiotherapy, kill cancer cells by inducing DNA damage and eliciting a death response. This is typically most effective in rapidly proliferating cells, that is cells which have lost their ability to respond to antigrowth signals and are self-sufficient in positive growth signals, two of the hallmarks of cancer [98]. There is considerable evidence to show that CSCs have a different DNA damage response when compared to the more differentiated tumour cells and thus are resistant to these treatments [99, 100]. For example, $CD133^+$ glioblastoma stem cells from primary samples and from xenografts had enhanced DNA repair relative to $CD133^-$ cells, most probably due to activation of a cell cycle checkpoint [101]. The resulting radioresistance could be reversed by the use of Chk1 and Chk2 inhibitors. Interestingly, it has also been shown that hypoxic conditions expand the stem cell population in glioblastoma [102]. In leukaemia stem cells it has been shown that a p21-mediated arrest of the cell cycle allows for DNA repair and survival of the stem cell [103]. Cell cycle stage and control over cell cycle checkpoints have a significant impact on treatment outcome. In addition, heterogeneity within stem cells gives an added layer of complexity as seen in haematopoietic stem cells, where there are both quiescent and proliferating stem cells. Both are protected from ionising radiation through DNA damage response and repair mechanisms, but whereas proliferating cells opt for the homologous recombination pathway when repairing double-strand breaks, the quiescent cells opt for the more error-prone non-homologous end-joining pathway [104]. The mechanism of radioresistance in breast CSCs is due to the acquisition of less DNA damage and reduced reactive oxygen species (ROS) as a result of an increase in ROS scavenging enzymes [105]. The BMI1 polycomb protein can also confer radioresistance in glioblastoma stem cells by promoting the recruitment of ATM kinase, a crucial mediator of the DNA damage response, to the sites of damage [106].

In summary, resistance to DNA damage occurs through different mechanisms including activation of cell cycle checkpoints, enhanced DNA repair and quenching of ROS. It has been shown that by teasing apart mechanisms of which pathways are affected, specific inhibitors can be used to interfere with the CSC resistance mechanisms. In addition, it has also become apparent that the best strategy is likely to use a traditional treatment such as radiotherapy combined with a specific drug that inhibits a DNA repair or checkpoint pathway.

9.2.4 Inhibition of Apoptosis

Tumours progress not just due to increased proliferation but also as a result of lack of apoptosis. Evasion of apoptosis is another of the hallmarks of cancer [98]. Therefore, it is not surprising that CSCs have mechanisms to suppress apoptosis thus ensuring their long-term survival [107]. The inhibitor of apoptosis family of proteins (IAPs) are caspase inhibitors that typically contain the baculovirus IAP repeat (BIR). Two members, XIAP and cIAP1 are present at higher levels in $CD133^+$ glioblastoma cells [108]. Survivin, another IAP family member, is involved in maintaining normal stem cell function as well as preventing apoptosis [107]. NF-κB is a transcriptional transactivator, which sits at the centre of many response pathways in mammalian cells. One such pathway is the activation of anti-apoptotic proteins and survival factors [109]. In the prostate, parthenolide, which blocks NF-κB signalling, resulted in killing of CSCs but no loss of viability in $CD133^+$ normal cells [53]. A similar outcome with parthenolide was seen in myeloid and lymphoid leukaemias [110]. This suggests that interfering with NF-κB signalling, which is upregulated in CSCs, can have a therapeutic effect, although its multiplicity of functions in many cell types might restrict its direct inhibition as a treatment strategy. The Bcl-2 family of proteins has anti-apoptotic and pro-apoptotic properties. Overexpression of Bcl-2 results in an increase in the number of haematopoietic stem cells and also protects these stem cells against chemotherapy [111]. A similar effect is seen in glioma stem cells, where an increase in Mcl-1 (Bcl-2 family member) conferred resistance to treatment [112]. Thus, it seems that CSCs have harnessed anti-apoptotic mechanisms to ensure their survival and interfering with these signalling pathways could be considered as a treatment with the potential to target CSCs.

9.2.5 Targeting Stem Cell Maintenance as a Therapeutic Strategy

There are other signalling pathways that are emerging as targets for CSC treatment. These include the Wnt/β-catenin, Notch and Hedgehog signalling pathways [113]. These pathways are also active in normal stem cells, where they are involved in self-renewal, proliferation and differentiation. There are now active clinical trials involving compounds that interfere with these pathways. The effect of Notch inhibition in breast cancer cell lines and primary cells was to prevent sphere formation, proliferation and colony formation, indicating that elevated Notch has a crucial role to play in maintaining stem cell properties [114]. This effect was replicated in glioblastoma, where blocking the Notch pathway with γ-secretase inhibitor treatment resulted in depletion and reduced proliferation of $CD133^+$ cells and reduced neurosphere formation. This effect in turn not only reduced the number of cells able to form xenografts but also slowed the growth of tumours [115]. There is evidence that the Wnt and β-catenin are involved in radioresistance in mouse mammary CSCs [116].

Since these developmental signalling pathways display extensive crosstalk, it is likely that targeting only one pathway may not be sufficient. For example, in a pancreatic cancer xenograft model only a combination of Hedgehog inhibition (cyclopamine) with mTOR blocking (rapamycin) alongside chemotherapy (gemcitabine) reduced the numbers of CSCs [117].

9.2.6 A Role for microRNAs in Therapy Resistance

A developing area of research is the role of microRNAs in cancer development, EMT transition and drug resistance [118], where it seems that microRNAs also have a role in stem cell maintenance [119], stem cell division [120, 121] and CSCs [122]. There are oncogenic and tumour suppressive microRNAs, both of which are deregulated in cancers. For example, loss of miR-34, which promotes p53-mediated apoptosis, has been observed in p53-mutant gastric cancers. When expression was restored in two p53-mutant pancreatic cancer cell models, this resulted in inhibition of growth, invasion and an increase in G1 and G2/M arrest [123]. As miR-34 is a downstream target of p53, the addition of the microRNA restores not only this pathway but it also results in downregulation of Bcl-2, thus increasing apoptosis. Restoration of miR-34 acted to sensitise cells to radiotherapy and chemotherapy. There are now active investigations into how to interfere with microRNAs, to induce differentiation of CSCs or to halt EMT [124, 125].

9.3 Clinical Relevance of Cancer Stem Cells

There is now a pressing need to specifically target novel treatments to CSCs for two reasons; firstly, to prevent secondary tumours if indeed tumour spread is initiated by CSCs, but secondly and most worrying, recent findings that chemotherapy and radiotherapy can actually result in an enrichment of CSCs. There is in vitro evidence for this in breast cancer and colon cancer. Primary colon cancer cells were serially transplanted in xenografts and following cyclophosphamide or irinotecan treatment residual tumours were enriched with colon CSCs. These cells were tumorigenic and had increased levels of ALDH1, which conferred chemoresistance [126]. In breast cancer cell lines $CD24^{-/low}/CD44^+$ tumour-initiating cells increased in numbers following a course of radiation treatment [127]. This correlates with clinical findings that there is a boost in repopulation of tumour cells between radiotherapy treatments. Also in breast cancer cell lines $CD24^{-/low}/CD44^+$ tumour-initiating cells increased following selection in doxorubicin [128]. Significantly, these findings in cell lines were supported in a study using patient samples [129]. Interestingly, the residual tumour-initiating cells that remained following endocrine therapy or chemotherapy are $CD24^{-/low}/CD44^+$ and also have mammosphere-forming ability and have an increase in mesenchymal markers. This association of mesenchymal markers indicating EMT with therapy resistance has been reported before and suggests that CSCs are not only resistant to therapy but undergo changes that allow them to invade and initiate metastasis. If this is also the case in the patient, then traditional therapy could provide some initial relief but ultimately cause more harm than good in the long term.

9.4 Conclusion

To sum up, research into the field of CSCs has accelerated significantly for the last ten years. But perhaps it is now appropriate to look in overview at the nature and the impact of the thousands of publications in this area. We need to ask the fundamental questions that still need to be answered in cancer research and determine if what we know now regarding the CSC model has answered them. What impact has this discovery had on patient treatment? If we eliminate CSCs will it really eliminate the tumour? What is the future potential based on our new knowledge? And, what new problems do we have to face? How can we overcome the mechanisms of resistance of CSCs? Future areas of investigation must also take into account the plasticity and adaptability of the stem cells as well as their heterogeneity. Alternative techniques including mathematical and computer modelling are already playing their part in this process and can help to visualise the process of cell differentiation, such as the model presented by Enver et al. illustrating a landscape picture of cell differentiation [130]. Ultimately the hope is that targeting CSCs as part of cancer treatment can be more patient-specific and tumour-targeted with no relapse.

Acknowledgements This work was funded by Yorkshire Cancer Research.

References

1. Cairns J (1975) Mutation selection and the natural history of cancer. Nature 255(5505):197–200
2. Chury Z, Tobiska J (1958) [Clinical findings & results of culture in a case of stem-cell leukemia with pluripotential properties of the stem cells.]. Neoplasma 5(3):220–231
3. Furth J, Kahn MC (1937) The transmission of leukaemia of mice with a single cell. Am J Cancer 31:276–282
4. Hamburger AW, Salmon SE (1977) Primary bioassay of human tumor stem cells. Science 197(4302): 461–463
5. Huntly BJ, Gilliland DG (2005) Leukaemia stem cells and the evolution of cancer-stem-cell research. Nat Rev Cancer 5(4):311–321
6. Houghton J et al (2007) Stem cells and cancer. Semin Cancer Biol 17(3):191–203
7. Lee JT, Herlyn M (2007) Old disease, new culprit: tumor stem cells in cancer. J Cell Physiol 213(3): 603–609
8. Polyak K, Hahn WC (2006) Roots and stems: stem cells in cancer. Nat Med 12(3):296–300
9. Wicha MS, Liu S, Dontu G (2006) Cancer stem cells: an old idea–a paradigm shift. Cancer Res 66(4):1883–1890, discussion 1895–1896
10. Farrell A et al (2006) Nature milestones: cancer. Nat 440: S7–S23
11. Lapidot T et al (1994) A cell initiating human acute myeloid leukaemia after transplantation into SCID mice. Nature 367(6464):645–648
12. Bonnet D, Dick JE (1997) Human acute myeloid leukemia is organized as a hierarchy that originates from a primitive hematopoietic cell. Nat Med 3(7): 730–737
13. Matsui W et al (2004) Characterization of clonogenic multiple myeloma cells. Blood 103(6): 2332–2336
14. Castor A et al (2005) Distinct patterns of hematopoietic stem cell involvement in acute lymphoblastic leukemia. Nat Med 11(6):630–637
15. Cox CV et al (2004) Characterization of acute lymphoblastic leukemia progenitor cells. Blood 104(9):2919–2925
16. Cox CV et al (2007) Characterization of a progenitor cell population in childhood T-cell acute lymphoblastic leukemia. Blood 109(2):674–682
17. Ricci-Vitiani L et al (2007) Identification and expansion of human colon-cancer-initiating cells. Nature 445(7123):111–115
18. Al-Hajj M et al (2004) Therapeutic implications of cancer stem cells. Curr Opin Genet Dev 14(1): 43–47
19. Collins AT et al (2005) Prospective identification of tumorigenic prostate cancer stem cells. Cancer Res 65(23):10946–10951
20. Singh SK et al (2003) Identification of a cancer stem cell in human brain tumors. Cancer Res 63(18): 5821–5828
21. Singh SK et al (2004) Identification of human brain tumour initiating cells. Nature 432(7015):396–401
22. Chan KS et al (2009) Identification, molecular characterization, clinical prognosis, and therapeutic targeting of human bladder tumor-initiating cells. Proc Natl Acad Sci U S A 106(33):14016–14021

23. Boiko AD et al (2010) Human melanoma-initiating cells express neural crest nerve growth factor receptor CD271. Nature 466(7302):133–137
24. Schatton T et al (2008) Identification of cells initiating human melanomas. Nature 451(7176):345–349
25. Chiou SH et al (2008) Positive correlations of Oct-4 and Nanog in oral cancer stem-like cells and high-grade oral squamous cell carcinoma. Clin Cancer Res 14(13):4085–4095
26. Prince ME et al (2007) Identification of a subpopulation of cells with cancer stem cell properties in head and neck squamous cell carcinoma. Proc Natl Acad Sci U S A 104(3):973–978
27. Ma S et al (2007) Identification and characterization of tumorigenic liver cancer stem/progenitor cells. Gastroenterology 132(7):2542–2556
28. Bussolati B et al (2008) Identification of a tumor-initiating stem cell population in human renal carcinomas. Faseb J 22(10):3696–3705
29. Li C et al (2007) Identification of pancreatic cancer stem cells. Cancer Res 67(3):1030–1037
30. Zhang S et al (2008) Identification and characterization of ovarian cancer-initiating cells from primary human tumors. Cancer Res 68(11):4311–4320
31. Eramo A et al (2008) Identification and expansion of the tumorigenic lung cancer stem cell population. Cell Death Differ 15(3):504–514
32. Rutella S et al (2009) Cells with characteristics of cancer stem/progenitor cells express the CD133 antigen in human endometrial tumors. Clin Cancer Res 15(13):4299–4311
33. Alison MR, Islam S (2009) Attributes of adult stem cells. J Pathol 217(2):144–160
34. Kuci S et al (2009) Adult stem cells as an alternative source of multipotential (pluripotential) cells in regenerative medicine. Curr Stem Cell Res Ther 4(2):107–117
35. Wang Y, Armstrong SA (2008) Cancer: inappropriate expression of stem cell programs? Cell Stem Cell 2(4):297–299
36. Zhang H, Wang ZZ (2008) Mechanisms that mediate stem cell self-renewal and differentiation. J Cell Biochem 103(3):709–718
37. Clarke MF et al (2006) Cancer stem cells–perspectives on current status and future directions: AACR Workshop on cancer stem cells. Cancer Res 66(19):9339–9344
38. Jordan CT (2009) Cancer stem cells: controversial or just misunderstood? Cell Stem Cell 4(3):203–205
39. Aasen T et al (2008) Efficient and rapid generation of induced pluripotent stem cells from human keratinocytes. Nat Biotechnol 26(11):1276–1284
40. Park IH et al (2008) Reprogramming of human somatic cells to pluripotency with defined factors. Nature 451(7175):141–146
41. Takahashi K et al (2007) Induction of pluripotent stem cells from adult human fibroblasts by defined factors. Cell 131(5):861–872
42. Yu J et al (2007) Induced pluripotent stem cell lines derived from human somatic cells. Science 318(5858):1917–1920
43. Maitland NJ, Collins AT (2010) Cancer stem cells – a therapeutic target? Curr Opin Mol Therap 12(6):662–673
44. Quintana E et al (2008) Efficient tumour formation by single human melanoma cells. Nature 456(7222):593–598
45. Ishizawa K et al (2010) Tumor-initiating cells are rare in many human tumors. Cell Stem Cell 7(3):279–282
46. Michor F et al (2005) Dynamics of chronic myeloid leukaemia. Nature 435(7046):1267–1270
47. Garvalov BK, Acker T (2011) Cancer stem cells: a new framework for the design of tumor therapies. J Mol Med 89:95–107. doi: 10.1007/s00109-010-0685-3
48. Hermann PC et al (2010) Cancer stem cells in solid tumors. Semin Cancer Biol 20(2):77–84
49. Visvader JE, Lindeman GJ (2008) Cancer stem cells in solid tumours: accumulating evidence and unresolved questions. Nat Rev Cancer 8(10):755–768
50. Park PC et al (2007) Stem cell enrichment approaches. Semin Cancer Biol 17(3):257–264
51. Zhao RC, Zhu YS, Shi Y (2008) New hope for cancer treatment: exploring the distinction between normal adult stem cells and cancer stem cells. Pharmacol Ther 119(1):74–82
52. Mimeault M et al (2008) Functions of normal and malignant prostatic stem/progenitor cells in tissue regeneration and cancer progression and novel targeting therapies. Endocr Rev 29(2):234–252
53. Birnie R et al (2008) Gene expression profiling of human prostate cancer stem cells reveals a pro-inflammatory phenotype and the importance of extracellular matrix interactions. Genome Biol 9(5):R83
54. Blum R et al (2009) Molecular signatures of prostate stem cells reveal novel signaling pathways and provide insights into prostate cancer. PLoS One 4(5):e5722
55. Sakariassen PO, Immervoll H, Chekenya M (2007) Cancer stem cells as mediators of treatment resistance in brain tumors: status and controversies. Neoplasia 9(11):882–892
56. Gupta PB, Chaffer CL, Weinberg RA (2009) Cancer stem cells: mirage or reality? Nat Med 15(9):1010–1012
57. Frame FM et al (2010) Development and limitations of lentivirus vectors as tools for tracking differentiation in prostate epithelial cells. Exp Cell Res 316(19):3161–3171
58. Hope KJ, Jin L, Dick JE (2004) Acute myeloid leukemia originates from a hierarchy of leukemic stem cell classes that differ in self-renewal capacity. Nat Immunol 5(7):738–743
59. Wang X et al (2009) A luminal epithelial stem cell that is a cell of origin for prostate cancer. Nature 461(7263):495–500
60. Leong KG et al (2008) Generation of a prostate from a single adult stem cell. Nature 456(7223):804–808
61. Robinson EJ, Neal DE, Collins AT (1998) Basal cells are progenitors of luminal cells in primary

cultures of differentiating human prostatic epithelium. Prostate 37(3):149–160
62. Richardson GD et al (2004) CD133, a novel marker for human prostatic epithelial stem cells. J Cell Sci 117(Pt 16):3539–3545
63. Trerotola M et al (2010) CD133, Trop-2 and alpha2-beta1 integrin surface receptors as markers of putative human prostate cancer stem cells. Am J Transl Res 2(2):135–144
64. Goldstein AS et al (2010) Identification of a cell of origin for human prostate cancer. Science 329(5991):568–571
65. Goldstein AS, Stoyanova T, Witte ON (2010) Primitive origins of prostate cancer: in vivo evidence for prostate-regenerating cells and prostate cancer-initiating cells. Mol Oncol 4(5):385–396
66. Lawson DA et al (2010) Basal epithelial stem cells are efficient targets for prostate cancer initiation. Proc Natl Acad Sci U S A 107(6):2610–2615
67. Patrawala L et al (2006) Highly purified CD44+ prostate cancer cells from xenograft human tumors are enriched in tumorigenic and metastatic progenitor cells. Oncogene 25(12):1696–1708
68. Patrawala L et al (2007) Hierarchical organization of prostate cancer cells in xenograft tumors: the CD44+alpha2beta1+ cell population is enriched in tumor-initiating cells. Cancer Res 67(14):6796–6805
69. Tang DG et al (2007) Prostate cancer stem/progenitor cells: identification, characterization, and implications. Mol Carcinog 46(1):1–14
70. Maitland NJ et al (2010) Prostate cancer stem cells: Do they have a basal or luminal phenotype? Horm Cancer 2(1):47–61
71. Loberg RD et al (2006) Development of the VCaP androgen-independent model of prostate cancer. Urol Oncol 24(2):161–168
72. Guo R et al (2011) Description of the CD133+ subpopulation of the human ovarian cancer cell line OVCAR3. Oncol Rep 25(1):141–146
73. Yeung TM et al (2010) Cancer stem cells from colorectal cancer-derived cell lines. Proc Natl Acad Sci U S A 107(8):3722–3727
74. Liu T et al (2010) Establishment and characterization of multi-drug resistant, prostate carcinoma-initiating stem-like cells from human prostate cancer cell lines 22RV1. Mol Cell Biochem 340(1–2):265–273
75. Maitland NJ et al (2010) Gene transfer vectors targeted to human prostate cancer: do we need better preclinical testing systems? Hum Gene Ther 21(7):815–827
76. Miki J et al (2007) Identification of putative stem cell markers, CD133 and CXCR4, in hTERT-immortalized primary nonmalignant and malignant tumor-derived human prostate epithelial cell lines and in prostate cancer specimens. Cancer Res 67(7):3153–3161
77. Swift SL, Burns JE, Maitland NJ (2010) Altered expression of neurotensin receptors is associated with the differentiation state of prostate cancer. Cancer Res 70(1):347–356
78. Hirschhaeuser F et al (2010) Multicellular tumor spheroids: an underestimated tool is catching up again. J Biotechnol 148(1):3–15
79. Lang SH et al (2001) Prostate epithelial cell lines form spheroids with evidence of glandular differentiation in three-dimensional Matrigel cultures. Br J Cancer 85(4):590–599
80. Lang SH et al (2001) Experimental prostate epithelial morphogenesis in response to stroma and three-dimensional matrigel culture. Cell Growth Differ 12(12):631–640
81. Lang SH et al (2010) Modeling the prostate stem cell niche: an evaluation of stem cell survival and expansion in vitro. Stem Cells Dev 19(4):537–546
82. Sneddon JB, Werb Z (2007) Location, location, location: the cancer stem cell niche. Cell Stem Cell 1(6):607–611
83. Li L, Neaves WB (2006) Normal stem cells and cancer stem cells: the niche matters. Cancer Res 66(9):4553–4557
84. Josson S et al (2010) Tumor-stroma co-evolution in prostate cancer progression and metastasis. Semin Cell Dev Biol 21(1):26–32
85. van der Pluijm G (2011) Epithelial plasticity, cancer stem cells and bone metastasis formation. Bone 48(1):37–43
86. Iwatsuki M et al (2010) Epithelial-mesenchymal transition in cancer development and its clinical significance. Cancer Sci 101(2):293–299
87. Micalizzi DS, Farabaugh SM, Ford HL (2010) Epithelial-mesenchymal transition in cancer: parallels between normal development and tumor progression. J Mammary Gland Biol Neoplasia 15(2):117–134
88. Elliott A, Adams J, Al-Hajj M (2010) The ABCs of cancer stem cell drug resistance. IDrugs 13(9):632–635
89. Scotto KW (2003) Transcriptional regulation of ABC drug transporters. Oncogene 22(47):7496–7511
90. Tanei T et al (2009) Association of breast cancer stem cells identified by aldehyde dehydrogenase 1 expression with resistance to sequential Paclitaxel and epirubicin-based chemotherapy for breast cancers. Clin Cancer Res 15(12):4234–4241
91. Dean M, Fojo T, Bates S (2005) Tumour stem cells and drug resistance. Nat Rev Cancer 5(4):275–284
92. Ding XW, Wu JH, Jiang CP (2010) ABCG2: a potential marker of stem cells and novel target in stem cell and cancer therapy. Life Sci 86(17–18):631–637
93. Ma I, Allan AL (2011) The role of human aldehyde dehydrogenase in normal and cancer stem cells. Stem Cell Rev 7:292–306. doi: 10.1007/s12015-010-9208-4
94. Ginestier C et al (2007) ALDH1 is a marker of normal and malignant human mammary stem cells and a predictor of poor clinical outcome. Cell Stem Cell 1(5):555–567
95. Gatti L et al (2009) ABC transporters as potential targets for modulation of drug resistance. Mini Rev Med Chem 9(9):1102–1112

96. Fletcher JI et al (2010) ABC transporters in cancer: more than just drug efflux pumps. Nat Rev Cancer 10(2):147–156
97. Wu CP, Calcagno AM, Ambudkar SV (2008) Reversal of ABC drug transporter-mediated multidrug resistance in cancer cells: evaluation of current strategies. Curr Mol Pharmacol 1(2):93–105
98. Hanahan D, Weinberg RA (2000) The hallmarks of cancer. Cell 100(1):57–70
99. Ishii H et al (2008) Cancer stem cells and chemoradiation resistance. Cancer Sci 99(10):1871–1877
100. Morrison R et al (2011) Targeting the mechanisms of resistance to chemotherapy and radiotherapy with the cancer stem cell hypothesis. J Oncol 2011:941876. doi: 10:1155/2011/941876
101. Bao S et al (2006) Glioma stem cells promote radioresistance by preferential activation of the DNA damage response. Nature 444(7120):756–760
102. Sheehan JP et al (2010) Improving the radiosensitivity of radioresistant and hypoxic glioblastoma. Future Oncol 6(10):1591–1601
103. Viale A et al (2009) Cell-cycle restriction limits DNA damage and maintains self-renewal of leukaemia stem cells. Nature 457(7225):51–56
104. Mohrin M et al (2010) Hematopoietic stem cell quiescence promotes error-prone DNA repair and mutagenesis. Cell Stem Cell 7(2):174–185
105. Diehn M et al (2009) Association of reactive oxygen species levels and radioresistance in cancer stem cells. Nature 458(7239):780–783
106. Facchino S et al (2010) BMI1 confers radioresistance to normal and cancerous neural stem cells through recruitment of the DNA damage response machinery. J Neurosci 30(30):10096–10111
107. Fulda S, Pervaiz S (2010) Apoptosis signaling in cancer stem cells. Int J Biochem Cell Biol 42(1):31–38
108. Liu G et al (2006) Analysis of gene expression and chemoresistance of CD133+ cancer stem cells in glioblastoma. Mol Cancer 5:67
109. Baud V, Karin M (2009) Is NF-kappaB a good target for cancer therapy? Hopes and pitfalls. Nat Rev Drug Discov 8(1):33–40
110. Guzman ML et al (2007) An orally bioavailable parthenolide analog selectively eradicates acute myelogenous leukemia stem and progenitor cells. Blood 110(13):4427–4435
111. Domen J, Cheshier SH, Weissman IL (2000) The role of apoptosis in the regulation of hematopoietic stem cells: Overexpression of Bcl-2 increases both their number and repopulation potential. J Exp Med 191(2):253–264
112. Tagscherer KE et al (2008) Apoptosis-based treatment of glioblastomas with ABT-737, a novel small molecule inhibitor of Bcl-2 family proteins. Oncogene 27(52):6646–6656
113. Takebe N et al (2011) Targeting cancer stem cells by inhibiting Wnt, Notch, and Hedgehog pathways. Nat Rev Clin Oncol 8:97–106. doi: 10.1038/nrclinonc.2010.196
114. Grudzien P et al (2010) Inhibition of Notch signaling reduces the stem-like population of breast cancer cells and prevents mammosphere formation. Anticancer Res 30(10):3853–3867
115. Fan X et al (2010) NOTCH pathway blockade depletes CD133-positive glioblastoma cells and inhibits growth of tumor neurospheres and xenografts. Stem Cells 28(1):5–16
116. Woodward WA et al (2007) WNT/beta-catenin mediates radiation resistance of mouse mammary progenitor cells. Proc Natl Acad Sci U S A 104(2): 618–623
117. Mueller MT et al (2009) Combined targeted treatment to eliminate tumorigenic cancer stem cells in human pancreatic cancer. Gastroenterology 137(3): 1102–1113
118. Sarkar FH et al (2010) Implication of microRNAs in drug resistance for designing novel cancer therapy. Drug Resist Updat 13(3):57–66
119. Hatfield S, Ruohola-Baker H (2008) microRNA and stem cell function. Cell Tissue Res 331(1):57–66
120. Shcherbata HR et al (2006) The MicroRNA pathway plays a regulatory role in stem cell division. Cell Cycle 5(2):172–175
121. Hatfield SD et al (2005) Stem cell division is regulated by the microRNA pathway. Nature 435(7044): 974–978
122. Ji Q et al (2010) No small matter: microRNAs – key regulators of cancer stem cells. Int J Clin Exp Med 3(1):84–87
123. Ji Q et al (2009) MicroRNA miR-34 inhibits human pancreatic cancer tumor-initiating cells. PLoS One 4(8):e6816
124. Kong D et al (2010) Epithelial to mesenchymal transition is mechanistically linked with stem cell signatures in prostate cancer cells. PLoS One 5(8): e12445
125. Starnes LM, Sorrentino A (2011) Regulatory circuitries coordinated by transcription factors and microRNAs at the cornerstone of hematopoietic stem cell self-renewal and differentiation. Curr Stem Cell Res Ther 6:142–161
126. Dylla SJ et al (2008) Colorectal cancer stem cells are enriched in xenogeneic tumors following chemotherapy. PLoS One 3(6):e2428
127. Phillips TM, McBride WH, Pajonk F (2006) The response of CD24(-/low)/CD44+ breast cancer-initiating cells to radiation. J Natl Cancer Inst 98(24):1777–1785
128. Calcagno AM et al (2010) Prolonged drug selection of breast cancer cells and enrichment of cancer stem cell characteristics. J Natl Cancer Inst 102(21): 1637–1652
129. Creighton CJ et al (2009) Residual breast cancers after conventional therapy display mesenchymal as well as tumor-initiating features. Proc Natl Acad Sci U S A 106(33):13820–13825
130. Enver T et al (2009) Stem cell states, fates, and the rules of attraction. Cell Stem Cell 4(5):387–397

Part III

Breast Cancer/Stem Cells

The Role of the Basal Stem Cell of the Human Breast in Normal Development and Cancer

10

Jose Russo and Irma H. Russo

Abstract
MCF-10F, an ERα negative human breast epithelial cell line derived from normal breast tissue, is able to form ductal structures in a tridimensional collagen matrix system. MCF-10F cells that are estrogen transformed (trMCF cells) progressively express phenotypes of in vitro cell transformation, including colony formation in agar methocel and loss of the ductulogenic capacity. Selection of these trMCF cells for invasiveness identified cells (bcMCF) that formed tumors in severe combined immunodeficient mice. The cell lines derived from those tumors (caMCF) were poorly differentiated ER, PR, and ERBB2 negative adenocarcinomas. These characteristics are similar to the human basal cell-like carcinomas. This in vitro–in vivo model demonstrates the importance of the basal cell type as a stem cell that reconstitutes the branching pattern of the breast and that is also target of a carcinogenic insult leading to transformation and cancer.

10.1 Introduction

Breast cancer is an invasive and ultimately fatal disease whose incidence in postmenopausal women has gradually increased in most Western societies over the last few decades, but has sharply increased in younger women [1]. Invasive breast cancer is a heterogeneous disease that encompasses a variety of pathological features that are associated with specific clinical behavior [2]. The classification and grading of the tumors, which are essential for selecting therapeutic approaches and the prediction of their biological behavior and patient prognosis, are currently based on the Nottingham modification of the Scarff-Bloom-Richardson system (NSBR) grading scheme. This system, however, is hindered by the subjectivity of the morphological assessment of nuclear grade, mitosis, and tubular formation [3]. The discovery that the morphological heterogeneity of breast cancer is also reflected at the transcriptome level has allowed the classification of breast cancer into five main groups: luminal A and B, normal breast like, ERBB2 (HER2), and basal-like breast carcinomas [4–6].

J. Russo (✉)
Breast Cancer Research Laboratory,
Women Cancer Program, Fox Chase Cancer Center,
Philadelphia, PA 19111, USA
e-mail: J_Russo@fccc.edu

J.S. Rhim and R. Kremer (eds.), *Human Cell Transformation: Role of Stem Cells and the Microenvironment*, Advances in Experimental Medicine and Biology 720,
DOI 10.1007/978-1-4614-0254-1_10, © Springer Science+Business Media, LLC 2011

The luminal-like subtypes display moderate to high expression of ER and luminal cytokeratins. The basal-like carcinomas, which have been reported to have a more aggressive clinical behavior, are the focus of this work. They are composed of cells that consistently express genes usually found in normal basal/myoepithelial cells of the breast, including basal cytokeratins (5/6, 14, and 17), E-cadherin, caveolin 1, and p53 [7, 8]. Molecular analyses of basal-like carcinomas have confirmed the frequent lack of expression of estrogen (ER) and progesterone (PR) receptors and HER2, high levels of expression of proliferation-related genes, and frequent mutation of the TP53 gene. Morphologically, basal-like breast carcinomas present with high histological grade, high mitotic indices, central necrotic zones, pushing borders, and a conspicuous lymphocytic infiltrate. In addition, metaplastic elements and medullary/atypical medullary features have been reported. Similarities have been found between basal-like tumors and breast carcinomas occurring in BRCA1 mutation carriers, in premenopausal African American women, Hispanic women, and in general, in the younger breast cancer patient population. Breast cancer is more aggressive in African American women [9]; this phenomenon has been partly explained by a later stage at diagnosis [10], larger size of the tumors that are more commonly high grade, and both ER and HER2 negative [11–14].

Basal-like breast cancer has been strongly associated with African American race and Hispanic ancestry [15–18]. Surveillance, Epidemiology, and End Results (SEER) data [19] also show a shift toward more aggressive subtypes in African American and Hispanic women that is consistent across several studies that include basal cell type with a 5-year relative survival of only 14% [20, 21]. Basal-like breast cancers are measured with specificity by adding positive markers such as cytokeratin 5/6 or epidermal growth factor receptor [22]. Nonetheless, specific markers for basal-like breast cancer are not presently available in most studies, even though distinguishing true basal-like from triple negative breast cancers has important implications for clinical prognosis [22, 23].

10.2 Biological and Molecular Understanding of Basal Breast Cancer and Epithelial–Mesenchymal Transition (EMT)

10.2.1 The In Vitro Model of Basal-Like Breast Cancer

Primary mammary epithelial cells grown in collagen matrix are able to form tree-like structures resembling in vivo ductulogenesis [24]. The human breast epithelial cells MCF-10F form tubules when grown in type I collagen. The advantage of an in vitro model of three-dimensional (3D) growth is that it reproduces the epithelial architecture of the breast (Fig. 10.1) [24–30]. Normal epithelial cells form duct-like structures, having apical–basal polarity and well-organized tubular structures with stable adherent junctions and cell-basement communications. The cell's neoplastic transformation is associated with the loss of apical–basal polarity and monolayer morphology and significant deviations from normal epithelial behavior in 3D cultures [24–30].

Our observations that ductal carcinomas originate in the Lob.1 of the immature breast (Fig. 10.1) [31], which are the structures with the highest proliferative activity, and the fact that the cells that do proliferate in culture are ER negative suggests that the stem cells that originate from normal ductal structures and cancer are the ER negative proliferating cells. This idea is supported by our observations that MCF-10F, a spontaneously immortalized ER negative human breast epithelial cell line derived from breast tissues containing Lob.1 and Lob.2, is able to form normal ductal structures in a tridimensional collagen matrix system (Fig. 10.2). The ductal structures are lined by a monolayer of well-polarized epithelial cells that become malignant after exposure to either the chemical carcinogen benz[a]pyrene [32] or the natural estrogen 17β-estradiol (E_2) [24, 29].

We have developed an in vitro–in vivo model of human breast epithelial cells transformation induced by estradiol (Fig. 10.3) [29, 30].

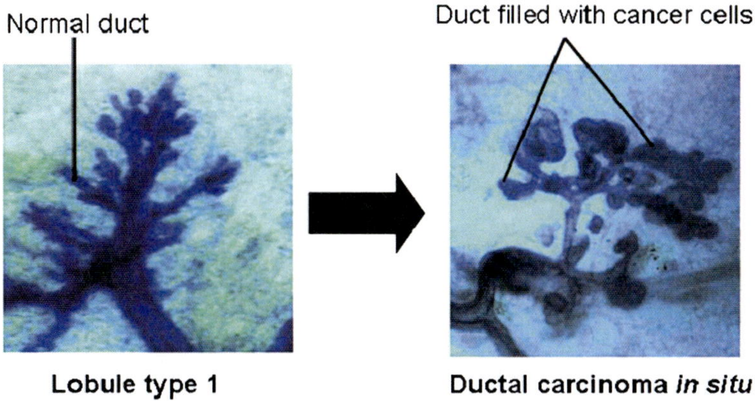

Fig. 10.1 Lobule type 1 or the terminal ductal lobular unit is the site of origin of ductal carcinoma in situ. Adapted from Russo et al. [76]

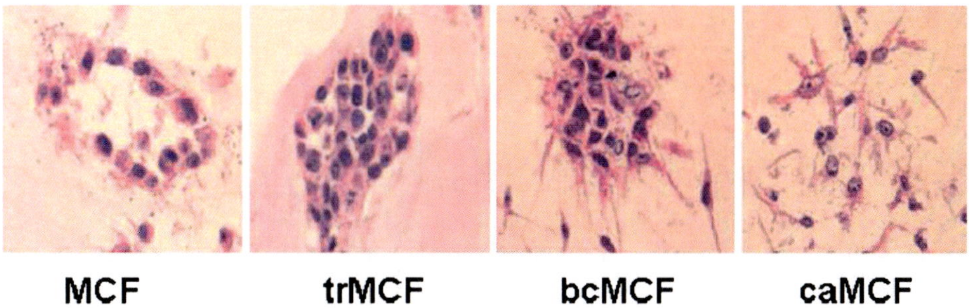

Fig. 10.2 In vitro–in vivo model of cell transformation. Different stages in the in vitro–in vivo model. The MCF basal cell type from normal ductal structures that loss the ductulogenic capacity forming the trMCF cells that evolves to bcMCF and caMCF cells. Adapted from: Russo et al. [29] and Tiezzi et al. [77]

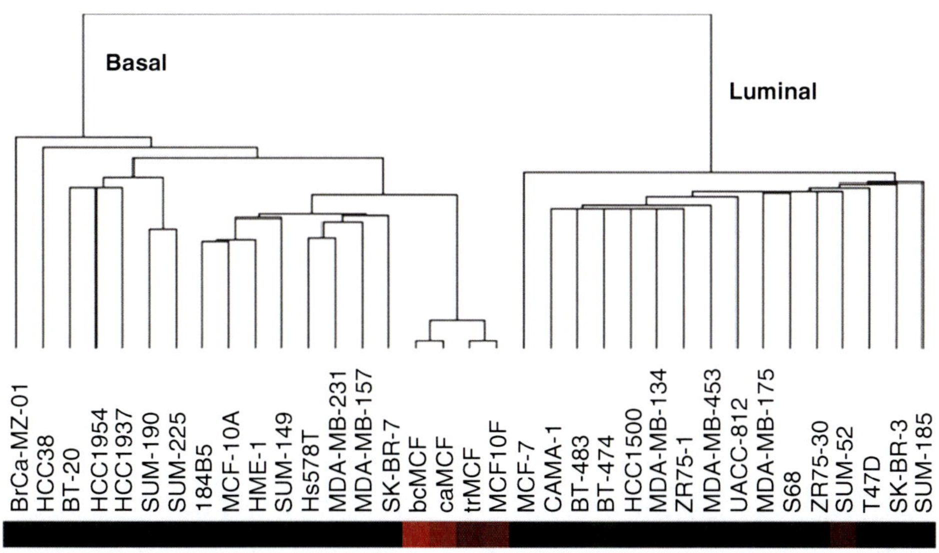

Fig. 10.3 MCF-10F and derived cell lines (trMCF, bcMCF, and caMCF) classify as basal cell lines. The expression data for these cell lines (26) was combined with the published expression values of the 364 gene set (14) and used to classify the breast cell lines by hierarchical clustering as described (14). The MCF-10F and derived cell lines occupy a distinct branch of the basal subtype (*left branch*) because the expression values were calculated independently from different batches of normalized genechips

In this model, the human breast epithelial cell line MCF-10F that is estrogen receptor α (ESR) negative was transformed by estradiol, and different cell lines that represent different stages of breast cancer progression were isolated [29, 30]. The MCF-10F progression model consists of four derived cell lines: (a) the spontaneously immortalized cell line MCF-10F, which does not show any characteristics of invasiveness or tumor formation and therefore is considered to be a normal-like breast epithelial cell line; (b) the transformed trMCF cells; (c) the invasive bsMCF cell line; and (d) the tumor cell lines, caMCFs, which show all characteristics of a fully malignant breast cancer cell types [29, 30] (Fig. 10.2). The bsMCF cells induced tumors in severe combined immunodeficient (SCID) mice that were poorly differentiated adenocarcinomas that were ESR, PR, and ERBB2 negatives and are also metastatic to the lung (Fig. 10.3).

Loss of the ductulogenic capacity is the earliest phenotype observed accompanied by increasing cell proliferation and the activation of genes related to DNA cell replication, inhibition of apoptosis, and the expression of genes related to cell polarity, cell positioning, and cellular architecture. Further selection of these trMCF cells for invasiveness in a Matrigel invasion system identified cells (bcMCF) that formed tumors in SCID mice. The cell lines derived from those tumors (caMCF) were poorly differentiated ER, PR, and ERBB2 negative adenocarcinomas [29, 30]. These characteristics are similar to the human basal cell-like carcinomas previously described [6]. To better understand the molecular events associated with the progressive phenotypic changes that were observed during estrogen-mediated malignant cell transformation, we analyzed chromosomal copy number (CN), loss of heterozygosity (LOH), and gene expression changes that occurred at different stages of cell transformation. By integrating these data we were able to identify associations between CN changes, LOH, transcript expression, and phenotypes of invasion and tumorigenicity, including a strong gene signature of epithelial to mesenchymal transition (EMT) that was confirmed by immunohistochemistry [30]. The bcMCF (invasive) and caMCF (tumor-derived) cells showed dramatic changes in morphology, losing epithelial characteristics of polarity and acquiring mesenchymal characteristics of a fibroblast-like spindle shape and increased migratory behavior, invasiveness, and metastatic capabilities. Changes in gene and protein expression were characteristic of EMT, namely, loss of intercellular adhesion (E-cadherin and occludins), downregulation of epithelial makers (cytokeratins), and upregulation of mesenchymal markers (vimentin and smooth muscle actins) [29, 30].

10.2.2 The Molecular Pathway

Immortal MCF-10F cells are nontransformed, nontumorigenic, and ER negative. Because malignant cell transformation of these cells produced poorly differentiated tumors characteristic of basal-like carcinomas, we chose to classify these cells relative to the breast cell lines described by Charafe-Jauffret et al. [33]. As shown in Fig. 10.4, these MCF-10F and derived cell lines clustered in the branch containing the basal breast cell lines. In our molecular characterization of malignant cell transformation [30], we identified the "intermediate filament" component enriched in gene ontology (GO) analysis, separating the nontumorigenic MCF-10F and trMCF cells from the tumorigenic bcMCF and caMCF cells. Numerous cytokeratins were suppressed or absent, whereas vimentin was strongly induced in bcMCF (7.0-fold) and caMCF (8.1-fold). Because of these findings, we generated a gene list from published literature for EMT markers and their regulators [30]. The 52 genes in this list were filtered by low stringency criteria of combined coefficient of variation >0.3 and "Present calls" in more than 30% of the samples. The 27 genes passing these criteria were used for sample and gene clustering (Fig. 10.5a). Two sample groups and two gene groups were identified. The nontumorigenic MCF-10F and trMCF cells were grouped into sample cluster κ, while the tumorigenic bcMCF and caMCF cells were grouped into cluster λ. On the other side, the genes were grouped into cluster α and β based on their expression pattern.

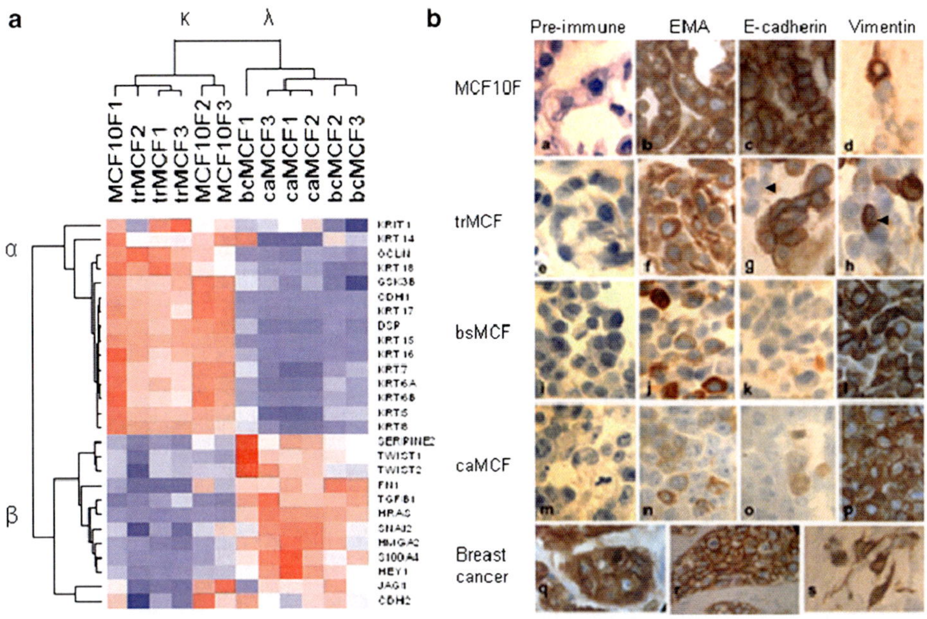

Fig. 10.4 Expression profile of EMT markers and their regulators during malignant cell transformation. (**a**) A list of EMT markers and promoting genes was generated a priory by literature search (26). Hierarchical clustering of cell lines and genes was performed using dChip software. Two sample clusters (κ and λ) and two gene clusters (α and β) were identified. The *red*, *white*, and *blue* colors represent level above, at, and below mean expression, respectively. (**b**) Detection of epithelial and mesenchymal markers by immunochytochemistry (100×). *a*: Histological sections of MCF-10F cells, reacted with preimmune mouse serum, were used as the negative control; *b–d*: MCF-10F reacted for EMA, E-Cadherin, vimentin, respectively; *e*: trMCF cells reacted with preimmune mouse serum used as negative control; *f–h*: trMCF cells reacted for EMA, E-cadherin, and vimentin, respectively; *i*: bsMCF cells reacted with preimmune mouse serum as a negative control; *j–l*: bsMCF cells reacted for EMA, E-cadherin, and vimentin, respectively; *m*: caMCF tumor cell line cells reacted with preimmune mouse serum used as negative control; *n–p*: caMCF tumor cell lines reacted for EMA, E-cadherin, and vimentin, respectively; *q* and *r*: invasive ductal carcinoma of the breast as positive control and immunoreacted for EMA and E-cadherin, respectively; *s*: histological section of an invasive adenocarcinoma immunoreacted for vimentin. From Huang et al. [30]

The epithelial markers *E-cadherin, occludin, desmoplakin,* and *cytokertins* were decreased, while the mesenchymal markers *fibronectin, vimentin,* and *N-cadherin* were increased in bcMCF and caMCF cells (Fig. 10.5a).

By real-time RT-PCR, it was confirmed that the expression of FN1, S100A4, SNAI2, HRAS, and TGFβ1 was increased, while CDH1 (E-cadherin) was decreased in bcMCF and caMCF cells (see [30]). Immunocytochemical analysis using antibodies against epithelial membrane antigen EMA (also called MUC1) and E-cadherin displayed significant loss of these epithelial markers and increased expression of the mesenchymal marker vimentin in tumorigenic cells (Fig. 10.5b). These findings confirmed the EMT phenotype revealed by gene expression profile in Fig. 10.5a.

In order to determine whether there is a relationship between our EMT gene signature (Fig. 10.5a) and the classification of the Basal A and Basal B breast cell lines reported in Neve et al. [34], we identified nine genes present in both our EMT gene signature and the 396 gene classifier set of Neve et al. [34]. By Prediction Analysis of Microarray (PAM), we have shown that the parent MCF-10F cells and trMCF cells can be classified as Basal A, whereas the bcMCF (invasive) and caMCF (tumor-derived) cells are classified as Basal B (Table 10.1). Based on these data, we hypothesize that both Basal subtypes A and B can arise from the same cell of origin, and may reflect differing degrees of EMT and invasive potential. To explore this hypothesis further, we extracted our EMT gene signature from the Genechip expression files of the 30 cell lines that

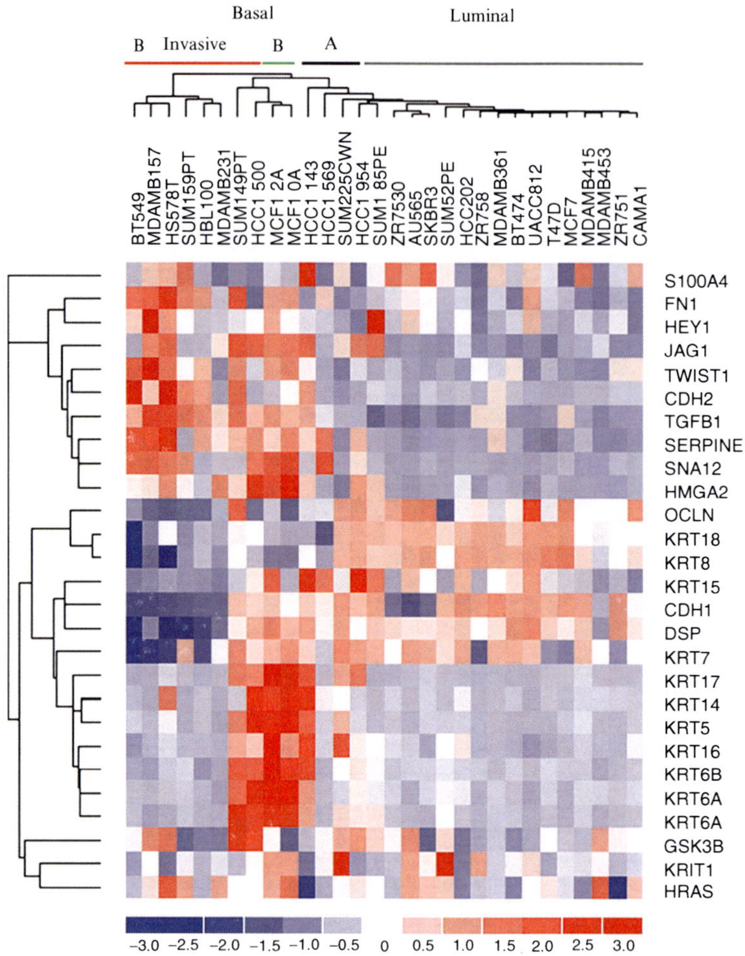

Fig. 10.5 Classification of breast cell lines using an EMT gene signature identifies subtype and invasiveness. Gene expression values were extracted from the CEL files of the 30 cell line subset of breast cancer cell lines that were previously characterized for invasive potential (see ref. [15], Fig. 10.4). The invasive Basal B cell lines are indicated by the *red bar*; noninvasive by a *green bar*

Table 10.1 Genes differentially expressed in bcMCF compared to MCF10F cells

Symbol	Gene name	Fold change
AZGP1	Alpha-2-glycoprotein 1, zinc-binding	–31.8
CDH1	Cadherin 1, type 1, E-cadherin (epithelial)	–497.8
EPB41L5	Erythrocyte membrane protein band 4.1 like 5	–17.7
FRMD3	FERM domain containing 3	–3.3
GRB14	Growth factor receptor-bound protein 14	–4.1
GPR56	G protein-coupled receptor 56	–27.8
KLF5	Kruppel-like factor 5 (intestinal)	–2.9
MEST	Mesoderm specific transcript homolog (mouse)	–4.3
MTUS1	Mitochondrial tumor suppressor 1	–2.4
RAB25	RAB25, member RAS oncogene family	–46.2
SFRP1	Secreted frizzled-related protein 1	–5.8
SFRP1	Secreted frizzled-related protein 1	–5.5
SFRP1	Secreted frizzled-related protein 1	–4.8
SRPX	Sushi-repeat-containing protein, X-linked	–6.4
ST14	Suppression of tumorigenicity 14 (colon carcinoma)	–5.8
ST14	Suppression of tumorigenicity 14 (colon carcinoma)	–4.2
S100A9	S100 calcium binding protein A9	–18.2
SIK1	Salt-inducible kinase 1	–6.9

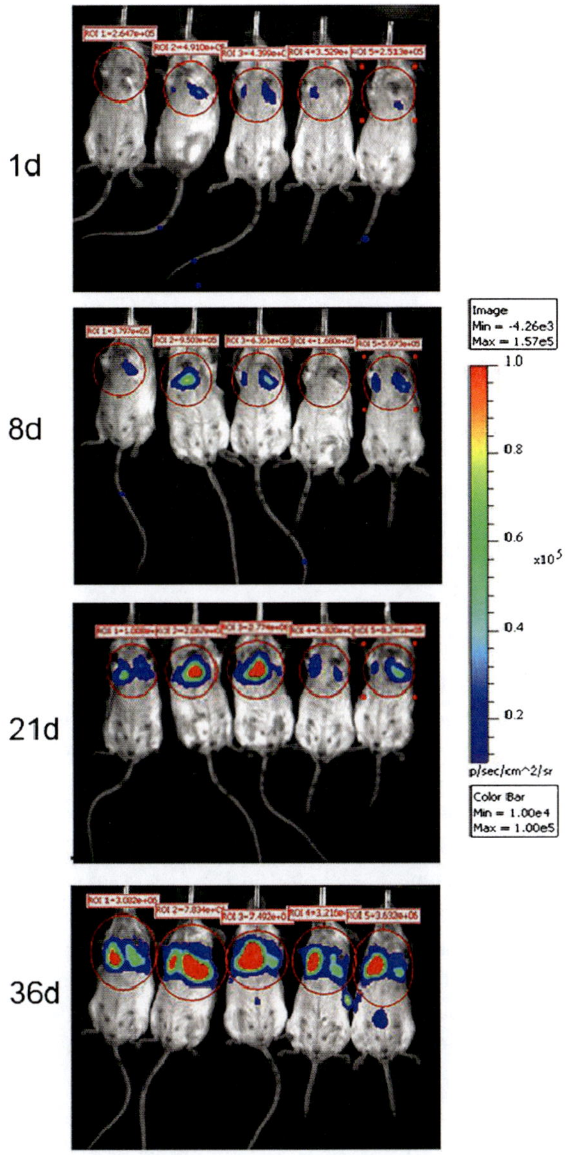

Fig. 10.6 bcMCF cells were transfected by Lipofectamine/Plus Reagent from Life Technologies. The plasmids that were used for the cotransfection were pGL3 control red (SV40-luc) from Promega/C. Contag, Standford University, and pSV40/Zeo from Invitrogen. After the cells were transfected, selection occurred over a period of 10–12 days using 1 μg/ml of Zeocin (Invitrogen). After selection, the cells were allowed a period of time to recover after which they were expanded. To ensure the presence of luciferase in the cells, a Luciferase Assay (Promega) was performed using the EnVision Workstation plate reader. Mice were placed in a plastic restraining device equipped with a hole from which we could access the tail. The tails of the mice were placed in warm water in order to dilate the lateral tail veins and to allow easy visualization. Two million (2×10^6) cells suspended in PBS were injected into the lateral tail vein using a 26-gauge needle. The animals were followed over a period of 5 weeks by Bioluminescence Imaging using the Caliper LS/Xenogen IVIS Spectrum System to determine the location of the bcMCF cells

were characterized for invasiveness in a modified Boyden chamber assay. As shown in Fig. 10.6, this EMT gene signature classified, with complete concordance to [34], the cell lines into luminal, Basal A and Basal B subtypes. Of importance to this proposal, the Basal B cell lines that scored as invasive grouped to the far left (Fig. 10.6). This result along with the recent report showing that EMT occurs more frequently in basal-like tumors [35] indicates the relevance of these breast cell lines for molecular analysis of the networks regulating EMT.

While the regulation of EMT is not fully understood, a network of several signaling pathways affecting the expression and/or function of a complex hierarchical network of transcription factors (TFs) has been partially elaborated [36, 37]. Known signaling pathways include multiple tyrosine kinase receptors leading to Ras-mediated activation of MAPK and PI3K pathways, TGF-β, Notch, and Wnt. From our studies we have evidence that enhanced TGF-β and Wnt signaling pathways are found in the EMT expressing bcMCF and caMCF cells [30]. TGF-β acting through Smad transcriptional complexes can repress expression of the Id TFs (Id1, Id2, Id3) and activate HMGA2, a DNA binding protein important for chromatin architecture [38]. Expression of HMGA2 is known to regulate several EMT controlling TFs including TWIST1, SNAI1, and SNAI2 (Slug) [30, 36, 37] (Figs. 10.5 and 10.6). TGF-β and Wnt signaling also affect the expression of several additional EMT-regulating TFs including ZEB1 (TCF8), TCF3 (E2A encoding E12 and E47), and LEF1 [37].

Analysis of the EMT expressing bcMCF cell line revealed the absence of expression of the secreted frizzled-related protein 1 (SFRP1), a repressor of Wnt signaling [30]. One allele of SFRP1 was deleted in these cells, with the remaining apparently silenced by methylation, accounting for the 28-fold reduction of this transcript. Loss and epigenetic inactivation of SFRP1 occurs often in invasive breast cancer and is associated with poor prognosis [39]. Inspection of the SFRP1 expression levels in Basal B cell lines [30] showed absent calls for 4 of the 8 invasive cell lines; and eightfold decrease in another three invasive cell lines relative to the noninvasive MCF-10A cells. Inspection of the expression files for bcMCF cells and the eight invasive Basal B cell lines [30] revealed that LEF1 was always absent, while TCF 3 and TCF 8 were expressed.

10.3 The Metastatic Phenotype

As indicated earlier, the bsMCF cells induced tumors in SCID mice that were poorly differentiated adenocarcinomas ER, PR, and ERBB2 negatives. bsMCF or caMCF (T4 or T5) cells inoculated in the tail of SCID mice induced metastatic foci in the lung (Fig. 10.3). This unique model allow us (a) to study the genomic and epigenomic changes present in the cell with metastatic capabilities, (b) to use metastatic and nonmetastatic cells from the same genetic background, and (c) to elucidate the functional role of each of the genes thus identified. We have found that more than 74 genes are either down or upregulated in the invasive and metastatic phenotype of the bsMCF (Fig. 10.7 and Table 10.2). Several genes controlling invasion and metastasis are significantly downregulated in bsMCF (C5) cells (Fig. 10.7 and Table 10.2). Many of these genes are silenced by methylation in specific GpC islands and the role in the process of invasion and metastasis has been already determined or suspected. For example, AZGP1 is known to be downregulated in malignant prostate epithelium [40, 41] but its precise role in metastasis is not clear. CLDN7 or Claudin 7 is significantly downregulated in bsMCF cells as well as in breast [42, 43] and other types of cancer [44]. Hypermethylation at the CLDN7 promoter was detected in 20% of colon cancer cells with low CLDN7 expression. EPB41L5 erythrocyte membrane protein band 4.1 like 5 is involved in cell polarity and in maintaining by separation of the apical and basolateral domains through specialized cell–cell junctions [45, 46] and could be an early marker of metastasis. GPR56 or G protein-coupled receptor 56 is downregulated in bsMCF cells and is markedly downregulated in the metastatic variants of melanoma. Functional studies have shown that overexpression of GPR56 suppresses tumor growth and metastasis, whereas reduced expression of GPR56 enhances tumor progression [47, 48]. KLF5 Kruppel-like factor 5 (Fig. 10.8) is downregulated also in bsMCF cells and reduced in expression in many types of human tumor [49].

FRMD3 is a member of the protein 4.1 superfamily and is a putative tumor suppressor [50] significantly downregulated in bsMCF cells. Grbl4, a growth factor receptor-bound protein 14 member of the Grb7 family of adapters, is an inhibitor of FGFR signaling. Grbl4 induces an

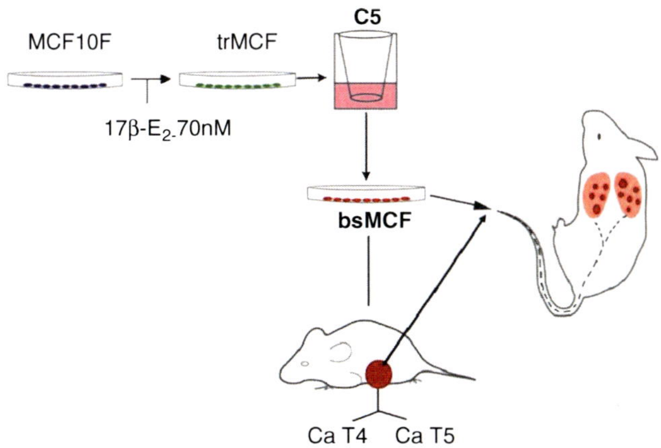

Fig. 10.7 Transformation of MCF-10F cells by 17β-estradiol treatment. Experimental protocol: MCF-10F cells treated with 70 nM 17β-estradiol (E_2) that expressed high colony efficiency (CE) and loss of ductulogenic capacity in collagen-matrix were classified as transformed (trMCF). Transformed cells that were invasive in a Matrigel Boyden type invasion chambers were selected (bsMCF) and plated at low density for cloning (bcMCF). MCF-10F, trMCF, bsMCF, and bcMCF were tested for carcinogenicity by injecting them into the mammary fat pad of 45-day-old female SCID mice. MCF-10F and trMCF cells did not induce tumors; bsMCF, formed solid tumors from which four cell lines, identified as caMCF, were derived and cells proven to be tumorigenic in SCID mice. bsMCF or caMCF when injected in the tail of SCID mice develop metastatic foci in the lung

Table 10.2 Upregulated genes differentially expressed genes in bcMCF compared to MCF10F

Symbol	Gene name	Fold change
FHL1	Four and a half domains 1	6.5
HEY1	Hairy/enhancer-of-split related with YRPW motif 1	5.4
ZEB1	Zinc finger E-box binding homeobox 1	12.0
ZEB2	Zinc finger E-box binding homeobox 2	11.5
COL6A3	Collagen, type VI, alpha 3	77.1
FN1	Fibronectin 1	4.9
FOSL1	FOS-like antigen 1	3.8
GNG11	Guanine nucleotide binding protein (G protein), gamma 11	10.5
HRAS	V-Ha-ras Harvey rat sarcoma viral oncogene homolog	6.0
NRP1	Neuropilin 1	6.9
RHOB	Ras homolog gene family, member B	5.7
S100A4	S100 calcium binding protein A4	18.8
TGFB1	Transforming growth factor, beta 1	3.1
TIMP1	TIMP metallopeptidase inhibitor 1	2.9
TNS1	Tensin 1	14.9

arrest of the signaling transduction cascades in the MDA-MB-231cells by blocking PLCy, ERK2, JNK1, and AKT [51, 52]. Another role of GRB14 is a binding partner of tankyrase 2. Tankyrase is an ankyrin repeat-containing poly[ADP-ribose] polymerase originally isolated as a binding partner for the telomeric protein TRF1. MEST or mesoderm specific transcript homolog is a hypermethylated gene that is highly enriched for targets of the PRC2 (Polycomb repressive complex 2) in embryonic stem cells. MTUS1 or microtubule associated tumor suppressor 1 is downregulated in bsMCF cells and is significantly downregulated in colon cancer and breast cancer [53–55]

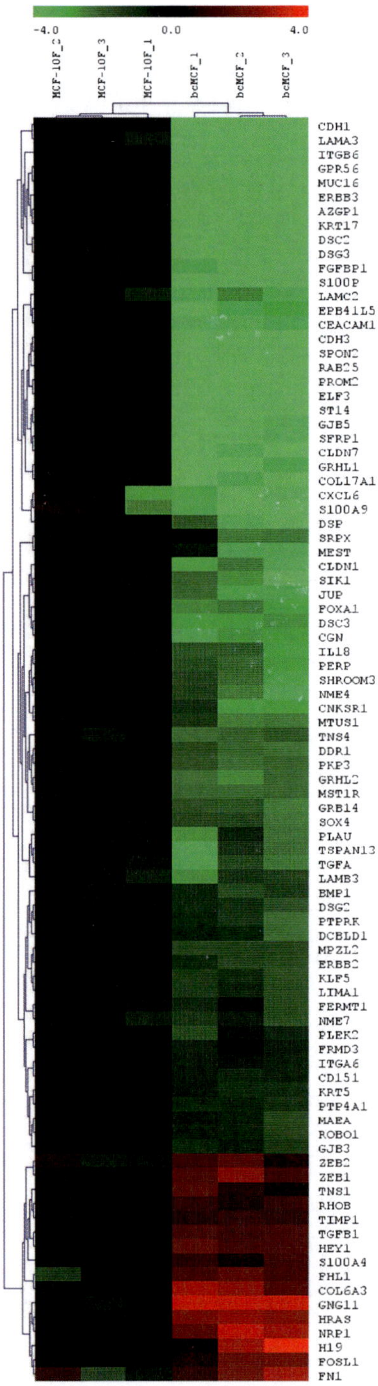

Fig. 10.8 Heat map of the bcMCF cells compared with the MCF10F cells

and in the triple negative (ER- PR- HER2-) breast carcinomas, a subgroup of highly proliferative tumors with poor outcome and no available targeted therapy. Functional studies indicate that silencing MTUS1 expression by siRNA increases cellular proliferation. Conversely, restoring endogenous levels of MTUS1 expression leads to reduced cancer cell proliferation, clonogenicity, anchorage-independent growth, and reduces the incidence and size of xenografts grown in vivo [56]. Loss of SFRP1 (secreted frizzled-related protein 1) expression is observed in breast, along with several other cancers [57–61], and is associated with poor patient prognosis. SFRP1 is significantly downregulated in bsMCF cells and it has been shown to be methylated in many pre and neoplastic breast cancer cell lines. SFRP1 antagonizes the Wnt/beta-catenin signaling pathway by competing with the Frizzled receptor for Wnt ligands resulting in an attenuation of the signal transduction cascade leading to the development of several human cancers, including breast cancer. SRPX sushi-repeat-containing protein, X-linked is downregulated in bsMCF cells and is also markedly reduced in carcinomas of colon, bladder, and ovary and closely linked to the progression of T-cell leukemia/lymphoma. The SRPX gene was originally isolated as a novel suppressor gene of v-src transformation and was shown to induce apoptosis in human cancer cells. It has been observed by us and others that TWIST is an important TF during embryonic development and has recently been found to promote the EMT phenomenon seen during the initial steps of tumor metastasis. TWIST is upregulated in bsMCF cells whereas SRPX is downregulated. Small interfering RNA (siRNA)-mediated depletion of TWIST with upregulation of SRPX. Indicating an important role of the SRPX gene in invasion and metastasis. Another gene downregulated in bsMCF cells is SNF1LK or the serine–threonine kinase SIK1 (salt-inducible kinase 1) as a regulator of p53-dependent anoikis. Inactivation of SIK1 compromised p53 function in anoikis and allowed cells to grow in an anchorage-independent manner. In vivo, SIK1 loss facilitated metastatic spread and survival of disseminated cells as micrometastases in lungs. The presence of functional SIK1 was required for the activity of the kinase LKB1 in promoting p53-dependent anoikis and suppressing anchorage-independent growth, matrigel invasion, and metastatic potential. Decreased expression of the gene encoding SIK1 closely correlated with development of distal metastases in breast cancers from three independent cohorts. Together, these findings indicate that SIK1 links LKB1 to p53-dependent anoikis and suppresses metastasis [62, 63]. SIK is an inducible gene target of TGFbeta/Smad signaling. Loss of endogenous SIK results in enhanced gene responses of the fibrotic and cytostatic programs of TGFbeta [64].

Among the upregulated genes in the bsMCF cells listed in Table 10.3 are FHL1, HEY1, ZEB1, ZEB2, FOSL1, and S100A4. FHL1 (four and a half LIM domains 1) may play an important role in ER signaling as well as breast cancer cell growth regulation [65, 66]. HEY1 (hairy/enhancer-of-split related with YRPW motif 1) and NOTCH3 are upregulated also in bsMCF cells and may be involved in the EMT process [66–68]. Transforming growth factor-beta (TGF-beta) is upregulated in bsMCF cells and is involved in the EMT through activation of Smad and non-Smad signaling pathways. EMT is the differentiation switch by which polarized epithelial cells differentiate into contractile and motile mesenchymal cells. Cell motility and invasive capacity are activated upon EMT. Multiple TFs, including deltaEF1/ZEB1, SIP1/ZEB2, and Snail/SNAI1, are induced by TGF-beta-Smad signaling and play critical roles in TGF-beta-induced EMT. In addition, both non-Smad signaling activated by TGF-beta and cross-talk with other signaling pathways play important roles in induction of EMT. Of these, Ras signaling synergizes with TGF-beta-Smad signaling, and plays an important role in the induction of EMT. FOSL1 is upregulated in bsMCF cells and has been shown to be overexpressed in MCF 7 cells after development of antiestrogen resistance. Fos is a component of the dimeric TF activator protein-1 (Ap-1), which is composed mainly of Fos (c-Fos, FosB, Fra-1, and Fra-2) and Jun proteins (c-Jun, JunB, and JunD). Unlike Fra-1 (encoded by Fosl1), c-Fos contains transactivation domains required for oncogenesis and cellular transformation [69]. The Fos related antigen-1 (Fra-1) is activated in multiple cancers and gene ablation can suppress the invasive phenotypes of many tumor cell lines [69–71]. S100A4 calcium binding protein A4 is upregulated in bsMCF cells and many other cancers [72, 73]. It promotes metastasis in several experimental animal models, and S100A4 protein expression is associated with patient outcome in a number of tumor types and possesses a wide range of biological functions, such as regulation of angiogenesis, cell survival, motility, and invasion [74, 75].

10.4 Summary and Conclusions

From these preliminary data we concluded that the EMT occurring in the breast basal cells depends predominately on TGF-β and Wnt signaling pathways, which increase the expression and function of transcription and chromatin organization factors that repress the epithelial and enhance the mesenchymal phenotype, thus favoring increased invasion and metastatic activity.

Acknowledgments Work supported by grant U01 ES/CA 12771 from the National Institute of Environmental Health Sciences (NIEHS) and the National Cancer Institute (NCI), NIH, DHHS. Its contents are solely the responsibility of the authors and do not necessarily represent the official views of the NIEHS or NCI, NIH.

References

1. Bouchardy C, Fioretta G, Verkooijen HM, Vlastos G, Schaefer P, Delaloye J-F, Neyroud-Caspar I, Balmer Majno S, Wespi Y, Forni M, Chappuis P, Sappino A-P, Rapiti E (2007) Recent increase of breast cancer incidence among women under the age of forty. Br J Cancer 96:1743–1746
2. Turner NC, Reis-Filho JS (2006) Basal-like breast cancer and the BRCA1 phenotype. Oncogene 25: 5846–5853
3. Tawfik O, Kimler BF, Davis M, Stasik C, Lai SM, Mayo MS, Fan F, Donahue JK, Damjanov I, Thomas P, Connor C, Jewell WR, Smith H, Fabian CJ (2007) Grading invasive ductal carcinoma of the breast: advantages of using automated proliferation index instead of mitotic count. Virchows Arch 450(6):627–636
4. Perou CM, Sørlie T, Eisen MB, van de Rijn M, Jeffrey SS, Rees CA, Pollack JR, Ross DT, Johnsen H, Akslen LA, Fluge O, PergamenschikovA WC, Zhu SX, Lønning PE, Børresen-Dale AL, Brown PO, Botstein D (2000) Molecular portraits of human breast tumours. Nature 406:747–752
5. Sorlie T, Perou CM, Tibshirani R et al (2001) Gene expression patterns of breast carcinomas distinguish

tumor subclasses with clinical implications. Proc Natl Acad Sci U S A 98:10869–10874
6. Sorlie T, Tibshirani R, Parker J et al (2003) Repeated observation of breast tumor subtypes in independent gene expression data sets. Proc Natl Acad Sci U S A 100:8418–8423
7. van de Rijn M, Perou CM, Tibshirani R et al (2002) Expression of cytokeratins 17 and 5 identifies a group of breast carcinomas with poor clinical outcome. Am J Pathol 161:1991–1996
8. Rakha EA, El-Sayed ME, Green AR, Lee AH, Robertson JF, Ellis IO (2007) Prognostic markers in triple-negative breast cancer. Cancer 109:25–32
9. Chlebowski RT, Chen Z, Anderson GL, Rohan T, Aragaki A, Lane D, Dolan NC, Paskett ED, McTiernan A, Hubbell FA, Adams-Campbell LL, Prentice R (2005) Ethnicity and breast cancer: factors influencing differences in incidence and outcome. J Natl Cancer Inst 97:439–448
10. Newman LA, Mason J, Cote D, Vin Y, Carolin K, Bouwman D, Colditz GA (2002) African-American ethnicity, socioeconomic status, and breast cancer survival: a meta-analysis of 14 studies involving over 10,000 African-American and 40,000 White American patients with carcinoma of the breast. Cancer 94:2844–2854
11. Li CI, Malone KE, Daling JR (2002) Differences in breast cancer hormone receptor status and histology by race and ethnicity among women 50 years of age and older. Cancer Epidemiol Biomarkers Prev 11:601–607
12. Joslyn SA (2002) Hormone receptors in breast cancer: racial differences in distribution and survival. Breast Cancer Res Treat 73:45–59
13. Lippman ME (1993) The development of biological therapies for breast cancer. Science 259:631–632
14. Cunningham JE, Butler WM (2004) Racial disparities in female breast cancer in South Carolina: clinical evidence for a biological basis. Breast Cancer Res Treat 88:161–176
15. Millikan RC, Newman B, Tse CK, Moorman PG, Conway K, Dressler LG, Smith LV, Labbok MH, Geradts J, Bensen JT, Jackson S, Nyante S, Livasy C, Carey L, Earp HS, Perou CM (2008) Epidemiology of basal-like breast cancer. Breast Cancer Res Treat 109:123–139
16. Vona-Davis L, Rose DP (2009) The influence of socioeconomic disparities on breast cancer tumor biology and prognosis: a review. J Womens Health (Larchmt) 18(6):883–893
17. Troester MA, Swift-Scanlan T (2009) Challenges in studying the etiology of breast cancer subtypes. Breast Cancer Res 11(3):104
18. Lin NU, Claus E, Sohl J, Razzak AR, Arnaout A, Winer EP (2008) Sites of distant recurrence and clinical outcomes in patients with metastatic triple-negative breast cancer: high incidence of central nervous system metastases. Cancer 113(10):2638–2645
19. Anderson WF, Jatoi I, Devesa SS (2005) Distinct breast cancer incidence and prognostic patterns in the NCI's SEER program: suggesting a possible link between etiology and outcome. Breast Cancer Res Treat 90:127–137
20. Wang X, Chao L, Li X, Ma G, Chen L, Zang Y, Zhou G (2010) Elevated expression of phosphorylated c-Jun NH2-terminal kinase in basal-like and "triple-negative" breast cancers. Hum Pathol 41:401–403
21. Bauer KR, Brown M, Cress RD, Parise CA, Caggiano V (2007) Descriptive analysis of estrogen receptor (ER)-negative, progesterone receptor (PR)- negative, and HER2-negative invasive breast cancer, the so-called triple-negative phenotype: a population-based study from the California cancer Registry. Cancer 109: 1721–1728
22. Cheang MC, Voduc D, Bajdik C, Leung S, McKinney S, Chia SK, Perou CM, Nielsen TO (2008) Basal-like breast cancer defined by five biomarkers has superior prognostic value than triple-negative phenotype. Clin Cancer Res 14:1368–1376
23. Phipps AI, Malone KE, Porter PL, Daling JR, Li CI (2008) Body size and risk of luminal, HER2-overexpressing, and triple-negative breast cancer in postmenopausal women. Cancer Epidemiol Biomarkers Prev 17:2078–2086
24. Russo IH, Russo J (2004) In vitro models for human breast cancer. In: Molecular basis of breast cancer prevention and treatment. Springer, Heidelberg, pp 227–80
25. Russo J, Hasan Lareef M, Balogh G, Guo S, Russo IH (2003) Estrogen and its metabolites are carcinogenic agents in human breast epithelial cells. J Steroid Biochem Mol Biol 87:1–25
26. Lareef MH, Garber J, Russo PA et al (2005) The estrogen antagonist ICI-182-780 does not inhibit the transformation phenotypes induced by 17-beta-estradiol and 4-OH estradiol in human breast epithelial cells. Int J Oncol 26:423–429
27. Fernandez SV, Russo IH, Russo J (2006) Estradiol and its metabolites 4-hydroxyestradiol and 2-hydroxyestradiol induce mutations in human breast epithelial cells. Int J Cancer 118:1862–1868
28. Russo J, Lareef MH, Tahin Q et al (2002) 17 Beta-estradiol is carcinogenic in human breast epithelial cells. J Steroid Biochem Mol Biol 80:149–162
29. Russo J, Fernandez SV, Russo PA et al (2006) 17--Beta-estradiol induces transformation and tumorigenesis in human breast epithelial cells. FASEB J 20:1622–1634
30. Huang Y, Fernandez SV, Goodwin S, Russo PA, Russo IH, Sutter TR, Russo J (2007) Epithelial to mesenchymal transition in human breast epithelial cells transformed by 17β-estradiol. Cancer Res 67: 11147–11157
31. Russo J, Gusterson BA, Rogers AE et al (1990) Comparative study of human and rat mammary tumorigenesis. Lab Invest 62:244–278

32. Calaf G, Zhang PL, Alvarado MV, Estrada S, Russo J (1995) C-Ha-ras enhances the neoplastic transformation of human breast epithelial cells treated with chemical carcinogens. Int J Oncol 6:5–11
33. Charafe-Jauffret E, Ginestier C, Monville F, Finetti P, Adelaide J, Cervera N, Fekairi S, Xerri L, Jacquemier J, Brinbaum D, Bertucci F (2006) Gene expression profiling of breast cell lines identifies potential new basal markers. Oncogene 25:2273–2284
34. Neve RM, Chin K, Fridlyand J, Yeh J, Baehner FL, Fevr T, Clark L et al (2006) A collection of breast cancer cell lines for the study of funtionally distinct cancer subtypes. Cancer Cell 10:515–527
35. Sarrio D, Rodriguez-Pinilla SM, Hardisson D, Cano A, Moreno-Bueno G, Palacios J (2008) Epithelial-mesenchymal transition in breast cancer relates to the basal-like phenotype. Cancer Res 68:989–997
36. Moustakas A, Heldin C-H (2007) Signaling networks guiding epithelial-mesenchymal transitions during embryogenesis and cancer progression. Cancer Sci 98:1512–1520
37. Peinado H, Olmeda D, Cano A (2007) Snail, ZEB and bHLH factors in tumor progression: an alliance against the epithelial phenotype? Nature Rev 7:415–428
38. Han H-J, Russo J, Kohwi Y, Kohwi-Shigematsu T (2008) SATB1 reprogrammes gene expression to promote breast tumour growth and metastasis. Nature 452:187–195
39. Klopocki E, Kristiansen G, Wild PJ, Klaman I, Castanos-Velez E, Singer G, Stohr R, Sauter G, Leibiger H, Essers L, Weber B, Hermann K, Rosenthal A, Hartmann A, Dahl E (2004) Loss of SFRP1 is associated with breast cancer progression and poor prognosis in early stage tumors. Int J Oncol 25:641–649
40. Henshall SM, Horvath LG, Quinn DI, Eggleton SA, Grygiel JJ, Stricker PD, Biankin AV, Kench JG, Sutherland RL (2006) Zinc-alpha2-glycoprotein expression as a predictor of metastatic prostate cancer following radical prostatectomy. J Natl Cancer Inst 98(19):1420–1424
41. Lapointe J, Li C, Higgins JP, van de Rijn M, Bair E, Montgomery K, Ferrari M, Egevad L, Rayford W, Bergerheim U, Ekman P, DeMarzo AM, Tibshirani R, Botstein D, Brown PO, Brooks JD, Pollack JR (2004) Gene expression profiling identifies clinically relevant subtypes of prostate cancer. Proc Natl Acad Sci U S A 101:811–816
42. Park D, Kåresen R, Axcrona U, Noren T, Sauer T (2007) Expression pattern of adhesion molecules (E-cadherin, alpha-, beta-, gamma-catenin and claudin-7), their influence on survival in primary breast carcinoma, and their corresponding axillary lymph node metastasis. APMIS 115:52–65
43. Sauer T, Pedersen MK, Ebeltoft K, Naess O (2005) Reduced expression of Claudin-7 in fine needle aspirates from breast carcinomas correlate with grading and metastatic disease. Cytopathology 16:193–198
44. Usami Y, Chiba H, Nakayama F, Ueda J, Matsuda Y, Sawada N, Komori T, Ito A, Yokozaki H (2006) Reduced expression of claudin-7 correlates with invasion and metastasis in squamous cell carcinoma of the esophagus. Hum Pathol 37:569–577
45. Usami Y, Satake S, Nakayama F, Matsumoto M, Ohnuma K, Komori T, Semba S, Ito A, Yokozaki H (2008) Snail-associated epithelial-mesenchymal transition promotes oesophageal squamous cell carcinoma motility and progression. J Pathol 215:330–339
46. Gosens I, Sessa A, den Hollander AI, Letteboer SJ, Belloni V, Arends ML, Le Bivic A, Cremers FP, Broccoli V, Roepman R (2007) FERM protein EPB4iL5 is a novel member of the mammalian CRB-MPP5 polarity complex. Exp Cell Res 313:3959–3970
47. Xu L, Begum S, Hearn JD, Hynes RO (2006) GPR56, an atypical G protein-coupled receptor, binds tissue transglutaminase, TG2, and inhibits melanoma tumor growth and metastasis. Proc Natl Acad Sci U S A 103: 9023–9028
48. Xu L, Hynes RO (2007) GRP56 and TG2: possible roles in suppression of tumor growth by the microenvironment. Cell Cycle 6:160–165
49. Kwak MK, Lee HJ, Hur K, Parkdo J, Lee HS, Kim WH, Lee KU, Choe KJ, Guilford P, Yang HK (2008) Expression of Krüppel-like factor 5 in human gastric carcinomas. J Cancer Res Clin Oncol 134:163–167
50. Haase D, Meister M, Muley T, Hess J, Teurich S, Schnabel P, Hartenstein B, Angel P (2007) FMRD3, a novel putative tumour suppressor in NSCLC. Oncogene 26:4464–4468
51. Kairouz R, Parmar J, Lyons RJ, Swarbrick A, Musgrove EA, Daly RJ (2005) Hormonal regulation of the Grb14 signal modulator and its role in cell cycle progression of MCF-7 human breast cancer cells. J Cell Physiol 203:85–93
52. Lyons RJ, Deane R, Lynch DK, Ye ZS, Sanderson GM, Eyre HJ, Sutherland GR, Daly RJ (2001) Identification of a novel human tankyrase through its interaction with the adaptor protein Grb14. J Biol Chem 276:17172–17180
53. Rodrigues-Ferreira S, Di Tommaso A, Dimitrov A, Cazaubon S, Gruel N, Colasson H, Nicolas A, Chaverot N, Molinié V, Reyal F, Sigal-Zafrani B, Terris B, Delattre O, Radvanyi F, Perez F, Vincent-Salomon A, Nahmias C (2009) 8p22 MTUS1 gene product ATIP3 is a novel anti-mitotic protein underexpressed in invasive breast carcinoma of poor prognosis. PLoS One 4:e7239
54. Frank B, Bermejo JL, Hemminki K, Sutter C, Wappenschmidt B, Meindl A, Kiechle-Bahat M, Bugert P, Schmutzler RK, Bartram CR, Burwinkel B (2007) Copy number variant in the candidate tumor suppressor gene MTUS1 and familial breast cancer risk. Carcinogenesis 28:1442–1445
55. Di Benedetto M, Bièche I, Deshayes F, Vacher S, Nouet S, Collura V, Seitz I, Louis S, Pineau P, Amsellem-Ouazana D, Couraud PO, Strosberg AD, Stoppa-Lyonnet D, Lidereau R, Nahmias C (2006) Structural organization and expression of human MTUS1, a candidate 8p22 tumor suppressor gene encoding a family of angiotensin II AT2 receptor-interacting proteins, ATIP. Gene 380:127–136

56. Zuern C, Heimrich J, Kaufmann R, Richter KK, Settmacher U, Wanner C, Galle J, Seibold S (2010) Down-regulation of MTUS1 in human colon tumors. Oncol Rep 23:183–189
57. Huang D, Yu B, Deng Y, Sheng W, Peng Z, Qin W, Du X (2010) SFRP4 was overexpressed in colorectal carcinoma. J Cancer Res Clin Oncol 136:395–401
58. Saini S, Liu J, Yamamura S, Majid S, Kawakami K, Hirata H, Dahiya R (2009) Functional significance of secreted Frizzled-related protein 1 in metastatic renal cell carcinomas. Cancer Res 69:6815–6822
59. Hu J, Dong A, Fernandez-Ruiz V, Shan J, Kawa M, Martínez-Ansó E, Prieto J, Qian C (2009) Blockade of Wnt signaling inhibits angiogenesis and tumor growth in hepatocellular carcinoma. Cancer Res 69:6951–6959
60. Lin YW, Chung MT, Lai HC, De Yan M, Shih YL, Chang CC, Yu MH (2009) Methylation analysis of SFRP genes family in cervical adenocarcinoma. J Cancer Res Clin Oncol 135:1665–1674
61. Gauger KJ, Hugh JM, Troester MA, Schneider SS (2009) Down-regulation of sfrp1 in a mammary epithelial cell line promotes the development of a cd44high/cd24low population which is invasive and resistant to anoikis. Cancer Cell Int 9:11
62. Cheng H, Liu P, Wang ZC, Zou L, Santiago S, Garbitt V, Gjoerup OV, Iglehart JD, Miron A, Richardson AL, Hahn WC, Zhao JJ (2009) SIK1 couples LKB1 to p53-dependent anoikis and suppresses metastasis. Sci Signal 2:ra35
63. Takemori H, Katoh Hashimoto Y, Nakae J, Olson EN, Okamoto M (2009) Inactivation of HDAC5 by SIK1 in AICAR-treated C2C12 myoblasts. Endocr J 56: 121–130
64. Kowanetz M, Lönn P, Vanlandewijck M, Kowanetz K, Heldin CH, Moustakas A (2008) TGFbeta induces SIK to negatively regulate type I receptor kinase signaling. J Cell Biol 182:655–662
65. Ding L, Niu C, Zheng Y, Xiong Z, Liu Y, Lin J, Sun H, Huang K, Yang W, Li X, Ye Q (2011) FHL1 interacts with estrogen receptors and regulates breast cancer cell growth. J Cell Mol Med 15:72–85
66. Lin J, Ding L, Jin R, Zhang H, Cheng L, Qin X, Chai J, Ye Q (2009) Four and a half LIM domains 1 and receptor interacting protein of 140kDa (RIP140) interact and cooperate in estrogen signaling. Int J Biochem Cell Biol 41:1613–1618
67. Engin F, Bertin T, Ma O, Jiang MM, Wang L, Sutton RE, Donehower LA, Lee B (2009) Notch signaling contributes to the pathogenesis of human osteosarcomas. Hum Mol Genet 18:1464–1470
68. Miyazono K (2009) Transforming growth factor-beta signaling in epithelial-mesenchymal transition and progression of cancer. Proc Jpn Acad Ser B Phys Biol Sci 85:314–323
69. Pennanen PT, Sarvilinna NS, Ylikomi TJ (2009) Gene expression changes during the development of estrogen-independent and antiestrogen-resistant growth in breast cancer cell culture models. Anticancer Drugs 20:51–58
70. Matsuo K, Owens JM, Tonko M, Elliott C, Chambers TJ, Wagner EF (2000) Fosl1 is a transcriptional target of c-Fos during osteoclast differentiation. Nat Genet 24:184–187
71. Young MR, Colburn NH (2006) Fra-1 a target for cancer prevention or intervention. Gene 379:1–11
72. Wang HY, Zhang JY, Cui JT, Tan XH, Li WM, Gu J, Lu YY (2010) Expression status of S100A14 and S100A4 correlates with metastatic potential and clinical outcome in colorectal cancer after surgery. Oncol Rep 23:45–52
73. Hua J, Chen D, Fu H, Zhang R, Shen W, Liu S, Sun K, Sun X (2010) Short hairpin RNA-mediated inhibition of S100A4 promotes apoptosis and suppresses proliferation of BGC823 gastric cancer cells in vitro and in vivo. Cancer Lett 292:41–47
74. Boye K, Mælandsmo GM (2010) S100A4 and metastasis: a small actor playing many roles. Am J Pathol 176:528–535
75. Ismail TM, Zhang S, Fernig DG, Gross S, Martin-Fernandez ML, See V, Tozawa K, Tynan CJ, Wang G, Wilkinson MC, Rudland PS, Barraclough R (2010) Self-association of calcium-binding protein S100A4 and metastasis. J Biol Chem 285:914–922
76. Russo J, Hu YF, Yang X, Russo IH (2000) Developmental, cellular and molecular basis of human breast cancer. J Natl Cancer Inst Monogr 27:17–38
77. Tiezzi DG, Fernandez SV, Russo J (2007) Epithelial to mesenchymal transition during breast cancer progression. Int J Oncol 31:823–827

11 Breast Cancer Subtypes: Two Decades of Journey from Cell Culture to Patients

Xiangshan Zhao, Channabasavaiah Basavaraju Gurumurthy, Gautam Malhotra, Sameer Mirza, Shakur Mohibi, Aditya Bele, Meghan G. Quinn, Hamid Band, and Vimla Band

Abstract

Recent molecular profiling has identified six major subtypes of breast cancers that exhibit different survival outcomes for patients. To address the origin of different subtypes of breast cancers, we have now identified, isolated, and immortalized (using hTERT) mammary stem/progenitor cells which maintain their stem/progenitor properties even after immortalization. Our decade long research has shown that these stem/progenitor cells are highly susceptible to oncogenesis. Given the emerging evidence that stem/progenitor cells are precursors of cancers and that distinct subtypes of breast cancer have different survival outcome, these cellular models provide novel tools to understand the oncogenic process leading to various subtypes of breast cancers and for future development of novel therapeutic strategies to treat different subtypes of breast cancers.

11.1 Introduction

Despite the enormous progress made in understanding the biology of cancer over the past two decades, breast cancer remains the second leading cause of cancer-related deaths among women in USA [1]. The major problem remains that although the majority of breast cancers initially respond to therapy, they subsequently develop resistance to chemo-, radio- as well as biological therapies [2]. Moreover, the recent evidence that breast cancer is not a single disease but rather a group of diseases is making oncologists re-think their therapeutic strategies for treating breast cancer. Recent advancement in technology has made it possible for researchers to perform molecular profiling, which led to the identification of five major subtypes of breast cancers: a basal epithelial-like group, an ErbB2-overexpressing group, a normal breast epithelial-like group and two luminal subtypes [3, 4]. The significance of these studies became evident by analyses that showed different subtypes of breast cancers respond differentially to therapy as basal subtype of breast cancer

V. Band (✉)
Department of Genetics, Cell Biology and Anatomy, University of Nebraska Medical Center, Nebraska Medical Center, Omaha, NE, USA

Eppley Institute for Cancer and Allied Diseases and UNMC-Eppley Cancer Center, University of Nebraska Medical Center, Nebraska Medical Center, Omaha, NE, USA
e-mail: vband@unmc.edu

J.S. Rhim and R. Kremer (eds.), *Human Cell Transformation: Role of Stem Cells and the Microenvironment*, Advances in Experimental Medicine and Biology 720,
DOI 10.1007/978-1-4614-0254-1_11,

presents as the most aggressive disease with poor prognosis and poor disease-free and overall survival of these patients. On the other hand, luminal A subtype of breast cancer presents a favorable outcome after conventional therapies [3, 4]. Although initially these studies were unexpected, particularly the nature of basal subtypes of breast cancers, our analyses of in vitro cell cultures of human mammary gland confirmed the presence of a population of basal subtype of cells in human breast tissue [5]. More importantly, in vitro studies have shown high susceptibility of basal subtype of breast cells to oncogenic transformation [6]. Together, these studies strongly support the idea that breast cancer subtypes either represent malignancies of biologically distinct cell subtypes producing distinct disease entities that may require different treatment strategies, or different subtypes of breast cancers may arise from a common precursor-based oncogene-driven reprogramming. With our newfound knowledge of breast cancer subtypes, there is increasing evidence that breast cancer may arise from breast cancer initiating or breast stem cells; and with the notion that these cells may account for therapy resistance, there is a new hope for cancer researchers to define stem cell based therapeutics.

Our laboratory has been investigating these questions for over two decades and has developed cellular models with the goal to (1) identify and define cancer initiating cells in human mammary gland that may represent the precursors of different subtypes of breast cancers and (2) identify novel cellular pathways whose disruption leads to transformation of cells. Below we will briefly focus on the work conducted in our laboratory over the past 20 years.

11.2 Evidence that Human Breast Tissue Contains Various Epithelial Cell Subtypes

The human mammary gland consists of fatty and fibrous tissue along with a branching ductal system ending as terminal ducts and acinar structures called terminal ductal-lobular units (TDLUs) [7]. Although, not clearly demonstrated, it is believed that most breast cancers arise in TDLUs. It is unclear if a linear model of breast cancer progression similar to colon cancers exists; however, different stages of benign breast cancers (such as typical and atypical hyperplasia), noninvasive cancers (such as carcinoma in situ - ductal or lobular), and invasive cancers (such as invasive ductal or lobular carcinomas) do exist. It is generally accepted that hyperplasia and/or carcinoma in situ represent earlier stages of invasive carcinomas. Additionally, multiple types of in situ carcinomas, such as solid, cribriform, papillary, or comedo types have been reported and it is possible that these represent tumors originating from different epithelial subtypes [7].

Histological examination of TDLU shows two major types of cells, luminal cells (that surround the lumen) and myoepithelial cells (that line the basement membrane) [7, 8]. The question we have been addressing is, if there are only two types of cells in breast, then what is the origin of the different subtypes of breast cancers? Recent studies from our laboratory and that of others have shown existence of stem/progenitor cells in human tissues, which begins to answer these questions (discussed below).

11.3 Isolation of Stem Cells from Normal and Tumor Breast Tissues

Stem cells have an enormous replicative potential with the characteristics of self-renewal and multilineage differentiation. Although still a controversial notion, many cancers are thought to originate from cancer stem cells [9]. This idea has also generated great interest in the field of breast cancer and researchers have begun to examine the existence of mammary stem cells. The cellular milieu of the mammary gland undergoes significant changes during pregnancy, lactation, and involution [7]. These include bursts of proliferation of existing cells during pregnancy, continued differentiation during lactation and apoptosis during involution at the end of the cycle. This cyclical behavior strongly supports the idea of a stem cell-like population in the mammary gland, which would meet the demand of a pregnancy cycle. The existence of adult mammary epithelial stem cells has therefore been proposed. Direct evidence for the existence of such cells has come from clear fat-pad transplantation, retroviral tagging, and X-chromosome inactivation studies in rodent model [8–14].

Recently, using various stem cell markers, attempts have been made to identify the human mammary epithelial stem cells [11–14]. Using immunomagnetic cell sorting based on surface antigen markers (Muc and ESA), Gudjonsson et al. were able to isolate Muc-/ESA+/K-19 cells that were able to self-renew and give rise to MUC+/ESA+cells, thus exhibiting characteristics of breast stem cells [13]. Dontu et al. isolated undifferentiated mammospheres from single cell suspensions of mammary epithelial cells (MECs) obtained by mechanical and enzymatic dissociations of mammary tissue [14]. Primary mammospheres can be further passaged to generate secondary mammospheres. Primary as well as secondary mammospheres were highly enriched in early progenitor or stem cells capable of differentiating along multiple lineages and of self-renewal [14].

Apart from normal mammary stem cells, the possible existence of a breast cancer stem cell has been reported in the literature [15, 16]. In a NOD/SCID xenotransplant model, Al-Hajj et al. used four cell surface markers, CD44, CD24, ESA, and B38.1 (a Breast/ovarian cancer specific marker) [16], and lineage markers to sort different populations of breast cells from breast tumor tissues. All mice injected with Lin-, CD44+, B38.1+, CD24–/low generated tumors while none of the mice injected with CD44-, B38.1- cells showed tumors. Lin-, CD44+, B38.1 fractions were further subdivided based on the expression of ESA. When used in xenotransplants, Lin-, ESA+, CD44+, CD24–/low cells generated tumors whereas other populations did not [16]. The presence of such a population in breast tumor tissues suggests the possible existence of breast cancer stem cells.

11.4 Establishment of Culture Conditions to Isolate and Propagate Human Mammary Progenitor/Stem Cells

For more than 20 years, our laboratory has developed techniques and culture conditions to establish normal human mammary epithelial cell (hMEC) [17]. In our laboratory, we defined a medium (DFCI-1) that allows proliferation of hMECs for 3–4 passages (about 15–20 population doublings) and then most of the cells senesce (unable to proliferate) [17, 18]. However, after about a month of nonproliferating heterogenous cell population, a more homogenous cell population emerges that continues to proliferate further for 50–60 population doublings, prior to senescence [17, 18]. This process of emergence of cells capable of proliferating for extended period is also known as self-selection; prior to selection, the cells are termed preselection cells whereas those that emerge after selection are called post-selection cells. Several years ago, using two-dimensional gel electrophoresis we had shown that hMECs cultured in DFCI-1 medium express keratin markers (K5, K6, K7, K14, K17, K18, and K19) that would characterize them as progenitors rather than a particular lineage, i.e., luminal (K8, K18, K19) or myoepithelial (K5, K14). Furthermore, these cells expressed markers of basal (K5 and K6) subtypes of cells [5]. Although at that point it seemed a curious finding, recent direct cDNA microarray profiling of human breast cancers has led to similar insights, i.e., identification of multiple subtypes of human breast cancers that present distinct outcomes. Importantly, phenotypic and gene expression characteristics of these breast cancer subtypes point to their possible origin from specific subtypes of MECs, such as basal or luminal cells [3, 4].

It is important to mention that hMECs exhibiting stem cell markers still underwent cellular senescence [5]. They entered into senescence either because (1) currently available in vitro culture conditions developed by us and others for hMECs do not allow mammary stem cells to self-renew indefinitely, or (2) these hMECs are not stem cells but are uncommitted progenitor cells and after several population doublings they lose their ability to self-renew. Unavailability of markers to distinguish stem vs. progenitor cells in mammary tissue makes it difficult for researchers to ascertain if hMECs cultured under these conditions are stem or progenitor cells.

In order to study these cells for further biological pathways, we have devised methods to isolate and immortalize (using hTERT-human telomerase reverse transcriptase) mammary stem/progenitor cells [19] and recently have been successful in defining culture conditions for their self-renewal and differentiation (discussed below) [20].

11.5 hTERT Immortal hMECs Exhibit Stem/Progenitor Cell Properties and Express Markers for Well-Known Stem Cell Signature Pathways

As described above, we have successfully isolated and immortalized stem cells that uniformly expressed various markers, such as basal (K5, K14, p63, vimentin) and luminal (K8, K18, E-cadherin) cell markers. One cell strain (70N) also expressed K19 (a known luminal cell marker), CD29, and CD49f (two known stem cell marker) [11–14], but did not express detectable levels of the differentiated luminal cell marker MUC1 [13, 20] or differentiated myoepithelial cell markers (such as α-smooth muscle actin (α-SMA), CD10 or Thy-1) [20]. Notably, despite expression of several luminal cell markers, the primary and hTERT-immortalized hMECs lacked the expression of estrogen receptor alpha (ER-alpha) providing further support to the idea that stem cells lack ER expression [21]. Immunofluorescence analyses confirmed that the same cells expressed different lineage markers (Fig. 11.1). We also demonstrated the presence of such progenitor/stem cells in normal reduction mammoplasty tissue (Fig. 11.2). Importantly, hMECs expressing hTERT exhibit the same sets of markers as their parental primary hMECs indicating that stem/progenitor characteristics were maintained through immortalization [20]. Based on these results we speculate that different subtypes of progenitor cells do exist in primary hMEC cultures and a particular subset of cells from this population may give rise to a particular type of breast cancer during transformation process. These results are consistent with microarray expression analyses of various breast cancer specimens [3, 4].

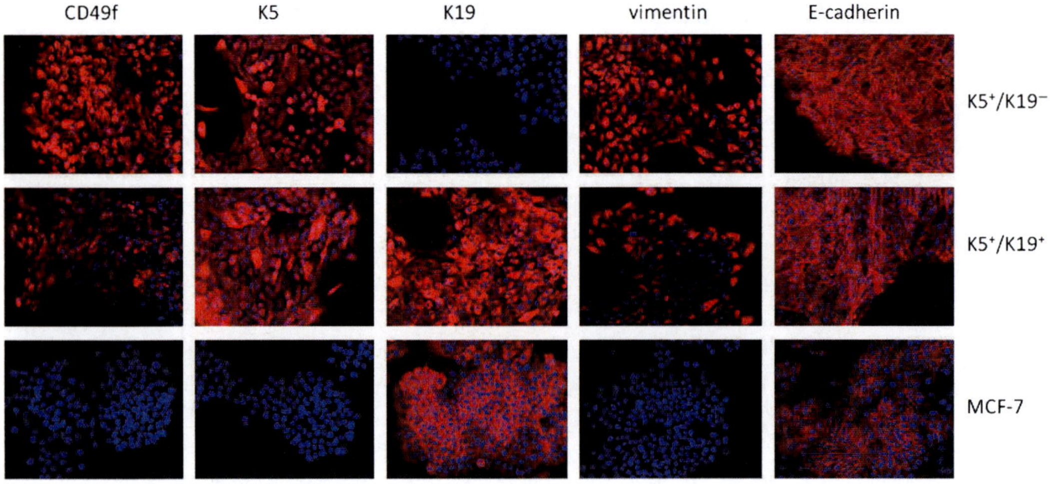

Fig. 11.1 Analyses of stem/progenitor cell markers in parental and hTERT immortalized human mammary epithelial cells. CD49f, K5, K19, Vimentin, E-cadherin (*red*); DAPI (*blue*)

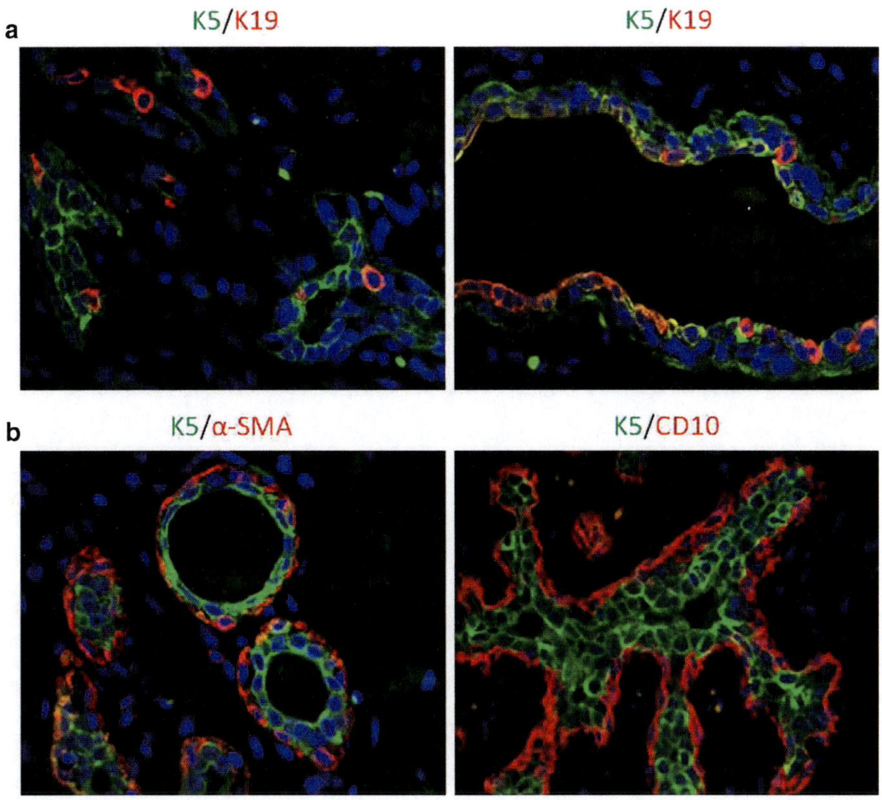

Fig. 11.2 Immunofluorescence staining of human mammary tissue specimen with anti-K5 (*green*), anti-K19 (*red*) double staining (**a**). Anti-K5 (*green*), co-staining with myoepithelial marker anti-α-SMA or CD10 (**b**)

Although we observed markers of stem/progenitor cells, the proof of principle of a stem/progenitor cell is in its ability to self-renew and differentiate. Thus, we devised a method to assess the ability of mammary progenitor cells to differentiate into myoepithelial and luminal cells in vitro [20]. Two types of stem/progenitor cells ($K5^+/K19^-$ and $K5^+/K19^+$) when cultured in MEGM medium [14, 20] on 2D plastic substratum proliferate as tightly packed epithelial colonies; after 1 month a proportion of cells near the periphery adopt a spindle shaped morphology (Fig. 11.3a). While cells in the center (undifferentiated progenitor cells) expressed the basal, luminal, and stem cell markers, the spindle-shaped cells forming the peripheral halo were K5-negative and acquired several well-known myoepithelial cell markers (α–SMA, CD10, and Thy-1) that were absent on the parental cells as well as in the central compact part of colonies grown in MEGM [20] (Fig. 11.3). Differential trypsinization and re-plating of the central part of cells showed their ability to regenerate similar colonies as well as to generate differentiated myoepithelial cells. In contrast, most of the differentiated myoepithelial cells failed to proliferate; however, if myoepithelial cells were transferred to DFCI-1 medium, some cells continued to self-renew as myoepithelial cells, suggesting the presence of myoepithelial progenitor cells. Furthermore, a subset of cells within the center of the colonies was MUC1-positive consistent with their differentiation along the luminal cell lineage ([13, 20] and references therein) (Fig. 11.3). Taken together these results support the conclusion that immortalized $K5^+/K19^-$ and $K5^+/K19^+$ cell types retain mammary stem or progenitor properties with the capability to self-renew and differentiate along luminal and myoepithelial lineages.

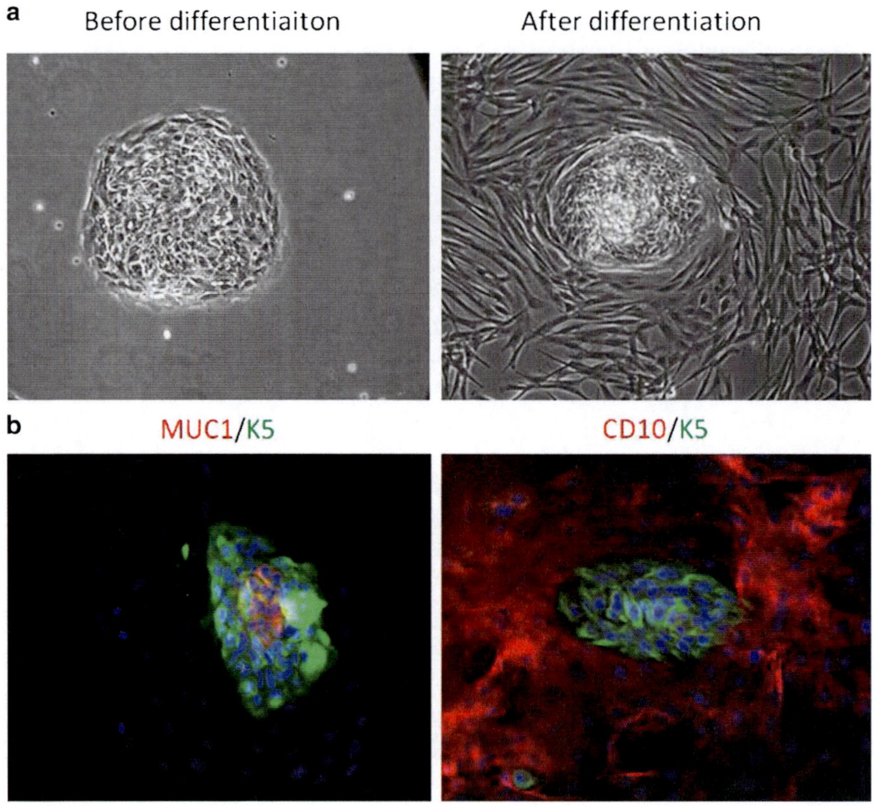

Fig. 11.3 In vitro self-renewal, luminal and myoepithelial cell differentiation of $K5^+/K19^-$ cell type in MEGM medium. (**a**) Morphology of cells before and after start of differentiation. (**b**) Immunofluorescence staining of cells after start of differentiation with K5 (*green*), luminal cell marker MUC1 (*red*), and myoepithelial cell marker CD10 (*red*)

11.6 Mammary Progenitor/Stem Cells Express Markers of Well-Defined Stem Cell Pathways

There is increasing evidence that several pathways, such as Notch, Hedgehog, Wnt, and Sox are altered in stem cell differentiation [22–25]. Thus, we performed Affymetrix-based genome-wide mRNA expression profiling of both $K5^+/K19^-$ and $K5^+/K19^+$ cell types and compared these with myoepithelial differentiated cells. These experiments showed significant alteration in known pathways in stem cell differentiation further substantiating our findings that $K5^+/K19^-$ and $K5^+/K19^+$ cell types represent stem/progenitor cells [20].

11.7 Mammary Progenitor Cells are Highly Susceptible to Preneoplastic Transformation

Over two decades, our studies have defined high susceptibility of mammary progenitor cells $K5^+/K19^-$ to oncogenic transformation. We have shown that introduction of various viral oncogenes, such as urogenital carcinoma-associated human papillomavirus (HPV) oncogenes E6 and E7, exposure to gamma-irradiation, as well as cellular genes (such as mutant p53, activated Rho GTPases and Bmi-1, a member of the polycomb group of transcriptional repressors) induce preneoplastic transformation, i.e., immortalization of progenitor cells [18, 19, 26–28].

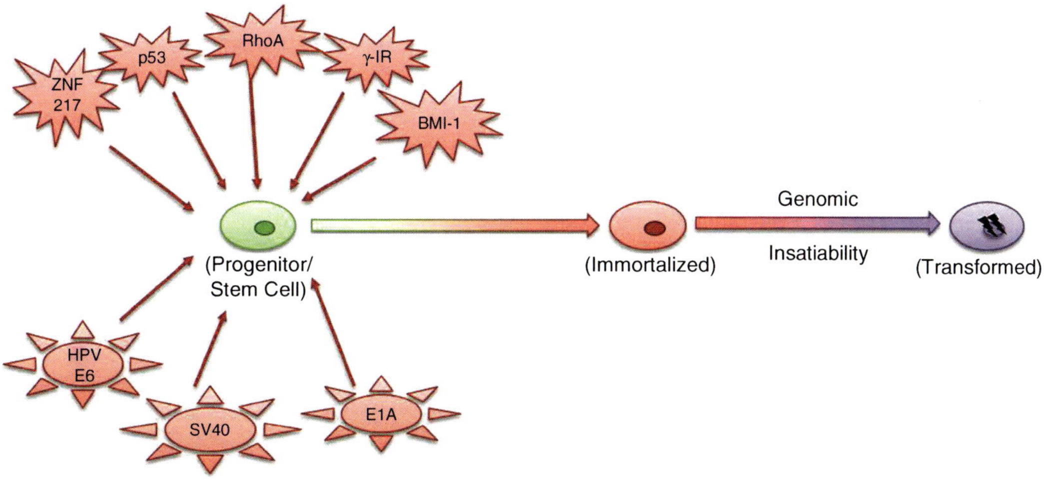

Fig. 11.4 Human mammary stem/progenitor cells are highly susceptible to preneoplastic transformation. Several viral oncogenes, such as HPV E6 and SV40, irradiation and cellular mutant/overexpressed genes (ZNF217, mutant p53, RhoA, and Bmi-1) induce immortalization/preneoplastic transformation. Genomic instability induced by additional oncogenes or by exposure to DNA damage further leads to full transformation of hMECs

Other study has shown that ZNF217, a zinc finger protein that is overexpressed in breast cancers, can promote the immortalization of post-selection human MECs [29] (Fig. 11.4).

11.8 Cellular Targets of E6 and Their Potential Role in hMEC Immortalization

Characterization of oncogenesis-related cellular targets of HPV E6 and E7 oncoproteins has been facilitated by their ability to dominantly immortalize primary human cells in vitro [18, 19, 30, 31]. The cellular protein that was first shown to interact with high-risk HPV E6 proteins was the tumor suppressor protein p53 via an ubiquitin ligase E6AP [32]. The p53 protein is a key regulator of DNA repair, cell cycle arrest, and apoptosis in response to genotoxic stress [31–33]. These responses are essential to prevent the emergence of cells with unstable genomes that are expected to be susceptible to oncogenic transformation. Indeed, mis-sense mutations that inactivate p53, deletion of its gene, or loss of its expression, occur in nearly half of all human cancers [31–33]. The reproducibility and relatively high efficiency with which E6 (in post-selection hMECs) or E6 and E7 combination can induce the immortalization of hMECs have therefore opened a practical approach to elucidate the biochemical mechanisms of hMEC immortalization. Using the HPV16 E6 as a bait in yeast two hybrid analyses of a hMEC cDNA library, we have identified several known as well as novel targets of the E6 oncogene that represent potential mediators of E6-mediated hMEC immortalization [34]. These include ADA3 (alteration/deficiency in activation 3), a novel coactivator of p53 and steroid receptors (estrogen receptor and retinoic acid receptor) that enhances transcription. shRNA-mediated knockdown of ADA3 dramatically inhibits function of p53, ER, and RXR proteins [35–37]; E6 targeted protein 1 (E6TP1), a novel GTPase activating for Rap small G protein; and protein kinase N (PKN), an effector for Rho small G protein [27, 34, 38]. Furthermore, MamL1, a human homologue of Drosophila mastermind gene and a known coactivator for Notch, also interacts with E6 [39, 40]. These studies have implicated the p53, Notch, ER, Rho, and Rap signaling pathways in early transformation of hMECs. Consistent with these analyses we have shown that expression of mutant p53 [26] or activated Rho [27] directly induces the immortalization of hMECs. Recently, we identified a human orthologue of drosophila *Ecdysoneless* (Ecd) protein as a E6 binding protein that also associates with p53 and induces p53-mediated cellular senescence in fibroblasts but not in epithelial cells [41]. In order to study the cellular function of Ecd in mammalian cells, we generated *Ecd*$^{lox/lox}$ mouse embryonic fibroblast cells from *Ecd* floxed mouse embryo. Cre-mediated deletion of Ecd in *Ecd*$^{lox/lox}$ MEFs led to a proliferative block due to a delay in G_1-S cell cycle progression; this defect was reversed by the introduction of human Ecd. Loss of Ecd led to marked downregulation of E2F target gene expression. Furthermore, Ecd directly binds to Rb at the pocket domain, and competes with E2F for binding to hypophosphorylated Rb. These results demonstrate that mammalian Ecd plays a role in cell cycle progression via Rb-E2F pathway [42, 43]. Recent studies from our laboratory show Ecd is overexpressed in several carcinomas, including breast cancers, and overexpression of Ecd in mammary cells enhances cellular entry to S-phase of synchronized cells upon growth factor stimulation. Importantly, overexpression of Ecd promotes oncogenic characteristics in human mammary epithelial cells (Bele et al., unpublished data). Taken together, these studies underscore importance of identification and characterization of E6 cellular targets in breast cells.

11.9 Conclusions

Our ability to culture and immortalize mammary stem/progenitor cells in vitro has provided the first reliable model of immortal mammary stem/progenitor cells and a wealth of knowledge about the behavior of mammary cells and the genes involved in normal cell growth and oncogenesis. Characterization of these cells has provided potential novel markers for early diagnosis and prognosis of breast cancers and given knowledge about potential precursor cells for breast cancers. In vitro transformation analyses have also proven important to understand the multistep nature of breast cancer. Lastly, using gene profiling we have begun to appreciate that breast cancers do not originate only from luminal cells but also from basal (and probably from myoepithelial progenitor) cells and that there are subtypes of breast cancers possibly originating from distinct normal precursors that have distinct clinical outcomes and may require different treatment strategies.

However, many critical questions remain unanswered: How do different subtypes of breast cancer cells originate? Do breast cancer stem cells originate from normal mammary stem cells? How to define cancer stem cell specific therapeutic strategy? In conclusion, experimental immortalization of breast progenitor cells has led to substantial progress in our understanding of the biology of breast cancer and future studies in these model systems should go a long way towards helping elucidate the nature of breast cancer heterogeneity and thus facilitate the development of more individualized stem cell based therapy for breast cancer patients.

Acknowledgments The authors thank the past and present members of their laboratories that have contributed to work published from our laboratories. Work in our laboratories was supported by the NIH Grants R01CA096844 and R01CA 144027 (VB), and R01CA099163, R01CA116552, R01CA105489, and R01CA087986 (HB) and department of defense breast cancer program W81XWH-07-1-0351 and W81XWH-11-1-0171 (VB), and W81XWH-11-1-0166 (HB) and Eppley Cancer Center Grant.

References

1. American Cancer Society. Cancer facts and figures 2011
2. Morrow PK, Hortobagyi GN (2009) Management of breast cancer in the genome era. Annu Rev Med 60: 153–165
3. Sorlie T, Tibshirani R, Parker J, Hastie T, Marron JS, Nobel A, Deng S, Johnsen H, Pesich R, Geisler S et al (2003) Repeated observation of breast tumor subtypes in independent gene expression data sets. Proc Natl Acad Sci USA 100:8418–8423
4. Hu Z, Fan C, Oh DS, Marron JS, He X, Qaqish BF, Livasy C, Carey LA, Reynolds E, Dressler L, Nobel A, Parker J, Ewend MG, Sawyer LR, Wu J, Liu Y, Nanda R, Tretiakova M, Ruiz Orrico A, Dreher D, Palazzo JP, Perreard L, Nelson E, Mone M, Hansen H, Mullins M, Quackenbush JF, Ellis MJ, Olopade OI, Bernard PS, Perou CM (2006) The molecular portraits of

breast tumors are conserved across microarray platforms. BMC Genomics 7:96
5. Wazer DE, Liu XL, Chu Q, Gao Q, Band V (1995) Immortalization of distinct human mammary epithelial cell types by human papilloma virus 16 E6 or E7. Proc Natl Acad Sci U S A 92(9):3687–3691
6. Dimri G, Band H, Band V (2005) Mammary epithelial cell transformation: insights from cell culture and mouse models. Breast Cancer Res 7(4):171–179, Review
7. Cardiff RD, Wellings SR (1999) The comparative pathology of human and mouse mammary glands. J Mammary Gland Biol Neoplasia 4:105–122
8. Kordon EC, Smith GH (1998) An entire functional mammary gland may comprise the progeny from a single cell. Development 125:1921–1930
9. Marx J (2003) Cancer research. Mutant stem cells may seed cancer. Science 301:1308–1310
10. Smith GH, Chepko G (2001) Mammary epithelial stem cells. Microsc Res Tech 52:190–203
11. Stingl J, Eaves CJ, Zandieh I, Emerman JT (2001) Characterization of bipotent mammary epithelial progenitor cells in normal adult human breast tissue. Breast Cancer Res Treat 67:93–109
12. Stingl J, Eaves CJ, Kuusk U, Emerman JT (1998) Phenotypic and functional characterization in vitro of a multipotent epithelial cell present in the normal adult human breast. Differentiation 63:201–213
13. Gudjonsson T, Villadsen R, Nielsen HL, Ronnov-Jessen L, Bissell MJ, Petersen OW (2002) Isolation, immortalization, and characterization of a human breast epithelial cell line with stem cell properties. Genes Dev 16:693–706
14. Dontu G, Abdallah WM, Foley JM, Jackson KW, Clarke MF, Kawamura MJ, Wicha MS (2003) In vitro propagation and transcriptional profiling of human mammary stem/progenitor cells. Genes Dev 17:1253–1270
15. Dick JE (2003) Breast cancer stem cells revealed. Proc Natl Acad Sci U S A 100:3547–3549
16. Al-Hajj M, Wicha MS, Benito-Hernandez A, Morrison SJ, Clarke MF (2003) Prospective identification of tumorigenic breast cancer cells. Proc Natl Acad Sci U S A 100:3983–3988
17. Band V, Sager R (1989) Distinctive traits of normal and tumor-derived human mammary epithelial cells expressed in a medium that supports long-term growth of both cell types. Proc Natl Acad Sci U S A 86(4): 1249–1253
18. Band V (2003) In vitro models of early neoplastic transformation of human mammary epithelial cells. Methods Mol Biol 223:237–248
19. Ratsch SB, Gao Q, Srinivasan S, Wazer DE, Band V (2001) Multiple genetic changes are required for efficient immortalization of different subtypes of normal human mammary epithelial cells. Radiat Res 155 (1 Pt 2):143–150, Review
20. Zhao X, Malhotra GK, Lele SM, Lele MS, West WW, Eudy JD, Band H, Band V (2010) Telomerase-immortalized human mammary stem/progenitor cells with ability to self-renew and differentiate. Proc Natl Acad Sci U S A 107(32):14146–14151
21. Sleeman KE, Kendrick H, Robertson D, Isacke CM, Ashworth A, Smalley MJ (2007) Dissociation of estrogen receptor expression and in vivo stem cell activity in the mammary gland. J Cell Biol 176(1): 19–26
22. Chiba S (2006) Notch signaling in stem cell systems. Stem Cells 24(11):2437–2447
23. Ruiz i Altaba A, Sanchez P, Dahmane N (2002) Gli and hedgehog in cancer: tumours, embryos and stem cells. Nat Rev Cancer 2(5):361–372
24. Nusse R (2008) Wnt signaling and stem cell control. Cell Res 18(5):523–527, Review
25. Chambers I, Smith A (2004) Self-renewal of teratocarcinoma and embryonic stem cells. Oncogene 23(43):7150–7160
26. Gao Q, Hauser SH, Liu XL, Wazer DE, Madoc-Jones H, Band V (1996) Mutant p53-induced immortalization of primary human mammary epithelial cells. Cancer Res 56(13):3129–3133
27. Zhao X, Lu L, Pokhriyal N, Ma H, Duan L, Lin S, Jafari N, Band H, Band V (2009) Overexpression of RhoA induces preneoplastic transformation of primary mammary epithelial cells. Cancer Res 69(2): 483–491
28. Dimri GP, Martinez JL, Jacobs JJ, Keblusek P, Itahana K, Van Lohuizen M, Campisi J, Wazer DE, Band V (2002) The Bmi-1 oncogene induces telomerase activity and immortalizes human mammary epithelial cells. Cancer Res 62(16):4736–4745
29. Nonet GH, Stampfer MR, Chin K, Gray JW, Collins CC, Yaswen P (2001) The ZNF217 gene amplified in breast cancers promotes immortalization of human mammary epithelial cells. Cancer Res 61(4): 1250–1254
30. Band V (1995) Preneoplastic transformation of human mammary epithelial cells. Semin Cancer Biol 6(3): 185–192, Review
31. Band V (1998) The role of retinoblastoma and p53 tumor suppressor pathways in human mammary epithelial cell immortalization. Int J Oncol 12(3): 499–507, Review
32. Huibregtse JM, Scheffner M, Howley PM (1991) A cellular protein mediates association of p53 with the E6 oncoprotein of human papillomavirus types 16 or 18. EMBO J 10(13):4129–4135
33. Münger K, Scheffner M, Huibregtse JM, Howley PM (1992) Interactions of HPV E6 and E7 oncoproteins with tumour suppressor gene products. Cancer Surv 12:197–217, Review
34. Gao Q, Srinivasan S, Boyer SN, Wazer DE, Band V (1999) The E6 oncoproteins of high-risk papillomaviruses bind to a novel putative GAP protein, E6TP1, and target it for degradation. Mol Cell Biol 19(1): 733–744
35. Kumar A, Zhao Y, Meng G, Zeng M, Srinivasan S, Delmolino LM, Gao Q, Dimri G, Weber GF, Wazer DE, Band H, Band V (2002) Human papillomavirus

oncoprotein E6 inactivates the transcriptional coactivator human ADA3. Mol Cell Biol 22:5801–5812
36. Zeng M, Kumar A, Meng G, Gao Q, Dimri G, Wazer D, Band H, Band V (2002) Human papilloma virus 16 E6 oncoprotein inhibits retinoic X receptor-mediated transactivation by targeting human ADA3 coactivator. J Biol Chem 277:45611–45618
37. Meng G, Zhao Y, Nag A, Zeng M, Dimri G, Gao Q, Wazer DE, Kumar R, Band H, Band V (2004) Human ADA3 binds to estrogen receptor (ER) and functions as a coactivator for ER-mediated transactivation. J Biol Chem 279:54230–54240
38. Gao Q, Kumar A, Srinivasan S, Singh L, Mukai H, Ono Y, Wazer DE, Band V (2000) PKN binds and phosphorylates human papillomavirus E6 oncoprotein. J Biol Chem 275(20):14824–14830
39. Wu L, Aster JC, Blacklow SC, Lake R, Artavanis-Tsakonas S, Griffin JD (2000) MAML1, a human homologue of Drosophila mastermind, is a transcriptional co-activator for NOTCH receptors. Nat Genet 26:484–489
40. Zhao Y, Katzman RB, Delmolino LM, Bhat I, Zhang Y, Gurumurthy CB, Germaniuk-Kurowska A, Reddi HV, Solomon A, Zeng MS, Kung A, Ma H, Gao Q, Dimri G, Stanculescu A, Miele L, Wu L, Griffin JD, Wazer DE, Band H, Band V (2007) The notch regulator MAML1 interacts with p53 and functions as a coactivator. J Biol Chem 282(16):11969–11981, Epub 2007 Feb 22
41. Zhang Y, Chen J, Gurumurthy CB, Kim J, Bhat I, Gao Q, Dimri G, Lee SW, Band H, Band V (2006) The human orthologue of Drosophila ecdysoneless protein interacts with p53 and regulates its function. Cancer Res 66(14):7167–7175
42. Kim JH, Gurumurthy CB, Naramura M, Zhang Y, Dudley AT, Doglio L, Band H, Band V (2009) Role of mammalian Ecdysoneless in cell cycle regulation. J Biol Chem 284(39):26402–26410
43. Kim JH, Gurumurthy CB, Band H, Band V (2010) Biochemical characterization of human Ecdysoneless reveals a role in transcriptional regulation. Biol Chem 391(1):9–19

Parathyroid Hormone Related Protein (PTHrP) in Tumor Progression*

12

Richard Kremer, Jiarong Li, Anne Camirand, and Andrew C. Karaplis

Abstract

Parathyroid hormone-related protein (PTHrP) is widely expressed in fetal and adult tissues and is a key regulator for cellular calcium transport and smooth muscle cell contractility, as well as a crucial control factor in cell proliferation, development and differentiation. PTHrP stimulates or inhibits apoptosis in an autocrine/paracrine and intracrine fashion, and is particularly important for hair follicle and bone development, mammary epithelial development and tooth eruption. PTHrP's dysregulated expression has traditionally been associated with oncogenic pathologies as the major causative agent of malignancy-associated hypercalcemia, but recent evidence revealed a driving role in skeletal metastasis progression. Here, we demonstrate that PTHrP is also closely involved in breast cancer initiation, growth and metastasis through mechanisms separate from its bone turnover action, and we suggest that PTHrP as a facilitator of oncogenes would be a novel target for therapeutic purposes.

12.1 PTHrP Background, Discovery, Gene Sequence and Protein Structure

12.1.1 Background

Malignancy-associated hypercalcemia (MAH) is a well-recognized syndrome that occurs in patients suffering from certain malignant cancers. The classic signs and symptoms of hypercalcemia are confusion, polydipsia, polyuria, constipation, nausea, vomiting and eventually coma. Hypercalcemia associated with malignancy of nonparathyroid tissues frequently occurs during

*This study was supported by grants from the Susan G. Komen Foundation (KG 100766) and the Canadian Institutes for Health Research (CIHR MOP 102555) to R. Kremer and AC Karaplis and the US Department of Defense Breast Cancer Research Program (BC023897) to AC Karaplis.

R. Kremer (✉)
Department of Medicine, McGill University Health Centre, 687 Pine Avenue West, Montréal, QC, Canada H3A1A1
e-mail: richard.kremer@mcgill.ca

J.S. Rhim and R. Kremer (eds.), *Human Cell Transformation: Role of Stem Cells and the Microenvironment*, Advances in Experimental Medicine and Biology 720,
DOI 10.1007/978-1-4614-0254-1_12,

bone invasion by tumor cells, where bone resorption is the direct result of osteoclast action. Osteoclastic bone-degrading activity is stimulated by several cytokines including MIP-1α, TNF-β, IL-1β, IL-3 and IL-6, which are produced locally by the tumor cells [1–4]. MAH is commonly associated with malignant tumors of the breast, lung, head and neck, esophagus, cervix, skin or kidney [5].

In 1936, Gutman conducted the first large group study of MAH; these patients suffered from myeloma and breast cancer, and most of them presented bone metastases [6]. Fuller Albright in 1941 was the first to propose that a PTH-like humoral factor was responsible for the hypercalcemia in patients with renal carcinoma which resolved after irradiation of bone metastasis. In the following years, the concept was accepted and the term "ectopic PTH syndrome" became widely used to describe patients with high circulating plasma calcium concentrations, low phosphorus, and few or no bone metastases [7–9]. In 1969, a thorough review was made of cases where patients suffered symptoms of hyperparathyroidism yet exhibited malignant tumors of nonparathyroid origin. The conclusions drawn from this work were that some nonparathyroid tumor cells could initiate transcription and translation of the *PTH* gene and secrete PTH [10]. Such cases of "ectopic PTH syndrome" [11] included kidney adenocarcinoma [12] and a malignant hepatoblastoma [13].

Rare cases were found where nonparathyroid tumors secreted PTH, and many studies demonstrated that the immunological properties of circulating of PTH-like material in hypercalcemic patients with nonparathyroid cancer or primary hyperparathyroidism were distinct from those of the immunoreactive PTH found in the serum of patients with primary hyperparathyroidism [14, 15]. In 1980, Stewart and associates established the first full biochemical characterization for 50 consecutive patients with cancer-associated hypercalcemia, with or without bone metastases. This study delineated characteristic laboratory findings that now define the PTH-like syndrome, and coined the term humoral hypercalcemia of malignancy (HHM): elevated nephrogenous cyclic AMP excretion levels, high serum calcium, low serum phosphorous, marked reduction in 1,25-dihydroxyvitamin D and low or suppressed reactivity with PTH antisera.

12.1.2 Discovery

Based on the biochemical characterization of humoral hypercalcemia of malignancy, much effort was devoted to the problem of identifying and isolating the tumor-secreted unknown factor that was responsible for this syndrome. In 1987, three independent groups simultaneously achieved: (1) purification of an active component with a molecular weight of 18 kDa from a human lung cancer cell line (BEN) [16], (2) purification of a 6-kDa active component from cultured human renal carcinoma cells [17], and (3) purification of a 17-kDa active component from a human breast tumor biopsy [18]. Most interesting was the fact that the N-terminal amino acid sequence of these adenylate cyclase-stimulating proteins revealed outstanding homology to PTH, identifying the existence of a PTH-like factor in those cancer cells.

Using the partial amino acid sequence information from these discoveries, oligonucleotide probes were synthesized and used to identify complementary DNAs (cDNAs) encoding BEN cell mRNA [19], mRNA from a human renal carcinoma [20] and mRNA from a renal carcinoma cell line [21]. Characterization of those clones revealed a gene and peptide sequence similar to that of human PTH. Eight of the initial 13 residues of the mature protein and the final 2 residues in the signal peptide were identical to those of human PTH. The term parathyroid hormone-related protein (PTHrP) derives from this high homology to the N-terminal sequence of PTH.

The elucidation of the PTHrP sequence allowed investigators to synthesize fragments of the peptide and to study their effects in different experimental conditions. Kemp et al. demonstrated that the N-terminal fragment of PTHrP (1–34) exhibited PTH-like activity in rat and chicken kidney and produced effects commonly seen in HHM such as increased excretion of

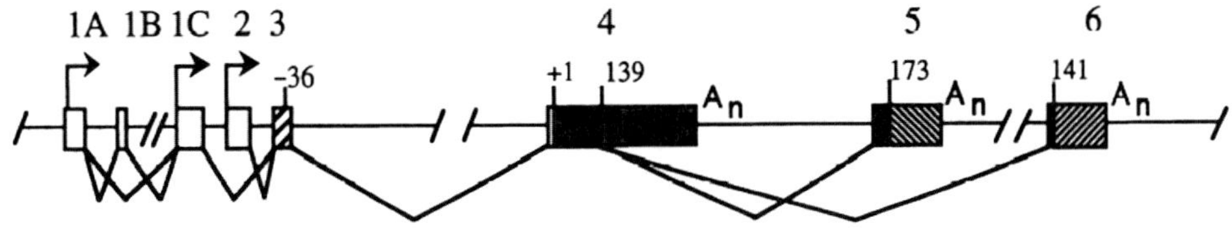

Fig. 12.1 Structure of human parathyroid hormone-related protein (PTHrP) gene (adapted from Philbrick et al. [35])

cAMP and phosphorus, and reduced urinary calcium excretion [22]. Using radioligand binding techniques, Nickols et al. were able to show that amino-terminal fragments of PTH and PTHrP interacted with identical receptors present on rabbit kidney microvessels and tubules [23]. Immunoassays were developed and indicated that PTHrP was circulating at high levels in patients with HHM [24, 25]. These findings provided strong evidence that circulating PTHrP produced by various cancers was the causal agent responsible for hypercalcemia.

12.1.3 Gene Sequence and Protein Structure

The *Pthrp* gene is localized to the short arm of chromosome 12 in humans [26] whereas the *PTH* gene is on the short arm of chromosome 11 [27]. The localization of *PTH* and *Pthrp* on these chromosomes and the similarity in sequence and organization of the two genes provide indirect evidence for a common evolutionary origin [26, 28].

The human *Pthrp* gene (Fig. 12.1) is a complex transcriptional unit which spans more than 15 kb of DNA, with 9 exons and at least 3 promoters. Alternative splicing of the premessenger RNAs produces three PTHrP isoforms differing at their carboxyl-terminal ends and containing 139, 141 or 173 amino acids. Exon 4 encodes a region common to all peptides, while exons 5 and 6 encode the unique carboxy termini of the other two peptides. *PTH* and *Pthrp* genes display an identical pattern of intron/exon organization in the region of exons 2 and 3 [19, 26, 28–30]. In contrast to the human organization, rat and mice *Pthrp* genes are relatively simple with a single promoter homologous to the downstream P3 promoter of the human gene. They produce mature peptides of 141 and 139 amino acids, respectively [31, 32]. The *Pthrp* gene in the chicken yields a single mature peptide of 139 amino acids [33].

The initial translational human PTHrP products undergo complex processing, including separation of the (−36 to −1) pre-pro sequence and endoproteolysis of the full-length (−36 to 139), (−36 to 141) and (−36 to 173) sequence at multibasic sites. All sequences present an N-terminal signal sequence for endoplasmic reticulum entry and a coding 1–139 peptide [34, 35]. The amino acid 35–111 region is dramatically conserved in human, rat, mouse and chicken sequences, with the human and rodent amino acid sequences differing in the 1–111 region by only two residues. This high evolutionary conservation suggests important physiological and biological functions. The 35–111 region is rich in putative proteolytic processing sites, with multiple dibasic amino acid groups [34, 35].

Region 107–111 (-Thr-Arg-Ser-Ala-Trp-) or TRSAW is evolutionally highly conserved, a suggesting that a peptide encompassing this region may be physiologically important [36].

12.2 PTHrP Physiology

12.2.1 PTHrP Functional Domains

Analysis of the *Pthrp* gene sequence reveals several functional domains (Fig. 12.2), including a signal peptide, a PTH-like N-terminal domain, a mid-region domain that begins at amino acid 38 and a unique carboxy-terminal domain named osteostatin [36, 37].

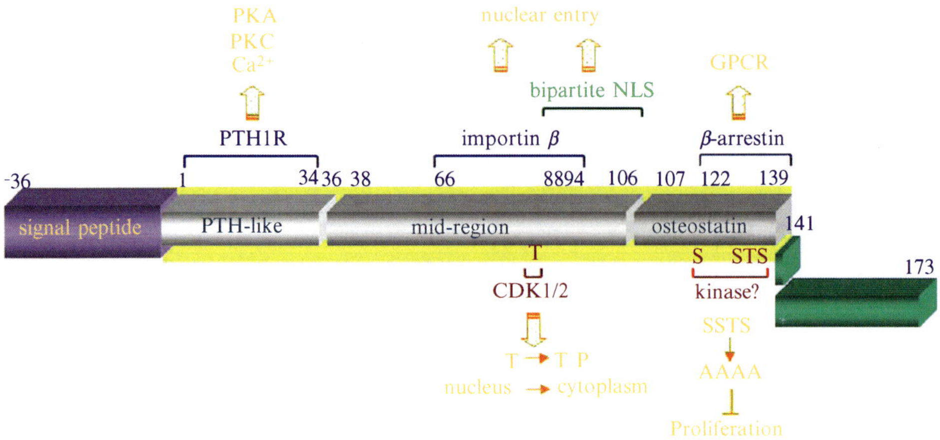

Fig. 12.2 The functional domains of the human PTHrP protein and its interaction partners. T-P denotes phosphorylation of Thr85. SSTS stands for residues Ser119, Ser130, Thr132 and Ser138 and AAAA denotes their replacements by alanines. *CDK* cyclin-dependent kinase, *GPCR* G-protein coupled receptor. *Green bars* show extension of the PTHrP protein in splicing variants –36/141 and –36/173 (adapted from Dittmer [138])

The N-terminal domain of PTHrP (1–36) shares with PTH a homologous sequence that interacts with the PTH/PTHrP type 1 receptor (PTH1R), a class II G-protein coupled receptor (GPCR). Upon binding PTHrP, the PTH1R receptor can either activate the formation of cyclic 3′,5′-adenosine monophosphate (cAMP) by stimulating adenylate cyclase (AC) through Gαs, or activate phospholipase Cβ (PLCβ) through Gαq/11. cAMP then activates protein kinase A (PKA), whereas activated PLCβ stimulates the formation of diacylglycerol (DAG) and 1,4,5-inositol triphosphate (IP3). In turn, DAG activates protein kinase C (PKC) and the production of IP3 leading to an increase in the intracellular free Ca^{2+} [38, 39]. The PTH1R receptor can also stimulate the influx of extra-cellular Ca^{2+} through regulation of calcium channels [40], and activate protein kinase C through a PLC-independent pathway [41–43]. The PTH1R receptor classical transduction pathways can lead to different biological effects in a cell type-specific manner.

The mid-region PTHrP comprises a bipartite nuclear localization sequence (NLS) consisting of residues 88–91 and 102–106. This NLS sequence may allow PTHrP to accumulate in the nucleus [44–46] and bind rRNA [47, 48]. This peptide likely enters the nucleus immediately after NLS binding to importin-β1 to form a complex that links to the microtubular trackway, where it is pulled by a dynien motor to a nuclear envelope pore and handed over to the pore channel transport machinery [49–52]. Although the NLS exact function is still obscure, it is likely to be of great significance since mice missing the NLS and the C-terminal region of the PTHrP gene present decreased cellular proliferative capacity and increased apoptosis in multiple tissues, retarded growth, early senescence and malnutrition leading to an early death [53]. In addition, in vitro studies indicate that the midregion sequence retains a CDK1/CDK2 phosphorylation site at Thr [85] [54], which suggests that translocation of PTHrP is associated with activation of the cell cycle [44, 55].

The PTHrP C-terminal domain 107–139 was named osteostatin because this peptide inhibits rat osteoclastic bone resorption with an incredibly low EC_{50} value of 10^{-15} M [36, 56, 57].

12.2.2 PTHrP Receptors

PTHrP binds to, and activates, the G-protein coupled receptor (GPCR) for PTH/PTHrP (PTH1R) which is expressed in PTH and PTHrP target cells such as osteoblasts in bone and renal tubular cells. PTH1R in humans and rodents is encoded by a multi-exon gene with potential for alternate splicing and alternate promoter usage [58]. The *PTH1R* gene is located on chromosome 3 in humans and possesses 14 exons.

PTH1R has seven transmembrane spanning domains [59] and has been cloned from opossum kidney, rat bone, and human bone and kidney [60–62]. The amino terminal region of PTH (1–34) and PTHrP (1–34) interacts with the J-domain, the functional portion of the receptor that contains the seven transmembrane-spanning helices and the connecting loops [63].

Following receptor–agonist interaction, PTH1R is activated and mediates not only the endocrine actions of PTH, but also the autocrine/paracrine actions of PTHrP.

12.2.3 Normal Physiological Functions of PTHrP

PTHrP is widely expressed in fetal and adult tissues, including cartilage, bone, breast, skin, skeletal heart and smooth muscle, uterus and placenta, as well as endocrine organs and the central nervous system [34, 35, 64]. PTHrP is a key regulator for cellular calcium transport and smooth muscle cell contractility, and possesses crucial roles in cell proliferation, development and differentiation [35]. It is important to note that the known biological properties of PTHrP are not only the results of its interaction with PTH1R and its subsequent signal transduction cascades, but also of PTHrP nuclear translocation [65, 66]. PTHrP is known to stimulate or inhibit apoptosis in various settings in a cell- or tissue-specific manner [67–69]. These actions are mostly performed in an autocrine/paracrine and intracrine fashion.

The biological actions of PTHrP are particularly important for bone development during endochondral bone formation. Targeted ablation of *Pthrp* results in homozygous$^{-/-}$ mice dying shortly after birth and presenting abnormalities in endochondral bone development [70]. In contrast, heterozygous *Pthrp*$^{+/-}$ animals are viable but demonstrate a reduction in trabecular bone volume and an early osteoporotic phenotype [71]. In addition, genetic mouse studies indicate PTHrP regulates hair follicle development, mammary epithelial development and tooth eruption [72–76].

12.2.4 PTHrP in Mammary Gland Development

During postnatal breast development in normal mice, PTHrP appears to be expressed by both luminal and myoepithelial cells of the mammary gland [77, 78], while myoepithelial cells and mammary stromal cells express the PTH/PTHrP receptor. Both stromal and myoepithelial cells are important in the branching growth of the mammary gland during sexual maturation and early pregnancy, and PTHrP ligand and receptor are appropriately positioned to participate in this process [77, 79, 80].

Pthrp-null mice die soon after birth of severe musculo-skeletal defects [70], but *Pthrp* rescue in the chondrocytes of these animals leads to a phenotype compatible with life. This strategy generates a mouse PTHrP-sufficient in chondrocytes, but PTHrP-null in all other sites including breast. These mice are characterized by the absence of normal epithelial-mesenchymal signaling cascade, failure to form mammary mesenchyme and the resorption of nascent mammary bud [73, 76].

12.3 PTHrP and Cancer Biology

12.3.1 Breast Cancer

Female breast cancer is a major medical problem with significant public health and societal ramifications. Although breast cancer death rates have declined in recent years, breast cancer remains the most commonly diagnosed cancer and the second leading cause of cancer death in women [81]. Since normal breast growth is regulated by many hormones, growth factors and receptors, it is not surprising that malignant cells derived from breast tissue also express the same hormones, growth factors and receptors. Numerous genes are controlled by complex regulatory networks and involved in the development and progression of breast cancer, and these genes are the key factors determining the characteristics of each tumor.

12.3.2 PTHrP and Cancer Development

PTHrP was originally discovered in patients with MAH [5]; approximately 80% of hypercalcemic patients with solid tumors have elevated PTHrP plasma concentrations as a result of increased secretion by the tumors [82]. However, PTHrP expression is present in many tumor types even in the absence of hypercalcemia. For instance, the great majority of breast cancer tumor samples show positive staining for PTHrP, and strong *Pthrp* gene activity in breast tumors is associated with onset of bone metastases independent of hypercalcemia [83–87].

Normal prostatic epithelial tissues express low levels of PTHrP, as determined by immunohistochemistry and in situ hybridization. In contrast, overexpression of PTHrP is common in prostate cancer [88, 89], and many prostate cancer cell lines in vitro and metastatic bone lesions in vivo express PTHrP. PTHrP stimulates primary prostate tumor growth and protects cells from apoptotic stimuli [90]. PTHrP expression was found in all major lung cancer cell types [91, 92] and was most common in squamous cell lung cancer [93]. While benign colorectal adenomas and non-neoplastic adjacent mucosal epithelia show no detectable PTHrP expression, about 95% of colorectal adenocarcinomas overexpress PTHrP mRNA and protein, and the expression level is higher in poorly differentiated than in well-differentiated adenocarcinomas [94, 95]. In a clinical study of 76 patients with various hematological malignancies, 50% of the 14 hypercalcaemic patients had significant elevation in plasma PTHrP concentrations [96]. Similar observations were made in stomach cancer where PTHrP is expressed in 77% of gastric adenocarcinomas without humoral hypercalcaemia. In contrast, only 5% of adenomas and none of the non-neoplastic epithelium showed PTHrP expression. Similarly, PTHrP expression was more common in moderately differentiated adenocarcinomas (95.5%) and poorly differentiated adenocarcinomas (100%) than in well-differentiated adenocarcinomas (43%). Furthermore, PTHrP expression was more intense in the deeply invasive portions than in the mucosal carcinomas [97].

It is well established that PTHrP is the major causative agent in MAH associated with a broad range of tumors. However, this is only one aspect of the multiple facets of PTHrP in cancer biology. Indeed, the complex growth factor-like properties of PTHrP have shed new light onto potential roles for this peptide in the regulation of tumor growth and invasion. In support of this, PTHrP expression has been shown to be under the control of numerous growth and angiogenic factors such as transforming growth factor (TGF-β,) epidermal growth factor (EGF), platelet-derived growth factor (PDGF) and vascular endothelial growth factor (VEGF) [98, 99]. Conversely, PTHrP stimulates the expression of many of these factors in various cell types and behaves as an angiogenic factor in endothelial cells [98, 100, 101]. More recently, PTHrP was shown to promote cell invasion in vitro [102, 103]. These observations clearly suggest a multifunctional role for PTHrP in cancer biology. Its functions include (1) regulation of tumor cells growth, differentiation, and progression; (2) induction of progression of osteolytic bone metastasis, (3) moderation of tumor cell survival factors and interference with apoptosis signaling pathway.

12.3.3 PTHrP in Breast Cancer Development

Breast cancer is frequently accompanied by PTHrP-induced hypercalcemia in advanced stages of the disease [25], and 50–60% of primary human breast cancer tumors overexpress PTHrP [85, 104, 105]. Several retrospective studies suggested that breast cancer patients with PTHrP-positive primary tumors were more likely to develop bone metastases compared to breast cancer patients with PTHrP-negative tumors

[85, 104–106]. PTHrP is expressed in 68% of surgically excised early breast cancers, compared with 100% of bone metastases [107], and 50% of cases of early breast cancer expressed the PTHrP receptor in contrast to 81% of cases of bone metastases. PTHrP expression without concomitant expression of its PTH1R receptor in primary tumors correlated with a reduced disease-free survival with a mortality rate of 6%, while co-expression of both predicted the worst clinical outcome at 5 years with a mortality rate of 32% [107]. A similar observation was noted regarding the differential expression of PTHrP isoforms in different stages of breast cancer. The levels of the 1–139 isoform mRNA was much higher in the tumors of patients who later developed metastases than in those of patients who developed no metastases. This mRNA isoform was also more abundant in breast tumors from patients who developed bone metastases than in those of patients who developed metastases in soft tissues. In contrast, the amounts of the 1–141 isoform mRNA in these groups of tumors were similar [87].

A prospective study conducted by Dr Kremer's team in patients with malignancy-associated hypercalcemia (including breast cancer patients) indicated that elevated circulating levels of PTHrP is an indicator of poor prognosis and is associated with reduced survival (Fig 12.3) [108]. Intriguingly, however, a retrospective clinical study of breast tumors collected at surgery suggests a better outcome and survival in patients whose primary tumor overexpresses PTHrP [109, 110].

To shed light on this controversy, Sato and co-workers treated animals presenting symptoms similar to those of HHM patients with a humanized anti-PTHrP antibody. The animals showed significant improvement in hypercalcemia and cachexia after antibody treatment [111]. Furthermore, administration of this antibody in nude mice injected intra-cardiac with the human breast cancer cell line MDA-MB-231 reduced the ability of these cells to form bone metastases [112]. Animal studies using mice transplanted with human tumors expressing PTHrP suggest that the humanized anti-PTHrP antibody could be an effective and beneficial agent for patients with HHM. In vitro, overexpression of PTHrP in the human breast cancer cell line MCF-7 caused an increase in mitogenesis, whereas the inhibition of endogenous PTHrP production resulted in decreased cell proliferation [113]. Enhanced bone tumor growth was also observed when the MCF-7 cells were transfected to overexpress PTHrP. Taken together, these results point to the pro-tumorigenic effects of PTHrP [94].

12.3.4 Role of PTHrP in Breast Cancer Metastasis to Bone

The bone matrix contains large amounts of growth factors such as type I collagen, insulin-like growth factors (IGFs), transforming growth factor α and β (TGF α and β), fibroblast growth factors (FGF-1 and -2), platelet derived growth factors (PDGFs) and bone morphogenic proteins (BMPs) [114].

The bone remodeling cycle involves a series of highly regulated steps which depend on the interactions of two cell types, the mesenchymal osteoblastic lineage and the hematopoietic osteoclastic lineage. Osteoblasts synthesize some of the growth factors stored in the matrix during bone formation. Most growth factors are released in active form into the marrow when bone is degraded during osteoclastic bone resorption. The resorption areas provide a fertile microenvironment for tumor cell colonization and proliferation [115, 116].

The inner structure of bone consists of bone marrow which is multicellular and contains hematopoietic stem cells and stromal cells. Hematopoietic stem cells differentiate into any type of blood and immune cells including macrophages, lymphocytes and osteoclasts. Stromal cells support the differentiation of the hematopoietic stem cells but importantly, have the ability to differentiate into osteoblasts [117]. In addition, physical factors within the bone microenvironment, including low oxygen levels, acidic pH and high extracellular calcium concentrations may also enhance tumor growth [118]. Furthermore,

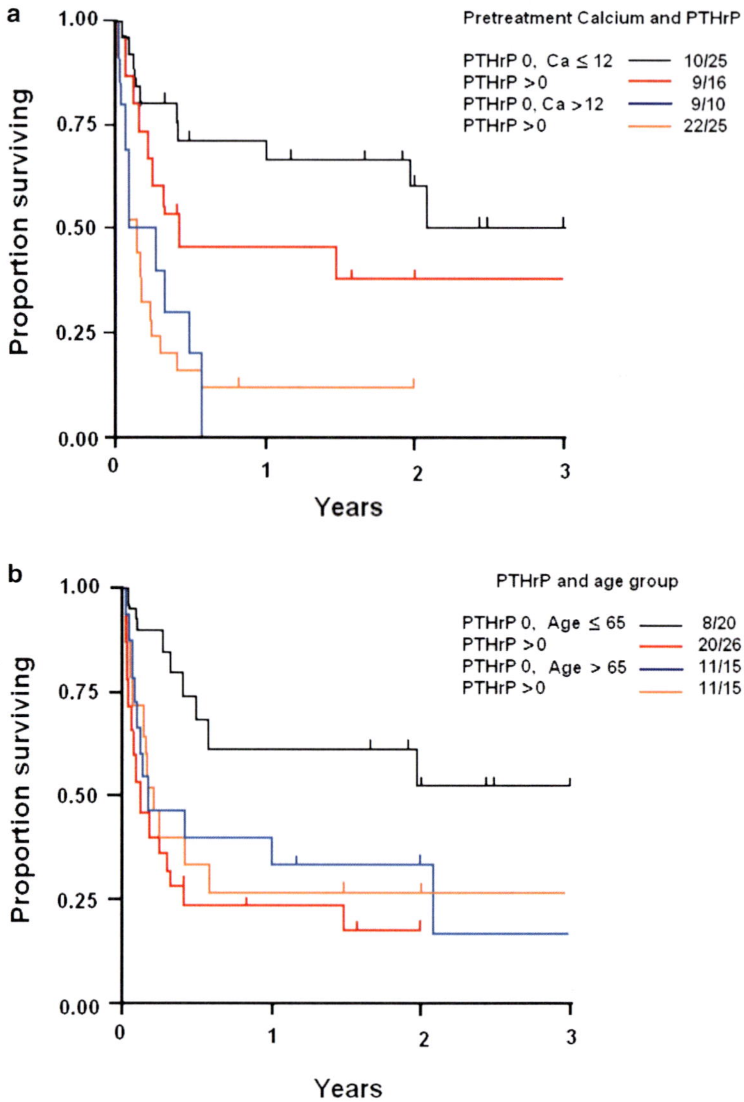

Fig. 12.3 Serum levels of PTHrP and survival of patients with MAH. (**a**) Survival in 76 hypercalcemic cancer patients, by PTHrP status and pretreatment calcium levels. Numbers shown in the inset are total numbers of deaths/number of patients at baseline. Numbers of patients at risk were 40 at 100 days, 22 at year 1, and 3 at year 3. (**b**) Survival in hypercalcemic cancer patients by PTHrP status and age group. The numbers of patients at risk were 41 at 100 days, 22 at year 1, and 3 at year 3. CA ≤ 12 = pretreatment serum calcium levels 10.3 to 12 mg/dl; CA > 12 = pretreatment calcium levels > 12 mg/dl; PTHrP 0 = PTHrP not elevated; PTHrP > 0 = PTHrP elevated. PTHrP and calcium levels are two independent prognostic factors for patient survival, and the effect of PTHrP is only significant in patients younger than 65

when the tumor cells stimulate osteoclastic bone resorption, the bone microenvironment becomes more enriched in bone-derived growth factors that enhance survival of the cancer and disrupt normal bone remodeling, resulting in bone destruction.

Metastasis is the spread and growth of tumor cells to distant organs, and represents the most devastating attribute of cancer. The common sites for metastatic spread of breast cancer are bone, lung and liver [119]. However, certain cancers will form metastases in specific organs or tissues at higher frequencies than predicted by blood flow patterns alone. For instance, breast, prostate, lung cancer and multiple myeloma frequently metastasize to bone. Bone metastases are often

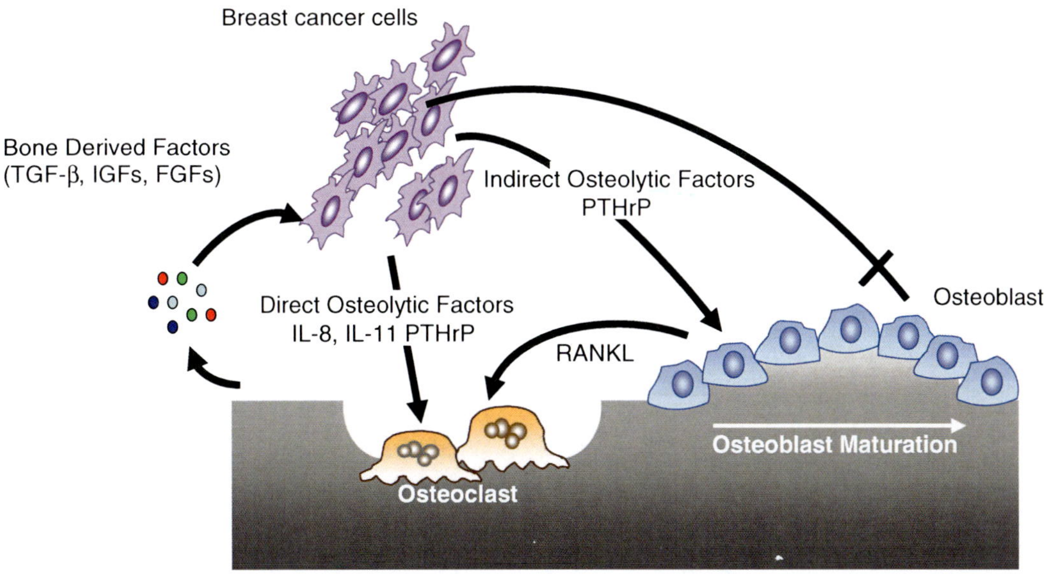

Fig 12.4 Interactions driving osteolytic metastases

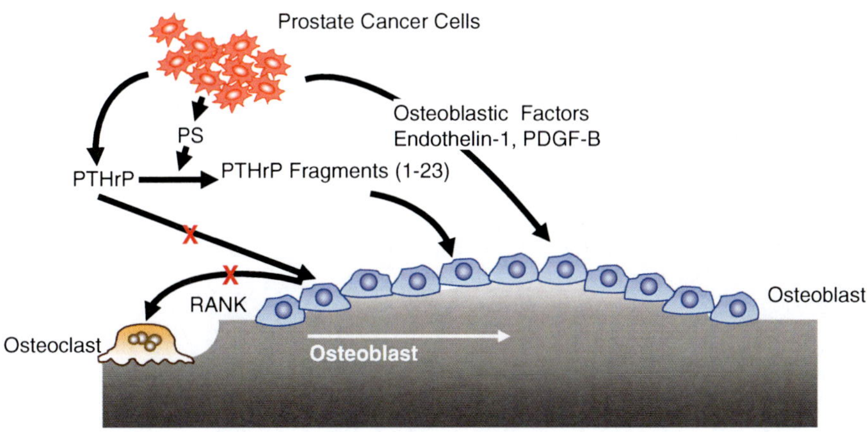

Fig 12.5 Interactions driving osteoblastic metastases

broadly classified as osteolytic (proceeding through bone destruction) (Fig. 12.4) or osteoblastic (proceeding through aberrant bone formation) (Fig. 12.5). Osteolytic bone lesions are typical of breast and lung carcinoma as well as hematological malignancies such as multiple myeloma. On the other hand, tumors in bone may stimulate new bone formation resulting in osteoblastic bone metastasis, a situation most often associated with prostate cancer although also encountered in breast cancer. The osteolytic-osteoblastic distinction is not absolute, as many patients with bone metastases have both types of lesions and individual metastatic lesions can contain both osteolytic and osteoblastic components. A dysregulation of the normal bone remodeling process is encountered in both types of lesions [120].

According to the Stanley Paget "seed and soil" hypothesis, tumor cells (seed) invading bone provide growth factors that activate the bone microenvironment (soil), which in turn produces growth factors that feed the tumor cells, creating a vicious cycle of destructive mutual cooperation [121].

PTHrP has been shown to play a key role in the osteoclastic bone resorption resulting from breast cancer metastasis to bone because it activates local bone turnover and consequently participates actively in the vicious cycle described above [117, 122, 123]. It has to be noted that PTHrP expression by breast cancer at metastatic sites differs dramatically from expression at primary sites. In fact, only 50% of primary breast cancers express PTHrP, whereas 92% of metastases of breast cancer to bone produce the peptide [124]. The increased local PTHrP stimulates RANKL expression and inhibits OPG secretion by osteoblasts. RANKL binds to its receptor RANK (expressed on osteoclasts) and enhances the differentiation and fusion of active osteoclasts in the presence of the macrophage-colony-stimulating factor. Bone-derived TGF-β, IGFs and FGFs released as a consequence of osteoclastic bone resorption stimulate tumor production of PTHrP via different receptors present in the cancer cells (PKC, MEK, P38 MAPK and Akt signaling pathway) [121, 125, 126]. Tumor cells might contribute to the vicious cycle by producing growth factors and cytokines which in turn sustain tumor growth [127]. It is, however, unclear whether the predilection of these cancers for spreading to bone results from the induction of PTHrP in the bone microenvironment, or whether tumors that produce PTHrP are more likely to metastasize to bone [124]. In contrast to its well-characterized role in bone metastasis development, the role of PTHrP in tumor progression outside the skeleton remains controversial.

12.4 *Pthrp* Gene Ablation in Mammary Epithelial Cells and Its Consequences on Tumor Initiation, Growth and Metastasis

12.4.1 Rationale

PTHrP is expressed in normal epithelial cells but its expression increases in breast cancer and becomes associated with multiple metastatic lesions and reduced survival. It is, however, still unknown whether PTHrP overexpression is simply a consequence of tumor progression, or whether it is mechanistically linked to the tumor progression process from initiation to metastasis. In order to shed light on this relationship, we ablated the *Pthrp* gene in mammary epithelial cells and determined the consequences of this ablation on tumor initiation, growth and metastasis.

We used a model in which PTHrP is specifically excised in mammary epithelial cells using the cre-loxPsystem. In this model, it is important to note that Cre is expressed shortly after birth (around 10 days) [128] but not during embryonic development [129], making the system suitable for the study of PTHrP roles in mammary development during puberty as well as in tumor initiation arising from the normal mammary epithelium.

12.4.2 MMTV-PyMT Breast Cancer Mouse Model

In the MMTV-PyMT transgenic mouse model [130], expression of the oncoprotein polyoma middle T antigen (PyMT) is under the control of the mouse mammary tumor virus long terminal repeat (LTR) and its expression is restricted to the mammary epithelium and absent from myoepithelial and surrounding stromal cells.

PyMT, a membrane-attached protein, is encoded by the small DNA polyoma virus. PyMT is not normally expressed in human breast tumor cells, but when overexpressed in the mammary epithelium of transgenic mice, it acts as a potent oncogene because its product (MT oncoprotein) binds various receptors and activates several signal transduction pathways, including those of the Src family kinase, ras and the PI3K pathways, which are frequently activated in human breast cancers [131]. Src phosphorylates Y315, which in turn directly activates the p85/p110/PI3K complex, which then activates PDK1 and 2 and results in AKT phosphorylation and activation [131]. Previous studies have shown that PTHrP can activate both ras [132] and AKT [133, 134] signaling pathways.

12.4.3 Tumor Progression from Hyperplasia to Metastastic Carcinoma

Mammary hyperplasia can be detected in this animal model as early as 4 weeks of age. The hyperplasia then progresses to adenoma in 6 weeks, to early carcinoma in 9 weeks and to late carcinoma 12 weeks, with pulmonary metastasis present in 100% of animals. The MMTV-PyMT mouse model of breast cancer is characterized by a high incidence of lung metastasis with highly reproducible progression kinetics. Increased metastatic potential has been shown to depend on the presence of macrophages in primary tumors and on the establishment of a chemoattractant paracrine loop of colony-stimulating factor-1 (CSF-1) and EGF ligands between macrophages and tumor cells [135]. Although PyMT transgenic mice do not develop bone metastasis per se, metastatic cells are found in the bone marrow relatively early and continue to grow in later stages without evidence of bone metastasis [136]

12.4.4 Conditional Knockout of *Pthrp* in Mammalian Epithelial Cells of the MMTV-PyVMT Transgenic Mouse Model

Disruption of the *Pthrp* gene in the mammary epithelium of the PyVMT transgenic mouse model produces mice that are homozygous (*PyVMT-Pthrp$^{flox/flox}$*) or heterozygous (*PyVMT-Pthrp$^{flox/+}$*) for the floxed *Pthrp* allele. Both groups of animals possess two active *Pthrp* alleles, whether flanked by flox sequences or not. These mice were crossed with a separate strain expressing Cre recombinase under the control of the MMTV promoter that targets Cre expression to the mammary epithelium. Excision of flox-bordered essential *Pthrp* sequences renders the gene nonfunctional. The resulting homozygous mice (Py*VMT-Pthrp$^{flox/flox}$; Cre^{+}*) therefore express no PTHrP in the mammary epithelium, while the heterozygous mice (*PyVMTPthrp$^{flox/+}$;Cre^{+}*) present lowered levels of PTHrP expression because they retain one active *Pthrp* allele.

A significant consequence of reduction or elimination of PTHrP expression in the mammary epithelium of the offspring is a marked delay in tumor onset. 100% of control animals (normal PTHrP levels) present tumors by day 55, while heterozygous (PTHrP haploinsufficiency) and homozygous animals (absent PTHrP) reach this percentage by days 77 and 85, respectively.

Metastatic spread to lungs was similarly reduced independent of tumor size, illustrating the crucial importance of ablating PTHrP signaling to prevent breast cancer progression and metastasis [137].

Other Cre/lox studies indicate ablation of floxed sequences in only 90% of the cells, leaving 10% of the cells to potentially express the knock-out protein. In order to overcome the problem of residual PTHrP expression, we isolated cells from *PyVMT Pthrp$^{flox/flox}$;Cre^{-}* tumors, transfected them with an adenoCre-GFP (or control adeno-GFP) construct, subcultured the cells and purified them by flow cytometry to obtain pure populations of Cre^{+} (or control Cre^{-}) cells with complete or no *Pthrp* ablation. When these cells were transplanted into the mammary fat pad of healthy syngeneic mice, tumor onset was significantly delayed post-tumor implantation in adeno-Cre animals compared to adeno-GFP controls [137]. Tumor growth was also significantly delayed in the adenoCre mice. Metastatic tumor cells were detectable in the bone marrow of adenoCre animals during killing, confirming that this model can be used to examine natural progression of breast cancer from the primary site to the skeleton. The ablation of *Pthrp* was also observed to inhibit G0/G1 to S transition in tumor cells, to enhance tumor cell apoptosis (increased TUNEL staining and decreased Bcl-2 expression) and to decrease Akt1 and increase Akt2 expression [137]. From these observations, we concluded that PTHrP is critical for the initiation of mammary tumorigenesis in vivo, and that ablation of PTHrP expression in mammary epithelial tumor cells inhibits tumor growth and metastasis by several mechanisms.

12.5 Summary and Conclusion

PTHrP has been associated with cancer as the causative agent for malignancy-associated hypercalcemia, and with progression of metastases in the skeletal environment. We show that PTHrP is involved in breast cancer initiation, growth and metastasis, and we suggest that PTHrP plays a role in a number of critical checkpoints for PyVMT, which points to a novel role as a facilitator of oncogenes and emphasizes the importance of attempting its targeting for therapeutic purposes.

References

1. Vanderschueren B et al (1994) Circulating concentrations of interleukin-6 in cancer patients and their pathogenic role in tumor-induced hypercalcemia. Cancer Immunol Immunother 39:286–290
2. Sugihara A et al (1998) Expression of cytokines enhancing the osteoclast activity, and parathyroid hormone-related protein in prostatic cancers before and after endocrine therapy: an immunohistochemical study. Oncol Rep 5:1389–1394
3. Shibahara T, Nomura T, Cui NH, Noma H (2005) A study of osteoclast-related cytokines in mandibular invasion by squamous cell carcinoma. Int J Oral Maxillofac Surg 34:789–793
4. Suzuki K, Yamada S (1994) Ascites sarcoma 180, a tumor associated with hypercalcemia, secretes potent bone-resorbing factors including transforming growth factor alpha, interleukin-1 alpha and interleukin-6. Bone Miner 27:219–233
5. Martin TJ, Suva LJ (1988) Parathyroid hormone-related protein: a novel gene product. Baillieres Clin Endocrinol Metab 2:1003–1029
6. Wysolmerski JJ, Broadus AE (1994) Hypercalcemia of malignancy: the central role of parathyroid hormone-related protein. Annu Rev Med 45:189–200
7. Grill V, Rankin W, Martin TJ (1998) Parathyroid hormone-related protein (PTHrP) and hypercalcaemia. Eur J Cancer 34:222–229
8. Svane S (1964) Hypercalcemia in malignant disease without evidence of bone destruction. A case simulating acute hyperparathyroidism. Acta Med Scand 175:353–357
9. Gellhorn A, Plimpton CH (1956) Hypercalcemia in malignant disease without evidence of bone destruction. Am J Med 21:750–759
10. Omenn GS, Roth SI, Baker WH (1969) Hyperparathyroidism associated with malignant tumors of nonparathyroid origin. Cancer 24:1004–1011
11. Buckle R (1974) Ectopic PTH syndrome, pseudohyperparathyroidism; hypercalcaemia of malignancy. Clin Endocrinol Metab 3:237–251
12. Buckle RM, McMillan M, Mallinson C (1970) Ectopic secretion of parathyroid hormone by a renal adenocarcinoma in a patient with hypercalcaemia. Br Med J 4:724–726
13. Tashjian AH Jr (1969) Animal cell cultures as a source of hormones. Biotechnol Bioeng 11:109–126
14. Riggs BL, Arnaud CD, Reynolds JC, Smith LH (1971) Immunologic differentiation of primary hyperparathyroidism from hyperparathyroidism due to nonparathyroid cancer. J Clin Invest 50:2079–2083
15. Benson RC Jr, Riggs BL, Pickard BM, Arnaud CD (1974) Immunoreactive forms of circulating parathyroid hormone in primary and ectopic hyperparathyroidism. J Clin Invest 54:175–181
16. Moseley JM et al (1987) Parathyroid hormone-related protein purified from a human lung cancer cell line. Proc Natl Acad Sci USA 84:5048–5052
17. Strewler GJ et al (1987) Parathyroid hormonelike protein from human renal carcinoma cells. Structural and functional homology with parathyroid hormone. J Clin Invest 80:1803–1807
18. Burtis WJ et al (1987) Identification of a novel 17,000-dalton parathyroid hormone-like adenylate cyclase-stimulating protein from a tumor associated with humoral hypercalcemia of malignancy. J Biol Chem 262:7151–7156
19. Suva LJ et al (1987) A parathyroid hormone-related protein implicated in malignant hypercalcemia: cloning and expression. Science 237:893–896
20. Mangin M, Ikeda K, Dreyer BE, Milstone L, Broadus AE (1988) Two distinct tumor-derived, parathyroid hormone-like peptides result from alternative ribonucleic acid splicing. Mol Endocrinol 2:1049–1055
21. Thiede MA, Strewler GJ, Nissenson RA, Rosenblatt M, Rodan GA (1988) Human renal carcinoma expresses two messages encoding a parathyroid hormone-like peptide: evidence for the alternative splicing of a single-copy gene. Proc Natl Acad Sci USA 85:4605–4609
22. Kemp BE et al (1987) Parathyroid hormone-related protein of malignancy: active synthetic fragments. Science 238:1568–1570

23. Nickols GA, Nickols MA, Helwig JJ (1990) Binding of parathyroid hormone and parathyroid hormone-related protein to vascular smooth muscle of rabbit renal microvessels. Endocrinology 126:721–727
24. Kao PC, Klee GG, Taylor RL, Heath H 3rd (1990) Parathyroid hormone-related peptide in plasma of patients with hypercalcemia and malignant lesions. Mayo Clin Proc 65:1399–1407
25. Burtis WJ et al (1990) Immunochemical characterization of circulating parathyroid hormone-related protein in patients with humoral hypercalcemia of cancer. N Engl J Med 322:1106–1112
26. Mangin M et al (1988) Identification of a cDNA encoding a parathyroid hormone-like peptide from a human tumor associated with humoral hypercalcemia of malignancy. Proc Natl Acad Sci USA 85:597–601
27. Mannens M et al (1987) Regional localization of DNA probes on the short arm of chromosome 11 using aniridia-Wilms' tumor-associated deletions. Hum Genet 75:180–187
28. Yasuda T, Banville D, Hendy GN, Goltzman D (1989) Characterization of the human parathyroid hormone-like peptide gene. Functional and evolutionary aspects. J Biol Chem 264:7720–7725
29. Mangin M, Ikeda K, Dreyer BE, Broadus AE (1989) Isolation and characterization of the human parathyroid hormone-like peptide gene. Proc Natl Acad Sci USA 86:2408–2412
30. Mangin M, Ikeda K, Dreyer BE, Broadus AE (1990) Identification of an up-stream promoter of the human parathyroid hormone-related peptide gene. Mol Endocrinol 4:851–858
31. Mangin M, Ikeda K, Broadus AE (1990) Structure of the mouse gene encoding parathyroid hormone-related peptide. Gene 95:195–202
32. Karaplis AC, Yasuda T, Hendy GN, Goltzman D, Banville D (1990) Gene-encoding parathyroid hormone-like peptide: nucleotide sequence of the rat gene and comparison with the human homologue. Mol Endocrinol 4:441–446
33. Thiede MA, Rutledge SJ (1990) Nucleotide sequence of a parathyroid hormone-related peptide expressed by the 10 day chicken embryo. Nucleic Acids Res 18:3062
34. Orloff JJ et al (1994) Parathyroid hormone-related protein as a prohormone: posttranslational processing and receptor interactions. Endocr Rev 15:40–60
35. Philbrick WM et al (1996) Defining the roles of parathyroid hormone-related protein in normal physiology. Physiol Rev 76:127–173
36. Fenton AJ et al (1991) A potent inhibitor of osteoclastic bone resorption within a highly conserved pentapeptide region of parathyroid hormone-related protein; PTHrP[107-111]. Endocrinology 129: 3424–3426
37. Cornish J, Callon KE, Nicholson GC, Reid IR (1997) Parathyroid hormone-related protein-(107-139) inhibits bone resorption in vivo. Endocrinology 138: 1299–1304
38. Blind E, Raue F, Knappe V, Schroth J, Ziegler R (1993) Cyclic AMP formation in rat bone and kidney cells is stimulated equally by parathyroid hormone-related protein (PTHrP) 1-34 and PTH 1-34. Exp Clin Endocrinol 101:150–155
39. Iida-Klein A et al (1997) Mutations in the second cytoplasmic loop of the rat parathyroid hormone (PTH)/PTH-related protein receptor result in selective loss of PTH-stimulated phospholipase C activity. J Biol Chem 272:6882–6889
40. Swarthout JT, D'Alonzo RC, Selvamurugan N, Partridge NC (2002) Parathyroid hormone-dependent signaling pathways regulating genes in bone cells. Gene 282:1–17
41. Takasu H, Guo J, Bringhurst FR (1999) Dual signaling and ligand selectivity of the human PTH/PTHrP receptor. J Bone Miner Res 14:11–20
42. Whitfield JF et al (2001) Stimulation of protein kinase C activity in cells expressing human parathyroid hormone receptors by C- and N-terminally truncated fragments of parathyroid hormone 1-34. J Bone Miner Res 16:441–447
43. Yang D, Guo J, Divieti P, Bringhurst FR (2006) Parathyroid hormone activates PKC-delta and regulates osteoblastic differentiation via a PLC-independent pathway. Bone 38:485–496
44. Massfelder T et al (1997) Opposing mitogenic and anti-mitogenic actions of parathyroid hormone-related protein in vascular smooth muscle cells: a critical role for nuclear targeting. Proc Natl Acad Sci USA 94:13630–13635
45. Henderson JE et al (1995) Nucleolar localization of parathyroid hormone-related peptide enhances survival of chondrocytes under conditions that promote apoptotic cell death. Mol Cell Biol 15:4064–4075
46. Nguyen M, He B, Karaplis A (2001) Nuclear forms of parathyroid hormone-related peptide are translated from non-AUG start sites downstream from the initiator methionine. Endocrinology 142:694–703
47. Aarts MM et al (1999) Parathyroid hormone-related protein interacts with RNA. J Biol Chem 274: 4832–4838
48. Aarts MM et al (2001) Parathyroid hormone-related protein promotes quiescence and survival of serum-deprived chondrocytes by inhibiting rRNA synthesis. J Biol Chem 276:37934–37943
49. Lam MH et al (1999) Importin beta recognizes parathyroid hormone-related protein with high affinity and mediates its nuclear import in the absence of importin alpha. J Biol Chem 274:7391–7398
50. Cingolani G, Bednenko J, Gillespie MT, Gerace L (2002) Molecular basis for the recognition of a nonclassical nuclear localization signal by importin beta. Mol Cell 10:1345–1353
51. Lam MH et al (2002) Nuclear transport of parathyroid hormone (PTH)-related protein is dependent on microtubules. Mol Endocrinol 16:390–401
52. Jans DA, Thomas RJ, Gillespie MT (2003) Parathyroid hormone-related protein (PTHrP): a nucleocytoplasmic shuttling protein with distinct

paracrine and intracrine roles. Vitam Horm 66: 345–384
53. Miao D et al (2008) Severe growth retardation and early lethality in mice lacking the nuclear localization sequence and C-terminus of PTH-related protein. Proc Natl Acad Sci USA 105:20309–20314
54. Lam MH et al (1999) Phosphorylation at the cyclin-dependent kinases site (Thr85) of parathyroid hormone-related protein negatively regulates its nuclear localization. J Biol Chem 274:18559–18566
55. Lam MH et al (1997) PTHrP and cell division: expression and localization of PTHrP in a keratinocyte cell line (HaCaT) during the cell cycle. J Cell Physiol 173:433–446
56. Fenton AJ et al (1991) A carboxyl-terminal peptide from the parathyroid hormone-related protein inhibits bone resorption by osteoclasts. Endocrinology 129:1762–1768
57. Fenton AJ, Martin TJ, Nicholson GC (1994) Carboxyl-terminal parathyroid hormone-related protein inhibits bone resorption by isolated chicken osteoclasts. J Bone Miner Res 9:515–519
58. Kong XF et al (1994) The rat, mouse, and human genes encoding the receptor for parathyroid hormone and parathyroid hormone-related peptide are highly homologous. Biochem Biophys Res Commun 201:1058
59. Datta NS, Abou-Samra AB (2009) PTH and PTHrP signaling in osteoblasts. Cell Signal 21:1245–1254
60. Abou-Samra AB et al (1992) Expression cloning of a common receptor for parathyroid hormone and parathyroid hormone-related peptide from rat osteoblast-like cells: a single receptor stimulates intracellular accumulation of both cAMP and inositol trisphosphates and increases intracellular free calcium. Proc Natl Acad Sci USA 89:2732–2736
61. Juppner H et al (1991) A G protein-linked receptor for parathyroid hormone and parathyroid hormone-related peptide. Science 254:1024–1026
62. Schipani E et al (1993) Identical complementary deoxyribonucleic acids encode a human renal and bone parathyroid hormone (PTH)/PTH-related peptide receptor. Endocrinology 132:2157–2165
63. Gardella TJ, Juppner H (2000) Interaction of PTH and PTHrP with their receptors. Rev Endocr Metab Disord 1:317–329
64. Strewler GJ (2000) The physiology of parathyroid hormone-related protein. N Engl J Med 342:177–185
65. Gensure RC, Gardella TJ, Juppner H (2005) Parathyroid hormone and parathyroid hormone-related peptide, and their receptors. Biochem Biophys Res Commun 328:666–678
66. Clemens TL et al (2001) Parathyroid hormone-related protein and its receptors: nuclear functions and roles in the renal and cardiovascular systems, the placental trophoblasts and the pancreatic islets. Br J Pharmacol 134:1113–1136
67. Amling M et al (1997) Bcl-2 lies downstream of parathyroid hormone-related peptide in a signaling pathway that regulates chondrocyte maturation during skeletal development. J Cell Biol 136:205–213
68. Hastings RH et al (2003) Parathyroid hormone-related protein ameliorates death receptor-mediated apoptosis in lung cancer cells. Am J Physiol Cell Physiol 285:C1429–C1436
69. Gagiannis S et al (2009) Parathyroid hormone-related protein confers chemoresistance by blocking apoptosis signaling via death receptors and mitochondria. Int J Cancer 125:1551–1557
70. Karaplis AC et al (1994) Lethal skeletal dysplasia from targeted disruption of the parathyroid hormone-related peptide gene. Genes Dev 8:277–289
71. Amizuka N et al (1996) Haploinsufficiency of parathyroid hormone-related peptide (PTHrP) results in abnormal postnatal bone development. Dev Biol 175:166–176
72. Philbrick WM, Dreyer BE, Nakchbandi IA, Karaplis AC (1998) Parathyroid hormone-related protein is required for tooth eruption. Proc Natl Acad Sci USA 95:11846–11851
73. Wysolmerski JJ et al (1998) Rescue of the parathyroid hormone-related protein knockout mouse demonstrates that parathyroid hormone-related protein is essential for mammary gland development. Development 125:1285–1294
74. Wysolmerski JJ, McCaughern-Carucci JF, Daifotis AG, Broadus AE, Philbrick WM (1995) Overexpression of parathyroid hormone-related protein or parathyroid hormone in transgenic mice impairs branching morphogenesis during mammary gland development. Development 121:3539–3547
75. Wysolmerski JJ et al (1994) Overexpression of parathyroid hormone-related protein in the skin of transgenic mice interferes with hair follicle development. Proc Natl Acad Sci USA 91:1133–1137
76. Foley J et al (2001) Parathyroid hormone-related protein maintains mammary epithelial fate and triggers nipple skin differentiation during embryonic breast development. Development 128: 513–525
77. Ferrari SL, Rizzoli R, Bonjour JP (1992) Parathyroid hormone-related protein production by primary cultures of mammary epithelial cells. J Cell Physiol 150:304–311
78. Seitz PK et al (1993) Parathyroid hormone-related peptide production and action in a myoepithelial cell line derived from normal human breast. Endocrinology 133:1116–1124
79. Wysolmerski JJ, Stewart AF (1998) The physiology of parathyroid hormone-related protein: an emerging role as a developmental factor. Annu Rev Physiol 60:431–460
80. Ferrari S, Rizzoli R, Chaponnier C, Gabbiani G, Bonjour JP (1993) Parathyroid hormone-related protein increases cAMP production in mammary epithelial cells. Am J Physiol 264:E471–E475
81. Jemal A et al (2009) Cancer statistics, 2009. CA Cancer J Clin 59:225–249
82. Schilling T et al (1993) Parathyroid hormone-related protein (PTHrP) does not regulate 1,25-dihydroxyvitamin D serum levels in hypercalcemia of malignancy. J Clin Endocrinol Metab 76:801–803

83. Bouizar Z, Spyratos F, Deytieux S, de Vernejoul MC, Jullienne A (1993) Polymerase chain reaction analysis of parathyroid hormone-related protein gene expression in breast cancer patients and occurrence of bone metastases. Cancer Res 53:5076–5078
84. Bucht E et al (1998) Parathyroid hormone-related protein in patients with primary breast cancer and eucalcemia. Cancer Res 58:4113–4116
85. Bundred NJ et al (1991) Parathyroid hormone related protein and hypercalcaemia in breast cancer. BMJ 303:1506–1509
86. Kohno N et al (1994) Parathyroid hormone-related protein in breast cancer tissues: relationship between primary and metastatic sites. Breast Cancer 1:43–49
87. Bouizar Z, Spyratos F, De vernejoul MC (1999) The parathyroid hormone-related protein (PTHrP) gene: use of downstream TATA promotor and PTHrP 1-139 coding pathways in primary breast cancers vary with the occurrence of bone metastasis. J Bone Miner Res **14**:406–414
88. Iwamura M et al (1993) Immunohistochemical localization of parathyroid hormone-related protein in human prostate cancer. Cancer Res 53:1724–1726
89. Kremer R et al (1997) Ras activation of human prostate epithelial cells induces overexpression of parathyroid hormone-related peptide. Clin Cancer Res 3:855–859
90. Dougherty KM et al (1999) Parathyroid hormone-related protein as a growth regulator of prostate carcinoma. Cancer Res 59:6015–6022
91. Brandt DW, Burton DW, Gazdar AF, Oie HE, Deftos LJ (1991) All major lung cancer cell types produce parathyroid hormone-like protein: heterogeneity assessed by high performance liquid chromatography. Endocrinology 129:2466–2470
92. Kitazawa S et al (1991) Immunohistologic evaluation of parathyroid hormone-related protein in human lung cancer and normal tissue with newly developed monoclonal antibody. Cancer 67: 984–989
93. Nishigaki Y, Ohsaki Y, Toyoshima E, Kikuchi K (1999) Increased serum and urinary levels of a parathyroid hormone-related protein COOH terminus in non-small cell lung cancer patients. Clin Cancer Res 5:1473–1481
94. Nishihara M, Kanematsu T, Taguchi T, Razzaque MS (2007) PTHrP and tumorigenesis: is there a role in prognosis? Ann N Y Acad Sci 1117:385–392
95. Nishihara M et al (1999) Clinicopathological implications of parathyroid hormone-related protein in human colorectal tumors. J Pathol 187:217–222
96. Kremer R, Shustik C, Tabak T, Papavasiliou V, Goltzman D (1996) Parathyroid-hormone-related peptide in hematologic malignancies. Am J Med 100:406–411
97. Alipov GK et al (1997) Expression of parathyroid hormone-related peptide (PTHrP) in gastric tumors. J Pathol 182:174–179
98. Goltzman D and Kremer R (2005) Target Genes: PTHrP. In Vitamin D 2nd edition.Feldman, Pike and Glorieux editors. Elsevier Academic Press 1:737–749
99. Kremer R et al (1991) Regulation of parathyroid hormone-like peptide in cultured normal human keratinocytes: effects of growth factors and 1,25-dihydroxyvitamin D on gene expression and secretion. J Clin Invest 87:884–893
100. Esbrit P et al (2000) C-terminal parathyroid hormone-related protein increases vascular endothelial growth factor in human osteoblastic cells. J Am Soc Nephrol 11:1085–1092
101. Akino K et al (2000) Parathyroid hormone-related peptide is a potent tumor angiogenic factor. Endocrinology 141:4313–4316
102. Luparello C, Burtis WJ, Raue F, Birch MA, Gallagher JA (1995) Parathyroid hormone-related peptide and 8701-BC breast cancer cell growth and invasion in vitro: evidence for growth-inhibiting and invasion-promoting effects. Mol Cell Endocrinol 111:225–232
103. Luparello C, Sirchia R, Pupello D (2003) PTHrP [67-86] regulates the expression of stress proteins in breast cancer cells inducing modifications in urokinase-plasminogen activator and MMP-1 expression. J Cell Sci 116:2421–2430
104. Bundred NJ et al (1992) Parathyroid hormone related protein and skeletal morbidity in breast cancer. Eur J Cancer 28:690–692
105. Southby J et al (1990) Immunohistochemical localization of parathyroid hormone-related protein in human breast cancer. Cancer Res 50:7710–7716
106. Kohno N et al (1994) The expression of parathyroid hormone-related protein in human breast cancer with skeletal metastases. Surg Today 24:215–220
107. Linforth R et al (2002) Coexpression of parathyroid hormone related protein and its receptor in early breast cancer predicts poor patient survival. Clin Cancer Res 8:3172–3177
108. Truong NU, deB Edwardes MD, Papavasiliou V, Goltzman D, Kremer R (2003) Parathyroid hormone-related peptide and survival of patients with cancer and hypercalcemia. Am J Med 115:115–121
109. Martin TJ, Moseley JM (2000) Mechanisms in the skeletal complications of breast cancer. Endocr Relat Cancer 7:271–284
110. Henderson MA et al (2006) Parathyroid hormone-related protein localization in breast cancers predict improved prognosis. Cancer Res 66:2250–2256
111. Sato K, Onuma E, Yocum RC, Ogata E (2003) Treatment of malignancy-associated hypercalcemia and cachexia with humanized anti-parathyroid hormone-related protein antibody. Semin Oncol 30: 167–173
112. Saito H et al (2005) Humanized monoclonal antibody against parathyroid hormone-related protein suppresses osteolytic bone metastasis of human breast cancer cells derived from MDA-MB-231. Anticancer Res 25:3817–3823
113. Falzon M, Du P (2000) Enhanced growth of MCF-7 breast cancer cells overexpressing parathyroid hormone-related peptide. Endocrinology 141: 1882–1892
114. Hauschka PV, Mavrakos AE, Iafrati MD, Doleman SE, Klagsbrun M (1986) Growth factors in bone

matrix. Isolation of multiple types by affinity chromatography on heparin-Sepharose. J Biol Chem 261:12665–12674
115. Yoneda T (2000) Cellular and molecular basis of preferential metastasis of breast cancer to bone. J Orthop Sci 5:75–81
116. Lipton A (2004) Pathophysiology of bone metastases: how this knowledge may lead to therapeutic intervention. J Support Oncol 2: 205–213; discussion 213–204, 216–207, 219–220
117. Akhtari M, Mansuri J, Newman KA, Guise TM, Seth P (2008) Biology of breast cancer bone metastasis. Cancer Biol Ther 7:3–9
118. Kingsley LA, Fournier PG, Chirgwin JM, Guise TA (2007) Molecular biology of bone metastasis. Mol Cancer Ther 6:2609–2617
119. Weigelt B, Peterse JL, van't Veer LJ (2005) Breast cancer metastasis: markers and models. Nat Rev Cancer **5**:591–602
120. Guise TA, Mundy GR (1998) Cancer and bone. Endocr Rev 19:18–54
121. Mundy GR (2002) Metastasis to bone: causes, consequences and therapeutic opportunities. Nat Rev Cancer 2:584–593
122. Guise TA et al (1996) Evidence for a causal role of parathyroid hormone-related protein in the pathogenesis of human breast cancer-mediated osteolysis. J Clin Invest 98:1544–1549
123. Kakonen SM, Mundy GR (2003) Mechanisms of osteolytic bone metastases in breast carcinoma. Cancer 97:834–839
124. Roodman GD (2004) Mechanisms of bone metastasis. N Engl J Med 350:1655–1664
125. Alokail MS (2005) Induction of erbB1 and erB2 receptors by PTHrP in breast cancer cells. AACR Meeting Abstracts 2005:1359
126. Chirgwin JM, Guise TA (2000) Molecular mechanisms of tumor-bone interactions in osteolytic metastases. Crit Rev Eukaryot Gene Expr 10:159–178
127. Peyruchaud O (2007) Mechanisms of bone metastasis formation. J Soc Biol 201:229–236
128. Wagner KU et al (2001) Spatial and temporal expression of the Cre gene under the control of the MMTV-LTR in different lines of transgenic mice. Transgenic Res 10:545–553
129. Andrechek ER et al (2000) Amplification of the neu/erbB-2 oncogene in a mouse model of mammary tumorigenesis. Proc Natl Acad Sci USA 97: 3444–3449
130. Guy CT, Cardiff RD, Muller WJ (1992) Induction of mammary tumors by expression of polyomavirus middle T oncogene: a transgenic mouse model for metastatic disease. Mol Cell Biol 12:954–961
131. Dilworth SM (2002) Polyoma virus middle T antigen and its role in identifying cancer-related molecules. Nat Rev Cancer 2:951–956
132. Aklilu F, Park M, Goltzman D, Rabbani SA (1997) Induction of parathyroid hormone-related peptide by the Ras oncogene: role of Ras farnesylation inhibitors as potential therapeutic agents for hypercalcemia of malignancy. Cancer Res 57:4517–4522
133. Shen X, Falzon M (2006) PTH-related protein upregulates integrin alpha6beta4 expression and activates Akt in breast cancer cells. Exp Cell Res 312:3822–3834
134. Agouni A et al (2007) Parathyroid hormone-related protein induces cell survival in human renal cell carcinoma through the PI3K Akt pathway: evidence for a critical role for integrin-linked kinase and nuclear factor kappa B. Carcinogenesis 28: 1893–1901
135. Fantozzi A, Christofori G (2006) Mouse models of breast cancer metastasis. Breast Cancer Res 8:212
136. Husemann Y et al (2008) Systemic spread is an early step in breast cancer. Cancer Cell 13:58–68
137. Kremer R et al (2010) Parathyroid hormone-related protein (PTHrP) gene ablation in breast cancer cells inhibits invasion in vitro and metastasis in vivo: role of AKT and CXCR4. In ASBMR 2010 Annual Meeting presentation 1229 (Toronto, ON)
138. Ditmer J (2004) Importance of PTHrP for cancer development. J Gene Ther Mol Biol 8:451–454

Part IV

Multistep/Genes

Mechanism of Radiation Carcinogenesis: Role of the *TGFBI* Gene and the Inflammatory Signaling Cascade

13

Tom K. Hei, Yongliang Zhao, Hongning Zhou, and Vladimir Ivanov

Abstract

Using an immortalized human bronchial epithelial cell line, we showed previously that the transforming growth factor beta-induced (*TGFBI*) gene was consistently downregulated by six- to sevenfold among radiation-induced tumorigenic human cells when compared with controls. Transfection of *TGFBI* gene into tumor cells resulted in a significant reduction in tumor growth as well as in vitro anchorage independent growth. The observations that *TGFBI* knock-out animals showed increased spontaneous tumor incidence and chemically induced tumors highlight the suppressive nature of the gene. There is evidence that extranuclear/extracellular targets are important in low-dose radiation response and that the cyclo-oxygenase-2 signaling pathway mediates the process. The involvement of NFκB-dependent cytokines and the resultant inflammatory response works in concert with in modulating radiation-induced bronchial carcinogenesis.

13.1 Introduction

Radiation is a two-sided sword; on the one hand, it is a well-established human carcinogen, and, on the other hand, it is an effective therapeutic regimen against a variety of human malignancies. Although ionizing radiation is a well-established human carcinogen and is known to induce tumors on various sites in human, the underlying cellular and molecular mechanism of radiation carcinogenesis remains unknown. The carcinogenic risk for human epithelial cells after exposure to low LET radiation has been estimated to be in the range of 10^{-12} per cell per Gy based on the breast cancer incidence among Japanese A-bomb survivors [1], an incidence too low to be reproduced under any laboratory setting. Immortalized human lung epithelial cell cultures offer the next best alternatives as useful models in assessing the

T.K. Hei (✉)
Department of Radiation Oncology,
Center for Radiological Research, Columbia University Medical Center, 630 West 168th Street,
New York, NY 10032, USA
e-mail: tkh1@Columbia.edu

J.S. Rhim and R. Kremer (eds.), *Human Cell Transformation: Role of Stem Cells and the Microenvironment*, Advances in Experimental Medicine and Biology 720,
DOI 10.1007/978-1-4614-0254-1_13,

various phenotypic and genotypic changes leading to malignancies after exposure to ionizing radiation. Although these cells are immortalized either by viral transduction [2] or by ectopic expression of telomerase [3], they are phenotypically normal and do not express any transformed characteristics such as anchorage independent growth and tumorigenicity in nude mice. After exposure to either 150 keV/μm α particles or 1 GeV/ nucleon ^{56}Fe ions, transformed cells arise through a series of sequential stages including altered growth pattern, resistance to serum-induced terminal differentiation, agar-positive growth, tumorigenicity, and metastasis [4, 5].

Previous studies have shown that a single 60 cGy dose of these high LET alpha particles or heavy ions irradiations can induce neoplastic transformation of the BEP2D cells in a step-wise fashion at a frequency of ~ 10^{-6} [4]. The immortalization step, therefore, increases the transformation yield of primary human epithelial cells by more than a million fold. It should be noted that, while the majority of agar-positive BEP2D clones are nontumorigenic, they all demonstrated the propensity to resist serum-induced terminal differentiation. In addition, each preceding stage represents a necessary, yet insufficient step toward the later, more malignant phase. Furthermore, tumorigenic BEP2D cells show no mutations in any of the *ras* oncogenes [4], a finding that is consistent with the observation that radiation is clastogenic that induces primarily multilocus deletions and chromosomal loss.

13.1.1 Changes in Gene Profiling in Radiation-Induced Tumorigenic Cells

The differential expression of known genes between tumorigenic and nontumorigenic control cells can be compared using a cDNA expression array. Briefly, total RNAs from both tumorigenic and control BEP2D cells are isolated and then treated with DNase I to remove any contaminating DNA from the RNA preparation. PolyA$^+$ RNA is then isolated, and ^{32}P-labeled complex cDNA probes are generated by reverse transcription in the presence of ^{32}P-labeled dATP. The cDNA probes can then be hybridized to the nylon membrane and the hybridization patterns can be analyzed by autoradiography. The intensity of the signal, which corresponds to the expression levels of the individual genes, can be easily quantified using phosphorimaging technique.

Using cDNA microarrays to compare differentially expressed genes in radiation-induced tumorigenic cell lines relative to parental BEP2D cells, there is evidence that *TGFBI* gene is downregulated by more than six- to seven fold in the tumorigenic cells [5]. TGFBI is a secreted protein induced by the transforming growth factor-β (TGF-β) in human adenocarcinoma cells as well as other human cell types [6]. *TGFBI* was first identified in a human lung adenocarcinoma cell line (A549) treated with TGF-β. This gene encodes a highly conserved 683 amino-acid protein that contains a secretary signal sequence and four internal homologous domains, the last of which contains an RGD (Arg-Gly-Asp) motif which can serve as a ligand recognition site for integrins [6]. TGFBI product has been shown to be a component of extracellular matrix (ECM) in lung and mediates cell adhesion and migration through interacting with integrin via integrin receptors: α3β1, αvβ3, and αvβ5 [7]. It is ubiquitously expressed in human normal tissues; however, its role in cancer appears to be cell type specific. Although downregulation or lost expression of this gene has been found in human tumor cell lines including lung, breast, and leukemia as well as in human primary lung and breast tumor specimens [5, 8], there are several reports that TGFBI is over-expressed in human colorectal, colon, esophageal, and pancreatic cancers [9, 10]. Furthermore, there is evidence that TGFBI mediates tumor metastasis and cancer cell extravasation [10]. For human lung cancer CpG island hypermethylation in the promoter region correlates with the silencing of *TGFBI* promoter and its subsequent down-expression [11]. In vitro studies have implicated its role in maintaining microtubule stability, and inhibiting tumorigenicity and tumor angiogenesis, suggesting a tumor suppressor function in vivo.

13.1.2 Ectopic Re-Expression of BigH3 Suppress Tumorigenic Phenotype

To ascertain the tumor suppressive effect of *TGFBI* gene in human lung cancer, it is necessary to re-express the lost or downregulated gene in tumor cells and determine the tumor outcome upon inoculation into nude mice. The approach was to ectopically express the gene in a highly malignant cell line (TL1) using pRc/CMV2 *TGFBI* expression vector. Thirty G418-resistant colonies (TL1–TL30) that expressed different levels of *TGFBI* gene were chosen for further studies [12]. From the northern and western blotting results, the parental TL1 cells and TL1-pRc/CMV2 cells expressed a low, detectable, and similar level of *TGFBI* gene. After transfection, the expression of *TGFBI* gene in TL1-clone 18 cells was restored to a level similar to that of control BEP2D cells, whereas in TL1-clone 28 cells it was fourfold higher than control. Upon inoculation into athymic nude mice, no tumor (0/8 mice) was detected in animals injected with the parental BEP2D cells after more than 20 weeks. However, 8/8 mice that were injected with either TL1 or TL1-pRc/CMV2 tumorigenic cells with vector control developed progressively growing tumors at 4 weeks, with average volumes of 1021.8 mm^3 and 970.6 mm^3, respectively. In contrast, only 9/16 mice injected with TL1-clone 18 formed small tumor nodules at 4 weeks, with an average volume of 86.7 mm^3 which was significantly smaller than that of the parental TL1 cells ($P<0.01$). The results of these studies provide unequivocal evidence of the *TGFBI* gene function in radiation carcinogenesis.

13.1.3 TGFBI Knock Out Mice Have Higher Spontaneous Tumor Incidence

To explore the physiological function of *TGFBI* and its role in tumorigenesis, *TGFBI*-deficient mice were generated by homologous recombination [13]. The correct targeting resulted in the replacement of exons 4–6 of *TGFBI* gene in mice with a neomycin-resistant gene and was identified by Southern analysis. *TGFBI*$^{-/-}$ mouse embryo fibroblasts (MEF) still expressed *TGFBI* mRNA, but the level was about sixfold lower than in wild-type MEFs. Moreover, the deletion of exons 4–6 in *TGFBI*$^{-/-}$ MEFs was demonstrated by RT-PCR, and the absence of TGFBI protein was revealed by western blot.

To determine the tumor-suppressor function of *TGFBI* in vivo, *TGFBI*$^{-/-}$, *TGFBI*$^{+/-}$, and wild-type *TGFBI*$^{+/+}$ animals were generated from back-crossing of heterozygous *TGFBI*$^{+/-}$ mice and the development of spontaneous malignancies for up to 20 months was examined [13]. From ages of 9 to 16 months, over 20% of *TGFBI*$^{-/-}$ mice died of systemic illness, whereas all *TGFBI*$^{+/+}$ mice were alive. To determine the cause of death, the moribund *TGFBI*$^{-/-}$ mice between ages of 9 and 16 months were sacrificed and subjected to detailed histopathological analysis. Four out of 12 mice developed malignancies including one invasive lung adenocarcinoma and three lymphomas, one of which was a highly disseminated lymphoma infiltrating liver and lung tissues. By the end of 20 months, 8.3% of *TGFBI*$^{+/+}$ mice, 13.3% of heterozygotes, and 37.04% (20/54) of *TGFBI*$^{-/-}$ mice had developed tumors. The tumor incidence in heterozygotes is higher than that in wild-type mice, but did not reach statistical significance ($P>0.05$, χ^2 test). Furthermore, topical treatment of *TGFBI*$^{-/-}$ animals with the chemical carcinogen 7,12-dimethylbenz(a)anthracene (DMBA) resulted in a tenfold higher skin tumors by 6 months of age when compared with wild-type animals. This data further confirmed the tumor suppressor function of the *TGFBI* gene.

13.1.4 *TGFBI*$^{-/-}$ MEFs Exhibit an Increased Frequency of Chromosomal Aberrations

To clarify whether disruption of *TGFBI* resulted in an increased frequency of chromosomal aberrations, *TGFBI*$^{-/-}$ and wild-type MEFs at passage 2 were treated with 0.05 μg/ml colcemid for 3–6 h, and chromosomal metaphases were prepared using previously published procedures.

Various types of chromosomal aberrations in *TGFBI*$^{-/-}$ MEFs were identified. Overall, 43.75% (7/17) of metaphases prepared from *TGFBI*$^{-/-}$ MEFs contained chromatid breaks, centric fragments, or chromosomal breaks, whereas only 13.3% (2/15) metaphases from wild-type MEFs contained only centric fragments [13]. In addition, frequency of micronuclei was also examined in the early passage of MEFs (P2). Twenty-four hours post-plating, cells were fixed with acid alcohol and stained with acridine orange in dark for 10 min. A total of 3,000 cells were counted for each experiment, and three independent assays were performed. Micronuclei frequency in *TGFBI*$^{-/-}$ MEFs was found to be 4.7-fold higher than that in wild-type MEFs.

13.1.5 Nontargeted Effects of Ionizing Radiation

Radiation is a well-established human carcinogen. Based principally on the cancer incidence found in survivors of the atomic bombs in Japan, the International Commission on Radiological Protection (ICRP) and the US National Council on Radiation Protection and Measurements (NCRP) have recommended that estimates of cancer risk for low-dose exposure be extrapolated from higher doses where data are available using a linear, no-threshold model. This recommendation is based on the dogma that the DNA of the nucleus is the main target for radiation-induced genotoxicity and, since fewer cells are directly damaged at lower doses, the deleterious effects of radiation decline proportionally. However, there is increasing evidence from a number of laboratories indicating that extranuclear targets/extracellular events may also play an important role in determining the biological responses of ionizing radiation, particularly, at low doses (reviewed in 14, 15). A major conceptual shift in our understanding on the target theory of ionizing radiation in the last decade has resulted from the discovery of the bystander effect.

The radiation-induced bystander effect is defined as the induction of biological effects in the cells that are NOT directly traversed by a charged particle, but are in close proximity to cells that are. The additional responding cells that received no radiation exposure were "bystanders" of either directly hit cells or resulted from agents released from the irradiated medium. Although circumstantial evidence in support of a bystander effect appears to be consistent, direct proof of such extranuclear/extracellular effects are most convincingly demonstrated using charged particle microbeams [16, 17].

Using microbeam technology, a variety of biological endpoints in both human and other mammalian cell lines have firmly established the presence of a bystander effect under either confluent or sparsely populated culture conditions. In general, as few as one cell in a population that is targeted by a single particle has been shown to be sufficient in eliciting a bystander response [17]. Furthermore, increasing the number of particle traversals per cell or the total dose delivered to the irradiated fraction does not increase the intensity of the bystander response. Thus, there is no evidence of a dose response in any of the biological endpoints examined thus far.

The apparent lack of a dose response in bystander effects observed in many in vitro studies conducted using either a microbeam or through medium-mediated approach could have significant, broader implication in radiation risk assessment. The dose–response concept and the linear no threshold model form the cornerstone in our current management of radiation risk, particularly at the low-dose region. However, at the low-dose range, there are other radiobiological effects come into play including adaptive response, genomic instability, and genetic susceptibility. Although the role of bystander effects in human health is not clear at the moment, there is evidence that it may modulate tumor incidence after heavy ion irradiation in mice [18].

13.1.6 Nature of the Signaling Molecules and Pathways Involved in Radiation-Induced Nontargeted Response

In our quest to identify the signaling pathways involved in radiation-induced bystander effect, we first focused on the genes that are differentially expressed among the bystander vs. control cells. Since the microbeam can only irradiate one cell at a time and a large number of cells are needed for gene array analyses, we employed a novel double mylar dish approach to define the bystander response. Briefly, two concentric stainless steel rings were fitted with mylar bottoms with the outer and inner rings covered by a 6- and 38-μm-thick mylar sheet, respectively. The thicker mylar sheet of the inner ring was sliced into strips. After sterilizing with 70% ethanol and air-dried, exponentially growing normal human lung fibroblasts (NHLF) were plated in the concentric strip dishes 3 days before irradiation to ensure a confluent state. Since the fibroblasts seeded on the 38-μm-thick mylar strips would not be irradiated due to the short penetrating distance of the α particles, these cells would effectively become the bystander cells, being seeded right next to the cells plated on the 6-μm mylar dishes that were directly irradiated.

Using a signal transduction pathway specific SuperArray, we compared the differentially expressed genes among the nonirradiated control NHLF cells and the bystander cells. Among the 96 genes represented on the platform, transcription level of one gene, cyclooxygenase-2 (*COX-2*), was found to be consistently upregulated by more than threefold, while the RNA level of insulin growth factor binding protein-3 (*IGFBP3*) was found to be consistently lower by more than sevenfold in multiple analyses of multiple bystander samples [19]. Semiquantitative reverse transcription (RT)-PCR was used to confirm the expression levels of these two genes using expression level of the glyceraldehyde 3-phosphate dehydrogenase (*GAPDH*) gene as an internal control. The expression of the COX-2 protein in the nonirradiated bystander cells was further confirmed by western blotting. Addition of the COX-2 inhibitor NS-398 (50 μM) suppressed COX-2 activity in NHLF cells and finally, after 24 h, reduced the COX-2 protein level in bystander cells to a nondetectable level [19]. These results indicated that expression of COX-2 is associated with the bystander effect.

13.1.7 Effects of COX-2 Inhibitor on the Nontargeted Effect

If the *COX-2* gene is causally linked to the bystander-signaling pathways, it should be possible to modulate the bystander response using a specific inhibitor of the COX-2 enzymatic activity. Experiments were conducted to show the effect of a noncytotoxic and nonmutagenic dose of NS-398, a specific inhibitor of COX-2 activity, on bystander mutagenesis at the hypoxanthine guanine phosphoribosyltransferase (*HPRT*) locus in NHLF cells irradiated with a 0.5-Gy dose of α particles using the track segment beam [19]. NHLF cells showed a bystander mutagenic yield of $\sim 4.2 \pm 1.2$ mutants per 10^6 survivors. By contrast, in cultures co-treated with NS-398 (50 μM) that did not increase the spontaneous mutant yield by itself, the bystander mutant fraction was reduced by more than six fold to a level of $\sim 0.7 \pm 0.2$ mutants per 10^6 survivors. Although NS-398 treatment was able to reduce the *HPRT* mutant fraction in the directly irradiated population as well, the magnitude of suppression, from 9.2 ± 3.5 to 5.9 ± 2.2 mutants per 10^6 survivors was only 36%.

13.1.8 Activation of MAPK Signaling Pathways in Bystander Cells

Insulin growth factor and other cytokines activate mitogen-activated protein kinase (MAPK) signaling cascade; and activation of extracellular signal-related kinase (ERK) by phosphorylation is a critical upstream event preceding *COX-2* expression. Our previous findings based on western blot analysis demonstrated a strong upregulation of phospho-ERK levels in both α-irradiated and bystander NHLF 4 h after treatment [19]. To further confirm the activation of ERK in bystander cells, PD98059 (50 μM), a specific inhibitor of MAP kinases (MEK) – extracellular signal-regulated kinase (ERK) – was added to cell cultures immediately after irradiation for a period of 4 h. In the presence of PD98059, the phosphorylated form of ERK and its activation were suppressed in both α particle irradiated and bystander cells. If activation of the MAPK signaling cascade and ERK phosphorylation are essential in mediating the bystander effect, it should be possible to mitigate the later response by using a specific inhibitor of the MEK–ERK signaling cascade. In fact, treatment of cells with a noncytotoxic dose of PD98059 (50 μM) completely suppressed bystander toxicity observed in NHLF cultures [19].

13.1.9 NF-κB Signaling Pathway in Modulating Radiation-Induced Nontargeted Response

Exposure of mammalian cells to ionizing radiation induces two completely different "information flows" that regulate cell response: from the nucleus to the cytoplasm and from the plasma membrane receptors via the cytoplasm to the nucleus. One of the universal and well-documented effects of ionizing radiation is the production of double-stranded breaks (DSB) in genomic DNA and DSB-induced signaling that targets and activates Ataxia telangiectasia mutated (ATM) kinase in the nucleus followed the initiation of the downstream ATM-mediated signaling pathways [20]. ATM-mediated phosphorylation and stabilization of p53 is one of the critical events in directly irradiated cells, which determines the fate of a cell whether to undergo growth arrest or to cell death via the mitochondrial apoptotic pathway. A general role of ATM/ATR in the regulation of radiation-induced bystander response was described several years ago and was confirmed by the subsequent investigation [21, 22]. However, ATM-p53 signaling pathway was not directly involved in the initiation of bystander response [19, 22]; furthermore, bystander response was observed for p53-null cells [23].

There is clear evidence that alpha particle irradiation upregulates NFκB binding activity in both directly irradiated and bystander cells, while Bay 11–7082, a pharmacological inhibitor of IKK/NFκB, efficiently suppresses this up-regulation and also reduces levels below the basal amount [24]. This inhibitor of NFκB activity also efficiently downregulates COX-2 and iNOS-expression levels in both directly irradiated and bystander fibroblasts. The NF-κB-dependent gene expression of *Interleukin 1 beta (IL1B), IL6, IL8, IL33, tumor necrosis factor (TNF)*, and *PTGS2/COX2*, in concert with other NF-κB-target genes, in directly irradiated human skin fibroblasts resulted in the production of other cytokines and their receptors, as well as COX-2-dependent prostaglandin E2 (PGE2), with autocrine/paracrine functions. These signaling molecules might further activate signaling pathways and target NF-κB-dependent gene expression in both directly irradiated and nontarget cells, using the plasma membrane receptor-initiated pathways through the cytoplasm into the nucleus.

13.1.10 Relevance of the Nontargeted Response in Radiation Carcinogenesis

Using the radiosensitive *Patched-1*$^{+/-}$ (*Ptch1*$^{+/-}$) mouse model system that has a defect in radiation-induced activation of the ATR-Chk1 checkpoint signaling pathway, there is evidence that irradiated of the lower half of the animal with a 3-Gy dose of X-rays resulted in the induction of medulloblastoma in the nonirradiated brain tissues [25]. A significant increase in medulloblastoma (39%) occurred in the partial body irradiated heterozygous mice compared to the sham-treated group.

On the basis of human serum analyses, there is clear evidence that plasma clastogenic factors are present many years after radiation exposure from the Japanese atomic bomb survivors, Chernobyl liquidators, and from patients who underwent radiotherapy [26]. To provide a better estimate of the frequency distribution of second primary tumor sites in relation to previous irradiation volumes, a cohort of 115 pediatric patients who developed such cancers was studied [27]. It was estimated that ~22% of these secondarily derived tumors arose from a distance of at least 5 cm from the irradiated site and ~6% arose from a distance that was > 10 cm away, i.e., from out-of-field locations. Although these findings are suggestive, nonetheless, the data highlight the potential of second tumor development outside the treatment field and at much lower dose level.

13.2 Conclusion

In vitro oncogenic transformation assays based on immortalized human bronchial epithelial cells are useful surrogate models in dissecting molecular mechanisms associated with each of the multiple steps involved in radiation carcinogenesis. In an attempt to identify the genes involved in the neoplastic process, cDNA arrays have been used to screen differentially expressed genes between control and radiation-induced tumorigenic cells. The identification of *TGFBI* as a tumor suppressor gene in tumorigenic human bronchial epithelial cells and human lung cancer samples is further confirmed by the availability of the TGFBI knock-out mouse model. In the mean time, advances in microbeam technology have resulted in the unequivocal demonstration of an extranuclear and an extracellular target in modulating radiobiological responses. The presence of COX-2 and various inflammatory cytokines in the nontargeted cells/tissues signified the critical role of the stress response in the process. The demonstration of a radiation-induced nontargeted effect in 3D human tissues in whole organism and that the progeny of nontargeted cells shows an increase in genomic instability ([28] for review) highlights the contributions of COX-2 and the inflammatory cascade in radiation carcinogenesis.

Acknowledgments Work supported by funding from the National Institutes of Health grants CA 49062, ES 11804, NIH Resource Center Grant RR 11623 and Environmental Center grant ES 09089.

References

1. Hei TK, Piao CQ, Sutter T, Willey JC, Suzuki K (1996) Cellular and molecular alterations in human epithelial cells transformed by high LET radiation. Adv Space Res 18:137–143
2. Willey JC, Broussoud A, Sleemi A, Bennett WP, Cerutti P, Curtis CC (1991) Immortalization of human bronchial epithelial cells by papillomaviruses 16 or 18. Cancer Res 51:5370–5376
3. Piao CQ, Liu L, Zhao YL, Balajee AS, Suzuki M, Hei TK (2005) Immortalization of human small airway epithelial cells by ectopic expression of telomerase. Carcinogenesis 26:725–731
4. Hei TK, Piao CQ, Willey JC, Sutters T, Hall EJ (1994) Malignant transformation of human bronchial epithelial cells by radon-simulated alpha particles. Carcinogenesis 15:431–437
5. Zhao YL, Shao G, Piao CQ, Berenguer J, Hei TK (2004) Down regulation of BigH3 gene is involved in tumorigenesis in human bronchial epithelial cells induced by heavy ion radiation. Rad Res 162: 655–659
6. Skonier J, Bennett K, Rothwell V, Kosowski S, Plowman G, Wallace P, Edelhoff S, Disteche C, Neubauer M, Purchio AF (1994) Betaig-h3: a transforming growth factor-beta-responsive gene encoding a secreted protein that inhibits cell attachment in vitro and suppresses the growth of CHO cells in nude mice. DNA Cell Biol 13(6):571–584
7. Kim JE, Jeong HW, Nam JO, Lee BH, Choi JY, Park RW, Park JY, Kim IS (2002) Identification of motifs in the fasciclin domains of the transforming growth factor-beta-induced matrix protein betaig-h3 that interact with the alphaVbeta5 integrin. J Biol Chem 277:46159–46165
8. Calaf GM, Echiburu-Chau C, Zhao YL, Hei TK (2008) Betaig-h3 protein expression as a marker for breast cancer. Int J Mol Med 21(5):561–568
9. Yamanaka M, Kimura F, Kagata Y, Kondoh N, Asano T, Yamamoto M, Hayakawa M (2008) BigH3 is overexpressed in clear cell renal cell carcinoma. Oncol Report 19:865–874
10. Ma, C., Rong, Y., Radiloff, D.R., Datto, M.B., Centeno, B., Bao, S. Cheng, A.W., Ln F., Jiang, S. Yeatman, T.J., Wang, X.F., Extracellular matrix protein TGFBI promotes metastasis of colon cancer by enhancing cell extravasation. Genes Develop. 22: 308–321, 2008.
11. Shao G, Berenguer J, Borczuk AC, Powell CA, Hei TK, Zhao Y (2006) Epigenetic inactivation of Betaig-h3 gene in human cancer cells. Cancer Res 66:4566–4573
12. Zhao YL, Piao CQ, Hei TK (2002) Overexpression of BigH3 gene downregulates integrin $\alpha 5\beta 1$ and suppresses tumorigenicity in radiation induced tumorigenic human bronchial epithelial cells. Brit J Cancer 86:1923–1928
13. Zhang Y, Wen GY, Shao G, Wang CD, Lin CS, Bhagat G, Balajee A, Hei TK, Zhao YL (2009) BigH3 deficiency induces genomic instability and predisposes mice to spontaneous tumor development. Cancer Res 69:37–44
14. Morgan WF (2003) Non-target and delayed effects of exposure to ionizing radiation: I Radiation induced genomic instability and bystander effects in vitro. Radiat Res 159:567–580
15. Hei TK, Zhou H, Ivanov VN, Mei H, Lieberman HB, Brenner DJ, Amundson SA, Geard CR (2008) Mechanism of radiation induced bystander effects: a unifying model. J Pharm Pharmacol 60:943–950
16. Zhou H, Randers-Pehrson G, Waldren CA, Vannais D, Hall EJ, Hei TK (2000) Induction of a bystander mutagenic effect of α particles in mammalian cells. Proc Natl Acad Sci USA 97:2099–2104

17. Shao C, Stewart V, Folkard M, Michael BD, Prise KM (2003) Nitric oxide-mediated signaling in the bystander response of individually targeted glioma cells. Cancer Res 63:8437–8442
18. Cucinotta FA, Chappell LJ (2010) Non-targeted effects and the dose response for heavy ion tumor induction. Mutat Res 687:49–53
19. Zhou HN, Ivanov VN, Gillespie J, Geard CR, Amundson SA, Brenner DJ, Yu ZL, Lieberman HB, Hei TK (2005) Mechanism of radiation-induced bystander effects: role of cyclooxygenase 2 signaling pathway. Proc Natl Acad Sci USA 102: 14641–146465
20. Lee JH, Paull TT (2007) Activation and regulation of ATM kinase activity in response to DNA doble-strand breaks. Oncogene 26:7741–7748
21. Burdak-Rothkamm S, Rothkamm K, Prise KM (2008) ATM acts downstream of ATR in the DNA damage response signaling of bystander cells. Cancer Res 68:7059–7065
22. Ivanov VN, Zhou H, Ghandhi SA, Karasic TB, Yaghoubian B, Amundson SA, Hei TK (2010) Radiation-induced bystander signaling pathways in human fibroblasts: a role for interleukin-33 in the signal transmission. Cell Signal 22:1076–1087
23. Zhang Y, Zhou J, Held KD, Redmond RW, Prise KM, Liber HL (2008) Deficiencies of double-strand break repair factors and effects on mutagenesis in directly gamma-irradiated and medium-mediated bystander human lymphoblastoid cells. Radiat Res 169:197–206
24. Zhou H, Ivanov VN, Lien YC, Davidson M, Hei TK (2008) Mitochondrial function and nuclear factor-kappaB-mediated signaling in radiation-induced bystander effects. Cancer Res 68:2233–2240
25. Mancuso M, Pasquali E, Leonardi S, Tanori M, Rebessi S, Di Majo V, Pazzaglia S, Toni MP, Pimpinella M, Covelli V, Saran A (2008) Oncogenic bystander radiation effects in Patched heterozygous mouse cerebellum. Proc Natl Acad Sci USA 105:12445–12450
26. Emerit I, Levy A, Cernjavski L, Arutyunyan R, Oganesyan N, Pogosian A, Mejlumian J, Sarkisian T, Gulkandanian M, Quastel M, Goldsmith J, Riklis E, Kordysh E, Poliak S, Merklin L (1994) Transferable clastogenic activities in plasma from person exposed as salvage personnel of the Chernobyl reactor. J Cancer Res Clin Oncol 120:558–561
27. Diallo I, Haddy N, Adjadj E, Samand A, Quiniou E, Chavaudra J, Alziar I, Perret N, Guerin S, Lefkopoulos D, de Vathaire F (2009) Frequency distribution of second solid cancer locations in relation to the irradiated volume among 115 patients treated for childhood cancer. Int J Radiat Oncol Biol Phys 74:876–883
28. Hei TK, Ballas LK, Brenner DJ, Geard CR (2009) Advances in radiobiological studies using a microbeam. J Radiation Res 50:A7–A12

Histone Deacetylase Inhibitor: Antineoplastic Agent and Radiation Modulator

14

Zacharoula Konsoula, Alfredo Velena, Rachel Lee, Anatoly Dritschilo, and Mira Jung

Abstract

Inhibitors of histone deacetylases (HDACs) have emerged as a new class of anticancer agents based on their actions in cancer cell growth and cell cycle arrest, terminal differentiation, and apoptosis. Previously, we rationally designed and developed a new class of hydroxamide- and mercaptoacetamide-bearing HDAC inhibitors. A subset of these inhibitors exhibited chemo-radiation sensitizing properties in various human cancer cells. Furthermore, some HDAC inhibitors protected normal cells from radiation-induced damage and extended the survival of mice following total body exposure to lethal dose radiation. Pathological analyses revealed that intestinal and bone marrow cellularities recovered significantly from radiation-induced damage by structural compartments restoration, suggesting the mechanism of action of these HDAC inhibitors. These findings support the hypothesis that epigenetic regulation may play a crucial role in the functional recovery of normal tissues from radiation injuries.

The epigenetic regulation governs gene expression, in part, through the balance of histone acetyltransferase (HAT) and histone deacetylase (HDAC) activities. HATs acetylate lysine groups at the amino terminal tails of nuclear histones to neutralize positive charges on the histones, yielding a more open, transcriptionally active chromatin structure [1]. By contrast, HDAC family members deacetylate lysine residues on histones and induce transcription repression through chromatin condensation. Increased acetylation of histones and nonhistone proteins leads to changes in the chromatin architecture and accessibility for key cellular proteins to specific target sites [2, 3]. As such, a shift in the relative activities of

M. Jung (✉)
Department of Radiation Medicine,
The Lombardi Comprehensive Cancer Center,
Georgetown University Medical Center,
3970 Reservoir Road, N.W, Box 571482,
Washington, DC 20057-1482, USA
e-mail: jungm@georgetown.edu

J.S. Rhim and R. Kremer (eds.), *Human Cell Transformation: Role of Stem Cells and the Microenvironment*, Advances in Experimental Medicine and Biology 720,
DOI 10.1007/978-1-4614-0254-1_14, © Springer Science+Business Media, LLC 2011

these enzymes profoundly influences biological processes, including DNA repair, replication, cell cycle checkpoint activation, and cellular differentiation [4].

14.1 Histone Deacetylase Family Members

HDACs are initially identified as components of large multiprotein complexes that bind to promoters and repress transcription, and are found in the nuclear and cytoplasmic compartments [5, 6]. Eleven human HDACs have been identified and divided into four classes based on structure, sequence homology, and domain organization. Class I consists of HDACs 1, 2, 3, and 8, which are predominantly nuclear and play roles in cell proliferation [5]. Class II HDACs are further subdivided into IIa (HDACs 4, 5, 7, and 9) and IIb (HDAC6 and 10) [7]. These enzymes are characterized by a large NH2-terminal domain or a second catalytic site and their expression is more restricted, suggesting roles in cellular differentiation and development [8]. Class III enzymes include the SIRTs (sirtuins), and are NAD-dependent deacetylases [9], which are not inhibited by TSA or other hydroxamates. HDAC 11 is characterized as class IV, based on a phylogenetic analysis, and whose function is least known [10].

14.2 Structural Classifications of HDAC Inhibitors (HDACIs)

Inhibitors of HDACs have been discovered and developed as promising anticancer drugs [11]. On the basis of structural divergence, HDAC inhibitors are classified: (1) short-chain fatty acids (i.e., sodium phenyl butyrate, valproic acid, AN-9), (2) hydroxamic acids (i.e., suberoylanilide hydroxamic acid (SAHA; vorinostat), oxamflatin, trichostatin A (TSA), *m*-carboxycinnamic acid bis-hydroxamide (CBHA), LBH-589, LAQ-824, PCI-24781), (3) cyclic peptides (i.e., FR901228 (depsipeptide; romidepsin), apicidin, cyclic hydroxamic acid-containing peptides (CHAPS), trapoxin), (4) benzamides (i.e., MS-275, CI-994), (5) ketones (i.e., trifluoromethyl ketone), and (6) miscellaneous (i.e., MGCD-0103). A subset of HDACIs is reportedly as pan-HDAC inhibitors, such as TSA and SAHA, while MS-275 and depsipeptide confer isomer or/and class-specificities. Some of which are currently in clinical trials [12], and two drugs, vorinostat (Zolinza®) and romidepsin (Istodax®), have recently been approved by FDA for the treatment of cutaneous T-cell lymphoma (CTCL) in patients ([13], www.hhs.gov).

14.3 Development of Second-Generation HDACIs

Despite a generally acceptable toxicity profile for HDACIs, the lack of target specificity and the presence of the toxicophore hydroxamide group may still cause unpredictable toxicity problems. To improve physicochemical and pharmacological properties of first-generation HDACIs, we rationally designed and developed several new classes of HDACIs, including ligands bearing either a hydoxamate or mercaptoacetamide group [14, 15]. A subset of these candidates exhibited 50% HDAC inhibition activity within nanomolar ranges in vitro and radiosensitizing effects on various cancer cells [14, 15]. The candidate inhibitors potently repressed the growth of cancer cells: the GI_{50}s for the hydroxamates ranged from 0.1 to >60 μM, while the mercaptoacetamides revealed weaker growth inhibitory activity against cancer cell lines. Table 14.1 presents the data demonstrating that the hydroxamates conferred potent cytotoxicity, while the mercaptoacetamides revealed less against all three cancer cell lines. However, normal human fibroblasts (NHP-5) and human primary skin fibroblasts (Hs-68) were found to be significantly resistant to these HDACIs, supporting the notion that candidate HDAC inhibitors preferentially suppress cancer cell proliferation.

Table 14.1 Effects of hydroxamate (H6CAHA) and mercaptoacetamide (6MAQH) HDAC inhibitors on pan-HDACs and the growth of cancer and normal cells

		Cell proliferation (GI_{50} μM)				
	50% pan-HDAC	Cancer cells			Normal cells	
Compound	activity inhibition (nM)	PC-3	SQ-20B	MCF-7	NHP-5	Hs-68
H6CAHA	176	1.77	2.88	3.98	>300	>500
6MAQH	44	40	30	23	>500	>500

Note: Cell lines were obtained from the Tissue Culture Shared Resources of the Lombardi Comprehensive Cancer Center, Georgetown University Medical Center. PC-3 is a prostate cancer cell line; SQ-20B is a head and neck squamous carcinoma cell line; MCF-7 is a breast cancer cell line; NHP-5 and Hs-68 are nonmalignant human epithelial cell lines. Cell proliferation was measured by the 3-(4,5-dimethylthiazol-2-yl)-2,5-diphenyltetrazolium bromide (MTT) assay after a 48-h incubation period, as previously described [16]. Growth inhibitory concentrations at 50% (GI_{50}) were calculated graphically and the mean GI_{50} values were obtained. Results are the mean of at least three determinations. Statistical comparisons between treatment groups were made using GraphPad Prism 4.0 software

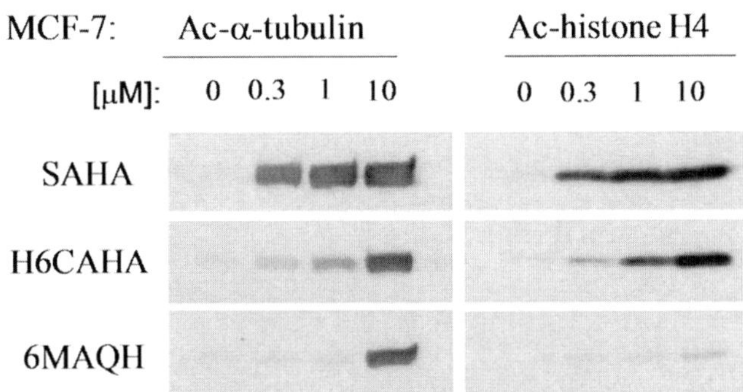

Fig. 14.1 Dose-dependent and differential effects of hydroxamate H6CAHA and mercaptoacetamide 6MAQH on histone H4 and α-tubulin acetylation levels in MCF7 cells were analyzed after 4-h treatment with indicated concentrations of HDAC inhibitors. Cells lysates were prepared in a lysis buffer containing 20-mM Tris–HCl (pH 7.4), 2-mM EDTA (pH 7.4), 2-mM EGTA (pH 7.4), 6-mM mercaptoethanol, 10 μg/ml leupeptin, 2 μg/ml aprotinin, and 1% NP40. The samples were subjected to SDS-polyacrylamide gel electrophoresis. The indicated proteins were probed with a polyclonal rabbit antibody against acetylated histone H4 or α-tubulin (Upstate, Inc). The immune-reactive bands were detected by enhanced chemiluminescence (Amersham Biosciences)

14.4 Cellular Target Specificity of HDACIs

Given that 11 HDAC isomers are present in mammalian cells, whether isomer-specific HDACIs are better drugs has been an outstanding question and is actively in ongoing investigation. In general, the determination of acetylated histone and α-tubulin levels has been used as a measure of HDACI efficacy in vitro and in vivo. Pan-HDACIs, such as SAHA and TSA, appear to target both histones H3/H4 and α-tubulin evidenced by their enhanced acetylation levels, in a dose-dependent manner. In contrast, other HDACIs, such as trapoxin B, sodium butyrate, and FK-228, do not exhibit these characteristics, suggesting their selectivity at the cellular level [17]. Others have reported that depsipeptide exerts modulation of p53, ErbB1, ErbB2, and Raf-1 expression in lung cancer cells [18]. Valproic acid, an anti-epileptic agent, has shown to be involved in the proteolysis of HDAC2 [19]. Hyperacetylations of α-tubulin and Hsp90 have also been tightly linked to class II HDAC6 inhibition. Our data, as shown in Fig. 14.1, demonstrate that mercaptoacetamide-based HDACI 6MAQH preferentially enhances acetylation of α-tubulin while hydroxamate HDACIs, SAHA and H6CAHA, enhance the acetylation levels of both histone H4 and α-tubulin in a dose-dependent manner, suggesting 6MAQH as a class II HDAC isomer-specific inhibitor. Therefore, understanding of the roles of various HDAC isoforms in cell processes and identifying HDACI-specific biomarkers in diseases may contribute to strategies for the identification of promising second-generation inhibitors that exhibit target specificity and that offer clinical relevance.

14.5 HDACs, DNA Damage Repair, and Radiation Sensitivity

Accumulated studies have shown that a particular subset of HDACIs induces radiosensitivity of various human cancer cells [14, 20–24]. Consistent with these findings, several recent reports have provided supporting evidence of specific HDAC isoforms' roles in cell cycle regulation and intrinsic radiation sensitivity [25–29].

A significant line of evidence has demonstrated that a subset of HDAC family members plays a critical role in DNA damage response and repair [27, 30–32] and that post-translational modifications (e.g., acetylation) of histones or nonhistone proteins flanking DSBs affect accessibility of repair and signaling proteins to the damaged regions of DNA [33, 34]. Moreover, changes in the chromatin architecture by HDACI have shown to enhance the activation of nonhistone proteins, including p53, pRB, and transcriptional factors that are involved in cell cycle regulation and affect cell proliferation and apoptosis, processes that are frequently aberrant in cancers [11, 35]. With their intrinsic anticancer properties, a subset of HDAC inhibitors has exhibited radiation sensitizing properties in certain cancer cell lines and in preclinical models [20–24, 36, 37]. However, underlying mechanisms are not fully understood.

Previously, we have shown that an HDAC inhibitor, trichostatin A (TSA), increased the amount of radiation-induced DNA damage and slowed the repair kinetics [31]. Gene expression profiling also revealed that a majority of genes that control cell cycle, DNA replication, and damage repair processes were down-regulated following TSA exposure. HDACI treatment modulated the radiation-induced DNA damage repair process, in part, by suppressing *BRCA1* gene expression [31]. Further studies have demonstrated prolonged appearance of γH2AX and Rad51 foci and suppression of DNA damage repair genes (*ATM, Brca1, and Brca2*). Functions of damage sensing and repair proteins, including ATM, MRE11, γ-H2AX, and 53BP1, are also affected by changes in chromatin structure conformation [38–40].

A sensor of DNA damage, ATM has been shown to interact with HDAC1, to play a role with HDAC4 in the DNA damage response pathway, and to be associated with ionizing radiation-induced changes in chromatin structure [25, 28]. Inactivation of HDAC3 affects S phase progression and enhances DNA damage [26]. HDAC4 co-localizes with 53BP1 (a protein which participates in the phosphorylation of p53 and Chk2 and in the maintenance of S and G_2 cell cycle checkpoints) in DNA damage-induced nuclear foci [25]. Finally, other studies have suggested that butyrates, TSA, and inhibitory HDAC RNAs are capable of sensitizing cancer cells to ionizing radiation [27, 30–32, 41].

14.6 HDACI and Radioprotection In Vitro

One of the hallmarks of HDACI action is to preferentially sensitize actively proliferating cancer cells [42–45]. Normal cells are considerably more resistant to HDAC inhibitors than are tumor cells [45]. To determine underlying mechanisms of biological discrepancy between cancer and normal cells, we examined the synergistic effect of HDACI on DNA damage response by scoring phosphorylated H2AX (γH2AX) foci, one of the first events occurring at DNA double-strand breaks (DSB). The data demonstrated that the foci formation in cancer cells (PC3, LNCaP, and DU145) was markedly increased within 0.5 h and gradually decreased by 3 h after irradiation (5 Gy) (Fig. 14.2). HDAC inhibitor treatment alone

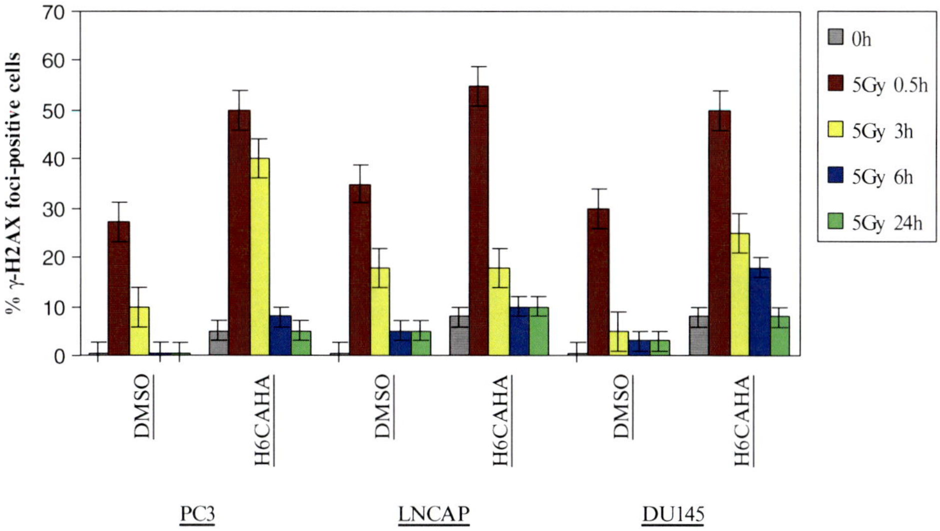

Fig. 14.2 Effect of H6CAHA on radiation-induced γH2AX foci; PC3, LNCaP, and DU145 cells were incubated with vehicle or 1-μM H6CAHA for 16 h and then exposed to 5 Gy of γ-radiation. The 0-h samples were not exposed to radiation. γH2AX foci were evaluated in 75 nuclei per treatment per experiment. Columns represent mean values from three independent experiments; *bars* represent SD. After staining, foci were visualized by Olympus FV300 equipped with FluoView software. Statistical comparisons between treatment groups were made using GraphPad Prism 4.0 software by Student's *t*-test. A probability value of $p<0.05$ was considered to be significant

resulted in the accumulation of γH2AX foci formation. However, after pretreatment with H6CAHA for 16 h followed by irradiation (5 Gy), γH2AX foci/cell growth increased significantly and sustained these levels up to 6 h and persisted for at least 24 h. This trend was not observed in normal cells (data not shown). These results indicate that disruption of the radiation-induced DSB damage repair mechanism may account for H6CAHA-induced radiosensitization of cancer cells and also indicate that H6CAHA may have the potential to facilitate DNA damage repair in normal cells.

14.7 HDACI and Radioprotection In Vivo

Given that the ability to selectively kill cancer cells while limiting damage of normal cells is the ultimate therapeutic goal and that HDACIs (phenylbutyrate, trichostatin A, and valproic acid) could suppress cutaneous radiation syndrome [42, 43], understanding the mechanisms of cancer cell radiation sensitization and normal cell protection by HDACIs is important.

To address how the candidate HDACIs affect normal cells in response to ionizing radiation in vivo, C57BL/6 mice were treated with an i.p. subtoxic dose (4 mg/kg) of either hydoxamate H6CAHA or mercaptoacetamide 6MAQH, 4 h prior to total-body irradiated (TBI) with 15 Gy of gamma radiation. The survival rate and body weight were monitored daily. The data demonstrated that 100% of animals undergoing TBI died within 6–7 days. The loss of body weight was evident, possibly indicating intestinal damage. However, HDACI treatment extended the survival of animals; 40% of animals treated with hydroxamide H6CAHA or mercaptoacetamide 6MAQH survived until the end of the study at 25 days. The data also showed that the duration of survival of animals pretreated with hydoxamate H6CAHA (median survival: 14 days) extended further than those treated with mercaptoacetamide 6MAQH (median survival: 9 days) (Fig. 14.3). These animals eventually gained

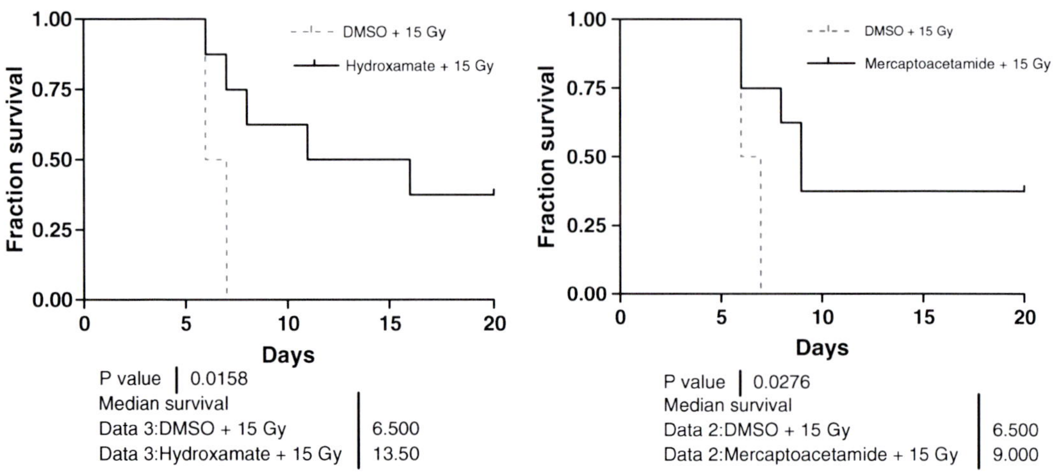

Fig. 14.3 The effects of each potential HDAC inhibitor, hydoxamate (H6CAHA) and mercaptoacetamide (6MAQH), on survival of mice undergoing total body irradiation of 15 Gy gamma. Each cohort (10 mice per group) received i.p. administered HDACI (4 mg/kg body weight) 4 h prior to irradiation. Survivals of animals were monitored and the body weights were measured daily for 25 days until the end of experiments. C57BL/6 mice (18–22 g) were treated with an intraperitoneal dose (4 mg/kg) of hydoxamate or mercaptoacetamide, 4 h prior to total-body irradiation (TBI). The animal protocol was reviewed and approved by the Georgetown University Animal Welfare Committee. The body weights of control and treated mice were measured at weekly intervals. Total body irradiation (TBI) was performed using a Mark-30 irradiator with a ^{137}Cs source at a fixed dose rate of 2.71 Gy/min. The Kaplan-Meier method was used to analyze survival data. All statistical analyses were performed on relative survival data (percentage change from initial survival). Student's *t*-tests were performed to determine if the differences between the treatments were significant. $p<0.05$ was considered statistically significant

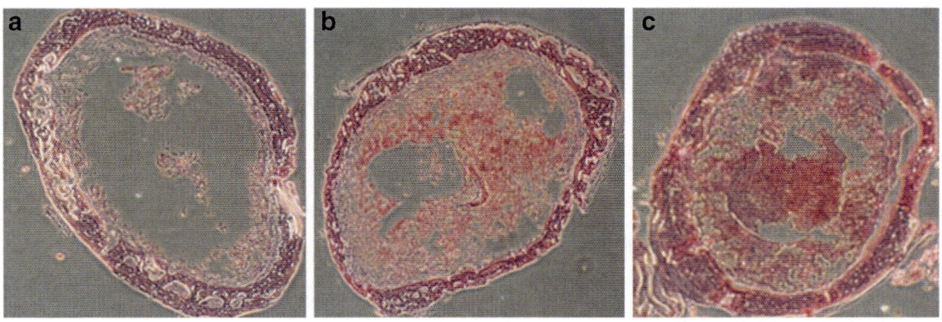

Fig. 14.4 Mouse bone marrow (BM) 2 days after exposure to 15 Gy of Cs-137 total body irradiation (IR). Femur cross sections: (**a**) DMSO (control); (**b**) 4 mg/kg of mercaptoacetamide 6MAQH; (**c**) 4 mg/kg of hydroxamate H6CAHA

weight and continued to survive until termination of the experiments.

To understand underlying mechanisms, histological analyses were performed using various tissues, including the pelvic bone marrow (BM) and the intestine. The data demonstrated that HDACI pretreatment resulted in less BM cell loss within 2 days compared with treatment with IR alone (Fig. 14.4). Structural recovery accompanied by an increase in small intestine crypts with villus height was also interpreted as an indicator of radioprotection (Fig. 14.5).

Although the underlying mechanisms responsible for survival remains unclear, these results

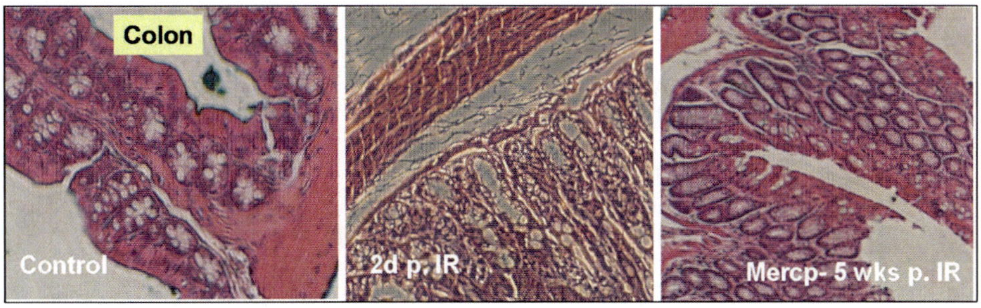

Fig. 14.5 Mouse intestine (colon) after exposure to 15 Gy of Cs-137 total body irradiation (IR). *Left*: DMSO control; *Middle*: 2 days post-IR; *Right*: merceptoacetamide 6MAQH, 5 weeks post-IR

suggest that the candidate HDACIs protect mice from BM and intestinal injuries after total-body irradiation (TBI), supporting the clinical potential of HDACI in protecting the damage of normal cells/tissues during radiation treatment. It is attempted to speculate that candidate HDAC inhibitors may trigger the activation of BM progenitors and promote their infiltration into the damaged sites and that suppression of cytokines by candidate compounds may contribute to reduced inflammation and/or cell death rates, thereby facilitating recovery from radiation-induced injuries.

14.8 Conclusion

Radiation-induced acute and late injuries often cause dysfunction of organs and generate secondary effects on functional impairment [46, 47]. Limitations and efficacy of radiation therapy in cancer treatment are contingent upon tolerances of normal tissues encompassed in the irradiated fields. Our novel and structurally distinct HDACIs, hydroxamide H6CAHA and merceptoacetamide 6MAQH, exhibited modulating effects on radiation responses by which preferentially killing cancer cells in vitro and in vivo, and enhancing the survival rate of mice following lethal doses of radiation. The data suggest that significant recovery of intestinal and BM cellularities and restoring structural compartments from radiation-induced damage is a mechanism of action of these HDAC inhibitors. Therefore, the use of HDAC inhibitors for radiosensitizing cancer cells and radioprotection of normal tissues identifies an entirely new category of drugs with clinical translation potential. Further studies are warranted for its precise evaluation in which these candidate compounds can be advanced as a treatment agent for normal cell protection from radiation-induced injury.

Acknowledgments We thank S. Lee and J. Tuturea for technical support. This work was supported in part by USMRC grants PC030471 (M. Jung) as well as the Lombardi Comprehensive Cancer Center Microscopy and Imaging Shared Resource, US Public Health Service Grant 2P30-CA-51008 and 1S10 RR15768-01.

References

1. Struhl K, Moqtaderi Z (1998) The TAFs in the HAT. Cell 94:1–4
2. Grunstein M (1997) Histone acetylation in chromatin structure and transcription. Nature 389:349–352
3. Katan-Khaykovich Y, Struhl K (2002) Dynamics of global histone acetylation and deacetylation in vivo: rapid restoration of normal histone acetylation status upon removal of activators and repressors. Genes Dev 16:743–752
4. Kouzarides T (1999) Histone acetylases and deacetylases in cell proliferation. Curr Opinion Genet Dev 9:40–48
5. De Ruijter AJ, Van Gennip AH, Caron HN, Kemp S, Van Kuilenburg AB (2003) Histone deacetylases (HDACs): characterization of the classical HDAC family. Biochem J 370:737–749
6. Kurdistani SK, Grunstein M (2003) Histone acetylation and deacetylation in yeast. Nat Rev Mol Cell Biol 4:276–284

7. Fischle W, Kiermer V, Dequiedt F, Verdin E (2001) The emerging role of class II histone deacetylases. Biochem Cell Biol 79:337–348
8. Verdin E, Dequiedt F, Kasler H (2003) Class II histone deacetylases: versatile regulators. Trends Genet 5:286–293
9. Michan S, Sinclair D (2007) Sirtulins in mammals: insight into their biological function. Biochem J 404:1–13
10. Gregoretti IV, Lee YM, Goodson HV (2004) Molecular evolution of the histone deacetylase family: functional implications of phylogenetic analysis. JMC 338:17–31
11. Marks P, Rifkind RA, Richon VM, Breslow R, Miller T, Kelly WK (2001) Histone deacetylases and cancer: causes and therapies. Nat Rev Cancer 1:194–202
12. Lane AA, Chabner BA (2009) Histone deacetylase inhibitors in cancer therapy. J Clin Oncol 27: 5459–5468
13. Mann BS, Johnson JR, Cohen MH, Justice R, Pazdur R (2007) FDA approval summary: vorinostat for treatment of advanced primary cutaneous T cell lymphoma. Oncologist 12:1247–1252
14. Jung M, Velena A, Chen B, Petukhov PA, Kozikowski AP, Dritschilo A (2005) Novel HDAC inhibitors with radiosensitizing properties. Radiat Res 163:488–493
15. Chen B, Petukhov PA, Jung M, Velena A, Eliseeva E, Dritschilo A, Kozikowski AP (2005) Chemistry and biology of mercaptoacetamides as novel histone deacetylase inhibitors. Bioorg Med Chem Let 15: 1389–1392
16. Konsoula Z, Cao H, Velena A, Jung M (2009) Pharmacokinetics-pharmacodynamics and antitumor activity of mercaptoacetamide-based histone deacetylase inhibitors. Mol Cancer Ther 8:2844–2851
17. Jung M, Yong KJ, Velena A, Lee SA (2009) Epigenetic Targets in drug discovery: cell-based assays for HDAC inhibitor hit validation. The Wiley-VCH series "Methods and Principles in Medicinal Chemistry" 42:119–137
18. Yu X, Guo ZS, Marcu MG et al (2002) Modulation of p53, ErbB1, ErbB2, and Raf-1 expression in lung cancer cells by depsipeptide FR901228. J Natl Cancer Inst 94:504–513
19. Kramer OH, Zhu P, Ostendorff HP et al (2003) the histone deacetylase inhibitor valproic acid selectively induces proteasomal degradation of HDA2. EMBO J 22:3411–3420
20. Biade S, Stobbe CC, Boyd JT, Chapman JD (2001) Chemical agents that promote chromatin compaction radiosensitize tumor cells. Int J Radiat Biol 77: 1033–1042
21. Karagiannis TC, El-Osta A (2006) Modulation of cellular radiation responses by histone deacetylase inhibitors. Oncogene 25:3885–3893
22. Arundel CM, Glicksman AS, Leith JT (1985) Enhancement of radiation injury in human colon tumor cells by the maturational agent Sodium Butyrate (NaB). Radiation Res 104:443–448
23. Kim JH, Shin JH, Kim IH (2004) Susceptibility and radiosensitization of human glioblastoma cells to trichostatin A, a histone deacetylase inhibitor. Int J Radiat Oncol Biol Phys 59:1174–1180
24. Zhang Y, Jung M, Dritschilo A, Jung M (2004) Enhancement of radiation sensitivity of human squamous carcinoma cells by histone deacetylase inhibitors. Radiat Res 161:667–674
25. Kao GD, McKenna WG, Guenther MG, Muschel RJ, Lazar MA, Yen TJ (2003) Histone deacetylase 4 interacts with 53BP1 to mediate the DNA damage response. JCB 160:1017–1027
26. Bhaskara S, Chyla BJ, Amann JM, Knutson SK et al (2008) Deletion of histone deacetylase 3 reveals critical roles in S phase progression and DNA damage control. Mol Cell 30:61–72
27. Miller KM, Tjeertes JV, Coates J, Legube G et al (2010) Human HDAC1 and HDAC2 function in the DNA-damage response to promote DNA nonhomologous end-joining. Nat Struct Mol Biol 17: 1144–1152
28. Kim GD, Choi YH, Dimtchev A, Jeong SJ, Dritschilo A, Jung M (1999) Sensing of ionizing radiation-induced DNA damage by ATM through interaction with histone deacetylase. J Biol Chem 274:31127–31130
29. Ju R, Muller MT (2003) Histone deacetylase inhibitors activate p21 (WAF1) expression via ATM. Cancer Res 63:2891–2897
30. Peterson CL, Cote J (2004) Cellular machineries for chromosomal DNA repair. Genes Dev 18:602–616
31. Zhang Y, Zhang Y, Carr T, Dimtchev A, Zaer N, Dritschilo A, Jung M (2007) Attenuated DNA damage repair by trichostatin A through BRCA1 suppression. Radiat Res 168:115–124
32. Goodarzi AA, Noon AT, Deckbar D, Ziv Y, Shiloh Y, Lo brich M, Jeggo PA (2008) ATM signaling facilitates repair of DNA double-strand breaks associated with heterochromatin. Mol Cell 31:167–177
33. Groth A, Rocha W, Verreault A, Almouzni G (2007) Chromatin challenges during DNA replication and repair. Cell 128:721–733
34. Kouzarides T (2007) Chromatin modifcations and their function. Cell 128:693–705
35. Johnstone RW (2002) Histone-deacetylase inhibitors: novel drugs for the treatment of cancer. Nat Rev Drug Discov 1:287–299
36. Camphausen K, Scott T, Sproull M, Tofilon PJ (2004) Enhancement of xenograft tumor radiosensitivity by the histone deacetylase inhibitor MS-275 and correlation with histone hyperacetylation. Clin Cancer Res 10:6066–6071
37. Folkvord S, Ree AH, Furre T et al (2009) Radiosensitization by SAHA in experimental colorectal carcinoma models-*in vivo* effects and relevance of histone acetylation status. Int J Radiat Oncol Biol Phys 74:546–552
38. Lavin MF (2008) Ataxia-telangiectasia: from a rare disorder to a paradigm for cell signalling and cancer. Nat Rev Mol Cell Biol 9:759–769

39. Keogh MC, Kim JA, Downey M, Fillingham J et al (2006) A phosphase complex that dephosphorylates gH2AX regulates DNA damage checkpoint recovery. Nature 439:497–501
40. Celeste A, Fernandez-Capetillo O, Kruhlak MJ, Pilch DR, Staudt DW, Lee A, Bonner RF, Bonner WM, Nusssenzwig A (2003) Histone H2Ax phoshporylation is dispensable for the initial recognition of DNA breaks. Nat Cell Biol 5:675–679
41. Glaser KB, Li J, Staver MJ, Wei RQ, Albert DH, Davidsen SK (2003) Role of class I and class II histone deacetylases in carcinoma cells using siRNA. Biochem Biophys Res Commun 310: 529–536
42. Chung YL, Wang AJ, Yao LF (2004) Antitumor histone deacetylase inhibitos suppress cutaneous radiation syndrome: implications for increasing therapeutic gain in cancer radiotherapy. Mol Cancer Ther 3: 317–325
43. Paoluzzi L, Figg WD (2004) Histone deacetylase inhibitors are potent radiation protectants. Cancer Biol Ther 3:612–613
44. Papeleu P, Vanhaecke T, Elaut G et al (2005) Differential effects of histone deacetylase inhibitors in tumor and normal cells-what is the toxicological relevance? Crit Rev Toxicol 35:363–378
45. Ungerstedt JS, Sowa Y, Xu WS et al (2005) Role of thioredoxin in the response of normal and transformed cells to histone deacetylase inhibitors. Proc Natl Acad Sci USA 102:673–678
46. Chang BK, Timmerman RB (2007) Stereotactic body radiation therapy: a comprehensive review. Am J Clin Onc 30:637–644
47. Timmerman R, McGarry R, Yiannoutsos C, Papiez L et al (2006) Excessive toxicity when treating central tumors in a Phase II study of stereotactic body radiation therapy for medically inoperable early-stage lung cancer. J Clin Onc 24:4833–4839

Human Fibroblasts for Large-Scale "Omics" Investigations of ATM Gene Function

15

Mira Jung, Olga Timofeeva, Amrita K. Cheema, Rency Varghese, Habtom Ressom, and Anatoly Dritschilo

Abstract

ATM (gene mutated in ataxia-telangiectasia) is a critical central component of the pleiotropic responses of cells to ionizing radiation-induced stress. To gain insight into molecular mechanisms and to enhance our understanding of ATM functions, we have advanced a human model cell system, derived from genetically defined immortal fibroblasts, and we have applied high-throughput genomic, proteomic and metabolomic technologies for a systems level analysis. The cellular characterizations reported here provide the background for application of a systems analysis to integrate transcription, post-translational modifications and metabolic activity induced by exposure of cells to ionizing radiation. We present here a summary of the derivation and characterization of cells comprising this model cell system and review applications of this model to systems analysis of ATM functions.

15.1 Introduction

Investigations of the human genetic disease, ataxia-telangiectasia (AT) have led to the discovery of ATM, the gene mutated in AT, as the central factor leading to a complex clinical syndrome marked by progressive neurological degeneration, as well as immunological deficiency and extreme sensitivity to ionizing radiation [1–3]. Established primary cells, lymphoid and fibroblasts derived from AT patients are available; however, growing such cells has proven difficult and most mechanistic studies utilize virus immortalized cell lines. We have used one such cell line, AT5BIVA to develop a model human fibroblast system for investigating the role of ATM in regulating gene expression, protein expression and post-translational modification, as well as metabolite generation. Here we characterize the cells and demonstrate feasibility for high-throughput analysis to globally define ATM mediated cellular

M. Jung (✉)
Department of Radiation Medicine,
Lombardi Comprehensive Cancer Center,
Georgetown University Medical Center,
3900 Reservoir Road NW, Washington, DC 20057, USA
e-mail: jungm@georgetown.edu

J.S. Rhim and R. Kremer (eds.), *Human Cell Transformation: Role of Stem Cells and the Microenvironment*, Advances in Experimental Medicine and Biology 720,
DOI 10.1007/978-1-4614-0254-1_15,

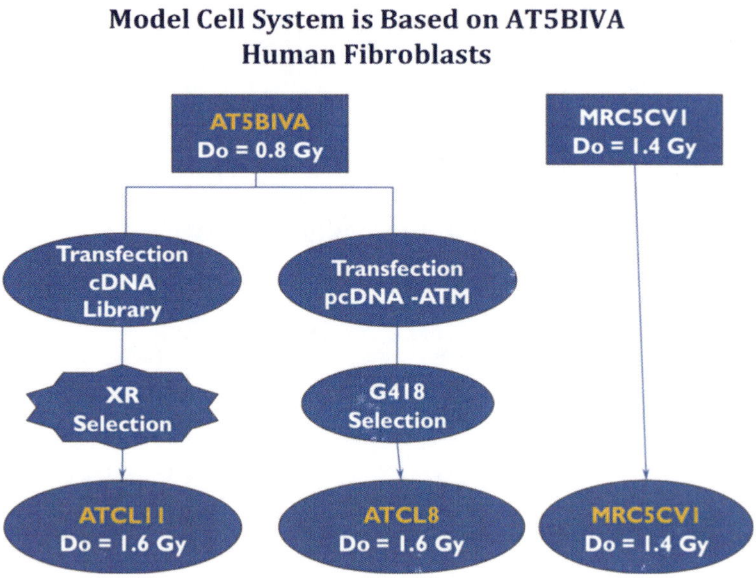

Fig. 15.1 Schematic diagram of cell model system

responses in the genetically defined model cell system. Bioinformatic integration of the genomic, proteomic and metabolomic analyses using commercially available software permits a systems view of cellular responses to radiation stress.

Although the clinical syndrome of AT is multifaceted, the disease is attributed to mutation in the single gene, ATM [4]. ATM spans more than 150 kb, consisting of 66 exons and transcribing a 13-kb transcript. The 3,056 amino acid gene product belong to the PI-3 kinase family of proteins and functions by phosphorylating and activating key molecules involved in cell cycle regulation, DNA repair, immune response, transcriptional regulation and genomic stability [4–6].

The activation of ATM in response to DNA damage results in phosphorylation of proteins involved in critical cellular processes, including cell cycle regulation and DNA repair. The phosphorylation cascade ultimately leads to transcriptional activation, and siRNA silencing of ATM has shown a significant impact on the transcriptional profile in the cell [7]. To our knowledge, there has been no comprehensive analysis of global gene expression changes in human cells in which ATM function has been restored. Therefore, our initial aim was to establish model cells suitable for investigating ATM-dependent and ATM-independent response to ionizing radiation exposure.

15.1.1 Establishment of the (ATM ±) Model Cell System

To establish a model cell system for gene expression analysis we selected AT human fibroblasts (AT5BIVA) with a known mutation in ATM, which leads to a truncated gene product. Introduction of the full-length *ATM* in a pcDNA3 expression vector resulted in a clonal cell line (ATCL8) with corrected radiation phenotype. A second important cell line was established following gene transfer and radiation selection experiments [8]. Cell line ATCL11 was found to have normal radiation response parameters in a background of mutant ATM. These cells have been previously reported and represent ATM-independent enhancement of cellular responses to radiation exposure attributed to the introduction of a mutated IκB-α, modifying cellular NF-κB regulation [8]. Figure 15.1 provides an overall schema of cell line derivation.

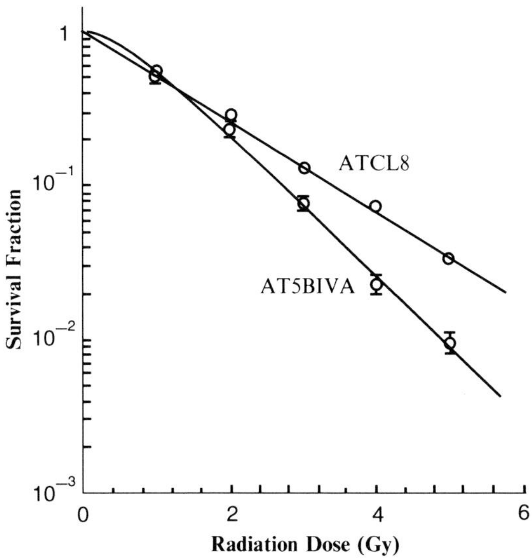

Fig. 15.2 Radiation clonogenic survival of AT cells. Logarithmically growing cells were exposed to graded doses of γ-radiation. Clonogenic survivors were measured and fit to the single-hit, multitarget and linear quadratic models as shown in a semi-logarithmic plot of the data. *Points* and *bars* represent means of SEMs from triplicate flasks

15.1.2 Characterization of Fibroblast Cell Lines

The radiation responses shown in Fig. 15.2 illustrate the survival of AT5BIVA cells to graded doses of γ-radiation exposure. Parameters derived from the single hit, multitarget model of cellular radiation survival, D_0 determinations provide a convenient assessment of radiation sensitivity as illustrated in Table 15.1 [9]. The smaller the value of D_0, the more sensitive are the cells to radiation killing. In our laboratories, cells with normal responses to ionizing radiation demonstrate D_0 values ranging from 1.2 to 1.7 Gy. Together, these several cell lines provide a unique, genetically defined model of the extremely sensitive fibroblasts of AT, AT5BIVA and the pcDNA transfected cells, as well as the radiation sensitivity complemented ATCL8 derived by the introduction of the "wild type" ATM. The cell line, ATCL11, contains a truncated IκB-α, a dominant negative factor which complements the radiation responses of AT5BIVA cells in an ATM-independent manner [8]. The MRC5CV1 cells are immortal human fibroblasts with "normal" radiation sensitivity ($D_0 \cong 1.4$ Gy) but in a different genetic background, serving as additional controls in radiation studies.

Table 15.1 Radiobiological parameters of model human fibroblasts

Cell line	Radiation category	D_0 (Gy)	$\tilde{N}$
AT derived cells and controls			
AT5BIVA	ERS	0.8	2.7
AT5BIVA/pcDNA	ERS	0.8	3.5
ATCL8	RS	1.6	1.2
ATCL11	RS	1.6	1.0
MRC5CV1	RS	1.4	5.2

ERS extreme radiation sensitivity; *RS* normal level of radiation sensitivity

15.1.3 Gene Expression Profiling in Human AT Fibroblasts

ATM has been implicated as a primary DNA damage sensing molecule in the cell [9]. To assess the effect of ATM on transcriptional regulation, we investigated gene expression patterns of the several AT5BIVA derived cells. A line graph of microarray analyses in Fig. 15.3 compares basal gene expression levels of cells in exponential growth showing the impact of ATM gene product, resulting in enhanced and suppressed gene "outliers." To assure reproducibility and quality of the data, experiments were performed in triplicate and samples were split prior to cRNA library preparation. This resulted in the analysis of six microarray chips per experimental point. Multidimensional scaling and gene-tree analysis of these samples from the genetically defined cell lines confirmed distinct separation by cell line, as reported elsewhere [10].

The expression differences demonstrated by microarray data were validated by quantitative Real-Time PCR (qRT-PCR) assays (Table 15.2). All samples were normalized to GAPDH controls. Overall, expression trends were remarkably consistent with data obtained by array analyses, albeit the more sensitive qRT-PCR generally showed higher expression levels.

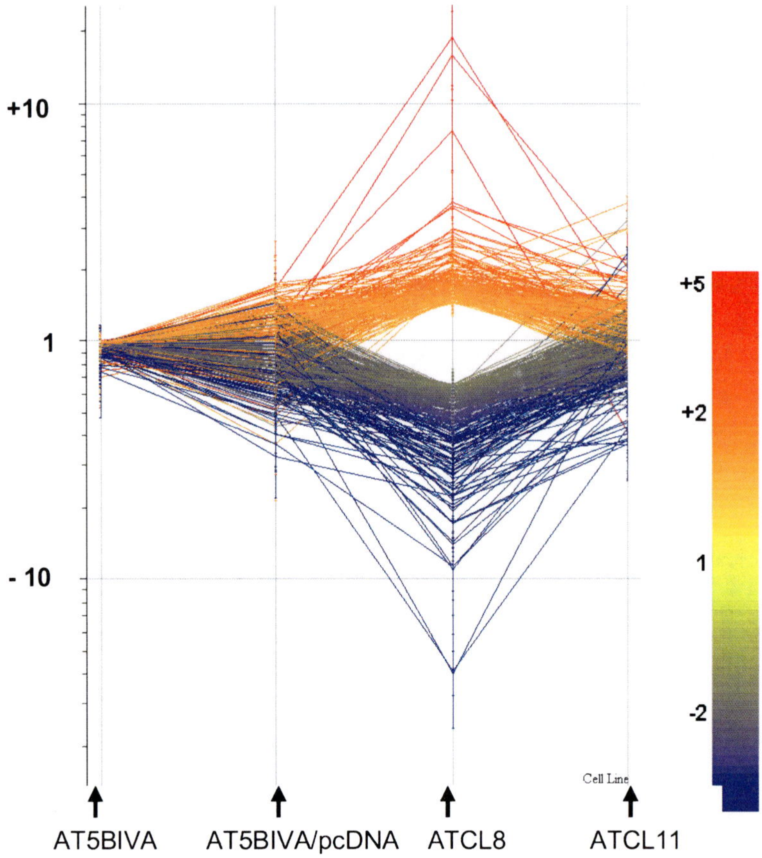

Fig. 15.3 *Line graph* of differential gene expression comparing AT5BIVA, vector control, ATCL8 and ATCL11 cells

Table 15.2 qRT-PCR validation of microarray determined expression differences comparing ATCL8 to AT5BIVA

Mean fold change (SEM)			
Accession number	Description	PCR	Microarray
MCL1	Myeloid cell leukemia sequence	−2.0 (0.183)	−2.1 (0.381)
RELA	Reticuloendotheliosis viral oncogene	1.52 (0.123)	1.88 (0.166)
GAS6	Growth arrest specific 6	−6.96 (1.64)	−10.24 (5.29)
CASP3	Caspase 3	−1.62 (0.258)	−1.55 (0.273)
CDH11	Cadherin 11	16.0 (2.39)	17.28 (1.71)
EPHB1	Ephrin B1	21.11 (5.87)	21.07 (3.29)
ATM	Ataxia telangiectasia mutated	34.30 (4.76)	20.17 (10.73)
SLC16A4	Solute carrier family 16	−13.93 (9.87)	−21.72 (16.83)
HDAC1	Histone deacetylase 1	2.30 (0.496)	1.47 (0.097)

15.1.4 Proteomic Studies of AT Cells

The product of the ATM gene triggers signaling cascades that regulate DNA damage and repair, cell cycle and apoptosis and stress response. These signaling pathways are governed by translational and post-translational events, leading to the need for analysis of differential protein levels. Since the primary focus has been on the radiation stress response, global changes in the proteome were determined by comparing 2-D gel patterns with and without radiation exposure [11].

A total of 435 and 630 differentially expressed proteins were identified for AT5BIVA and ATCL8 cell lines, respectively, across the time course study. We selected proteins with a high confidence

Table 15.3 Summary of selected proteins, expressed differentially in ATCL8 cells determined at indicated intervals following experiments 5 Gy γ-radiation

Protein name (transcription factors)	30 min	1 h	3 h	24 h
(Q99497) DJ-1 protein	ND	↓	↓	ND
(P67809) Nuclease sensitive element binding protein	ND	↑	↑	ND
(O60675) Transcription factor MafK	ND	↑	ND	ND
(Q86WZ6) Zinc finger protein 227	ND		ND	ND
(Q99676) Zinc finger protein 184	ND		ND	ND
(Q9Y265) RuvB-like 1	ND	ND		ND

Protein name (stress response)	30 min	1 h	3 h	24 h
(P07900) Heat shock protein – 90 alpha	↑	↑	↑	↑
(P08107) Heat shock protein 70	↓	↓	↓	↓
(P11021) 78 kDa glucose regulated protein	ND	↓	↑	↑

Protein name (DNA/RNA/protein processing)	30 min	1 h	3 h	24 h
(P24534) Elongation factor 1	↓	↓	ND	ND
(P31942) Heterogeneous nuclear riboprotein	↓	ND	ND	ND
(P17987) T-complex protein 1	↑	ND	ND	↓
(P49411) Elongation factor Tu	ND	↑	↓	↓
(P 31948) Stress-induced phosphoprotein	ND	↓	↓	↓
(P19338) Nucleolin	ND	↓	ND	ND
(Q01105) SET protein	ND	ND	↓	ND
(P35232) Prohibitin	ND	ND	↓	ND

Protein name (cell cycle/proliferation)	30 min	1 h	3 h	24 h
(Q05682) Caldesmon	ND	ND	↑	ND
(O60861) Growth arrest specific protein	ND	↓	ND	ND
(P12004) Proliferating cell nuclear antigen	ND	↓	ND	ND
(P11802) Cell division protein kinase 4	ND	ND	ND	↓
(P82979) Nuclear protein Hcc-1	ND	ND	ND	↑
(Q9HCT0) Fibroblast growth factor-22	ND	ND	ND	↑

score (>95%) for gene ontology analysis which showed a predominance of proteins involved in signaling, transcription, cell cycle and cytoskeletal structure and regulation. As an example, Table 15.3 summarizes selected differential protein levels as a function of time after exposure to ionizing radiation. Since 2-D gel analysis reflects extent of post-translational modifications, as well as changes in expression levels, the interpretation of upregulated and downregulated events are less relevant than the observations of changes in levels of a given protein or modified status.

Previously, we reported the effects of ATM on differential protein levels and/or post-translational modifications under basal conditions, by resolving whole cell protein lysates on 2-D gel electrophoresis [12]. A predominant representation of proteins was involved in pathways regulating cancer progression, cell growth and proliferation, cell death, small molecule biochemistry and amino acid metabolism. The pathways for purine, pyrimidine and amino acid metabolism were significantly altered in ATCL8 as compared to AT5BIVA [12]. Moreover, the integrated analysis identified significant ATM impact on RRM2, a potential constriction point in purine metabolism [12].

15.1.5 Metabolomic Studies of AT Cells

To determine effects of ATM restoration on levels of metabolic products, we performed ultra-performance liquid chromatography coupled with time of flight (UPLC TOF) mass spectrometry. The resultant features were analyzed by multivariate data analysis to yield ion rankings

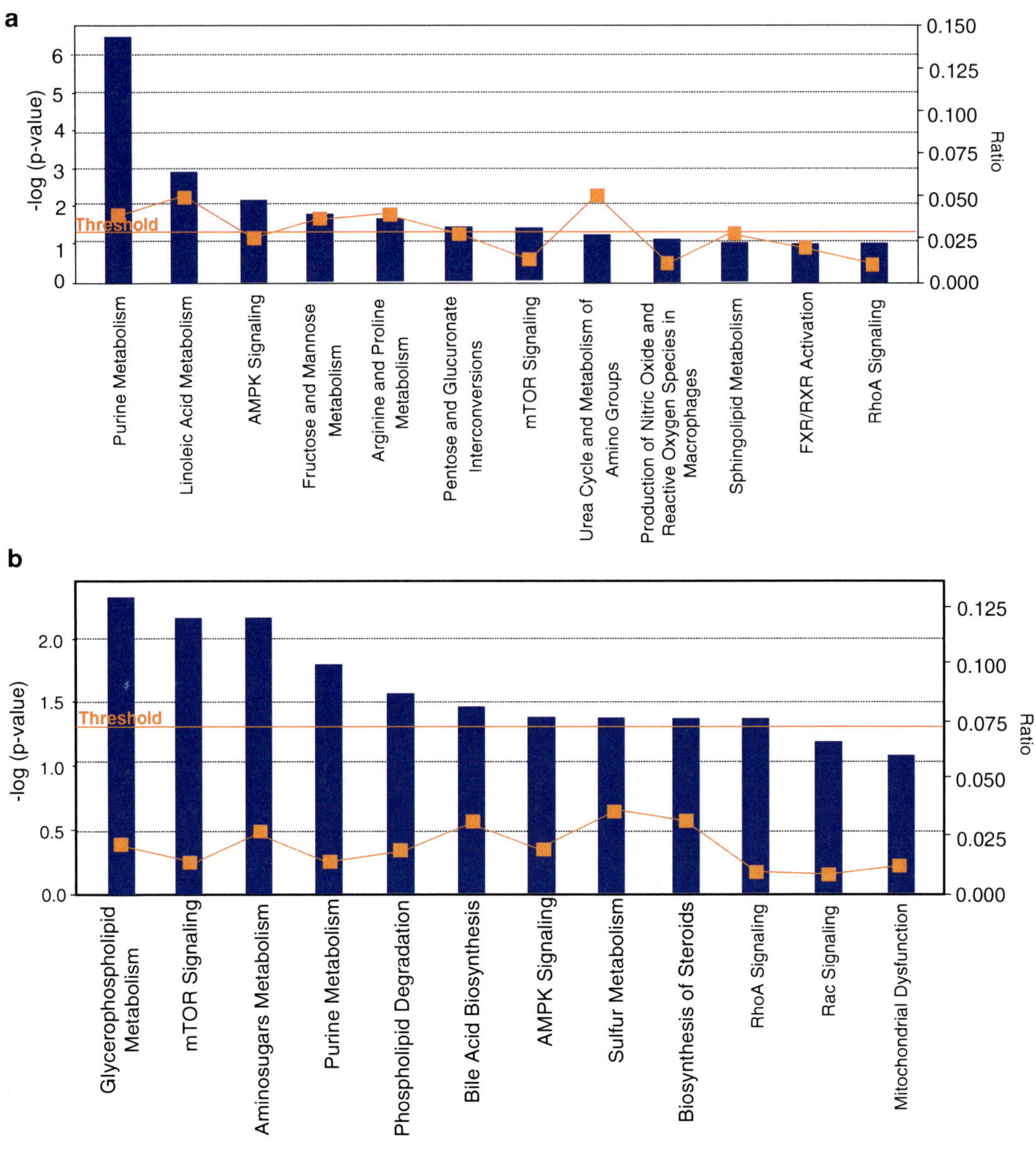

Fig. 15.4 Canonical pathways associated to the metabolites that showed significant changes in expression in response to radiation exposure of ATCL8 (**a**) and AT5BIVA (**b**) cells, respectively

for putative identification of biomarkers showing relative changes in metabolomic profiles. These ions comprised the dataset for mass based chemical identification using the Madison Metabolomic Consortium Database (MMCD). KEGG (Kyoto Encyclopedia of Genes and Genomes) IDs associated with metabolites showing significant fold changes and *p*-values for functional pathway analysis using "Ingenuity" software (ingenuity pathway analysis, IPA). Pathways showing significant perturbations included purine, pyrimidine and aminosugars metabolism (Fig. 15.4) [11].

15.1.6 Integration of Databases

To integrate the several functional pathways, the proteomics and microarray datasets were scored for differentially expressed proteins or genes known to participate in the common metabolic pathways using the IPA tool. As an example, Table 15.4

Table 15.4 Functional pathway analysis for network integration of "3-omics" data

	Genes			Proteins				Metabolites		
Pathway	Symbol	Description	ATCL8/ AT5BIVA	Symbol	Description	Acc ID	ATCL8/ AT5BIVA	Name	KEGG ID	ATCL8/ AT5BIVA
Purine metabolism	ABCC1	ATP-binding cassette	↓	ACTC1	Actin	P68032	↓	Adenine	C00147	↑
	ADSL	Adenylosuccinate lyase	↓	ATP5B	ATP synthase	P06576	↓	Adenosine	C00212	↑
	ATP1B1	ATPase	↓	BAT1	HLA-B	Q13838	↑	AMP	C00020	↑
	ENPP2	Phosphodiesterase 2	↓	DDX39	DEAD	O00148	↑	GMP	C00144	↑
	POLA2	DNA polymerase	↑	MYH1	Myosin	P12882	↓	Guanine	C00242	↑
	POLR3C	RNA polymerase III	↓	POLRMT	RNA polymerase	O00411	↓	Xanthine	C00385	↓
	SMARCA5	SWI/SNF related	↓	PSMC4	Proteasome	P43686	↓			
	TRAP1	TNF receptor-associated	↓	RUVBL2	RuvB-like 2	Q9Y230	↑			

illustrates pathway integration across all three levels of expression for purine metabolism. The selected ions representing small molecule metabolites were validated by mass spectrometry and empirical formulae by comparing MS/MS fragmentation patterns and retention times of the target ion and the standard metabolite [11].

The work summarized here utilized the iProXpress proteomic analysis system, the commercial software, Ingenuity (IPA) and the KEGG (http://www.genome.ad.jp/kegg/) pathway, one of the most widely used databases with curated pathway maps for metabolism and other cellular processes.

15.2 Discussion

We have described a genetically defined model cell system, established by transferring genes into the mutant AT5BIVA fibroblasts and selecting for clones exhibiting correction of the radiation sensitive phenotype. A full-length ATM expressing vector was used to derive ATCL8 and Western analysis has demonstrated characteristic ATM phosphorylation at serine 1981 1 h following exposure of cells to ionizing radiation in ATCL8, but not in AT5BIVA, confirming the presence of an activated ATM in ATCL8. Further characterization of the ATM competent and ATM deficient cells has included transcriptomic, proteomic and metabolomic studies, followed by an integrated analysis.

Collectively, our data show that ATM expression results in profound changes in the radiation phenotype and in gene expression. Comparisons of gene expression patterns of AT5BIVA to cells expressing a functional ATM following radiation exposure, demonstrate an attenuated, less robust response in AT5BIVA. Differences were observed in 15 functional categories, including those expected; such as apoptosis regulation, cell cycle and growth regulation. Therefore, these cells should permit detailed evaluation of ATM-dependent and ATM-independent radiation responses of cells.

The integrated systems approach to the global studies of cellular processes identifies common links at the mRNA, protein and small molecule metabolite levels. Purine metabolism was found to be a major ATM mediated pathway at all three levels of analysis.

Finally, this report shows the feasibility of performing integrated "omics" analysis for studying global changes that result from complex perturbations in a single gene. The technical aspects of the approach extending from genes to metabolites by integration within metabolic pathways have been demonstrated. The data highlight potential biomarker discovery and hypothesis generation in studies of complex phenotypes within the ATM model fibroblast system.

15.3 Materials and Methods

15.3.1 Cell Cultures

AT5BIVA and MRC5CV1 cells were obtained from the National Institute of General Medical Sciences (NIGMS). Cells were maintained in modified Eagle's medium with 20% fetal bovine serum, 100 U/ penicillin, and 100 pg/ml streptomycin. Cells were grown to 80% confluence and serum starved for 24 h for experiments. The ATCL8 cell line was established by transfecting AT5BIVA cells with the wild type, full-length "ATM" in a pCDNA expression vector and selected by screening for correction of radiation sensitivity.

15.3.2 Affymetrix Microarray Analysis

Total RNA was extracted using an RNeasy kit (Qiagen, Valencia, CA, USA). RNA labeling and hybridization were performed according to the Affymetrix protocol for one-cycle target labeling. For each experiment, fragmented cRNA was hybridized in triplicates to Affymetrix GeneChip HG-U95 arrays (Affymetrix, Santa Clara, CA). Affymetrix data analysis included pre-processing of the probe-level Affymetrix data (CEL files). We applied RMA for background adjustment, quantile method for normalization and the "median polish" for summarization. The triplicate arrays representing the same subject were averaged. Probe sets were considered statistically significant if their p-values were less than 0.001. Pathway analysis was performed with Database for Annotation, Visualization and Integrated Discovery (DAVID).

15.3.3 Quantitative RT-PCR

Total RNA was reverse-transcribed using the High Capacity cDNA Reverse Transcription Kit (Applied Biosystems). qRT-PCR was performed in triplicate using TaqMan Gene Expression Assays (Applied Biosystems) on the Applied Biosystems 7900HT Fast Real-time PCR System. Amplification of 18 S rRNA provided endogenous control to standardize the amount of sample added to the reaction. The comparative cycle threshold (CT) method was used to analyze the data by generating relative values of the amount of target cDNA (Applied Biosystems). The statistical analyses of these data were performed with a two-sided t test since the expression data showed normal distribution.

15.3.4 Two-Dimensional Protein Gel Electrophoresis

Five hundred microgram of each whole cell lysate were focused on the Protean IEF Cell from Bio-Rad. Proteins were resolved on a second dimension on 12% SDS-PAGE gels, fixed and stained with coomassie blue G250. Protein patterns were compared using Dimension imaging software (version 1.5, Syngene).

15.3.5 In-Gel Tryptic Digestion And Protein Identification by Mass Spectrometry

The protein spots of interest were manually excised from 2-D-gels and processed as previously described [13]. Tryptic peptides were extracted in acetonitrile and mass spectra were recorded with a matrix assisted laser desorption/ionization-time of flight, time of flight (MALDI-TOF-TOF) spectrometer (4700 Proteomics Analyzer, ABI, USA).

Peptide masses were compared with the theoretical masses derived from the sequences contained in Swiss-Prot databases using MASCOT. A subset of predicted proteins was validated using western blotting and the trend with respect to fold change was observed to be consistent with the experimental findings. For a time course comparisons of molecular networks, 30 min and 3 h samples were selected as representative of early and late responses following radiation exposure. Functional pathways were determined using the ingenuity pathway analysis (IPA).

15.3.6 Mass Spectrometry for Metabolomic Analyses

Samples were prepared and subjected to ultra-performance liquid chromatography-time of flight mass spectrometry (UPLC TOF MS). All reference chemicals were purchased from Sigma, St. Louis, MO. All solvents were LCMS grade (Fisher Scientific, USA– Optima grade).

15.3.7 Western Analyses

Protein extracts of AT5BIVA and ATCL8 cells were resolved by 1-D gel electrophoresis and transferred onto polyvinylidene fluoride (PVDF) membranes (Millipore Corporation, Bedford MA). Antibodies were purchased from Santa Cruz Biotechnology (USA). Antiphosphorylated serine 1981 ATM antibody was procured from Rockland Chemicals, USA.

15.3.8 Functional Pathway Analyses

IPA software was used to generate the pathways that are related to transcripts, proteins and metabolites. Network analysis generated functional association networks based on curated literature information of protein interaction, co-expression, and genetic regulation.

Acknowledgments This work was supported by NIH grant P01CA74175 and the CCSG grant NIH P30 CA51008 to the Lombardi Comprehensive Cancer Center supporting the Proteomics and Metabolomics Shared Resource.

References

1. Boder E, Sedgwick RP (1958) A familial syndrome of progressive cerebellar ataxia, oculocutaneous telangiectasia, and frequent pulmonary infection. Pediatrics 21:526–554
2. Taylor AM, Harnden DG, Arlett CF, Harcourt SA, Lehmann AR, Stevens S, Bridges BA (1975) Ataxia telangiectasia: a human mutation with abnormal radiation sensitivity. Nature 258:427–429
3. Chun HH, Gatti RA (2004) Ataxia-telangiectasia, an evolving phenotype. DNA Repair 3:1187–1196
4. Savitsky K, Bar-Shira A, Gilad S, Rotman G, Ziv Y, Vanagaite L, Tagle DA, Smith S, Uziel T, Sfez S et al (1995) A single ataxia telangiectasia gene with a product similar to PI-3 kinase. Science 268:1749–1753
5. Shiloh Y (2003) ATM and related protein kinases: safeguarding genome integrity. Nat Rev Cancer Rev 3:155–168
6. Lee SA, Dritschilo A, Jung M (1988) Impaired ionizing radiation-induced activation of a nuclear signal essential for phosphorylation of c-jun by dually phosphorylated c-Jun amino-terminal kinases in ataxia-telangiectasia fibroblasts. J Biol Chem 273: 32889–32894
7. Chen S, Wang G, Makrigiorgos GM, Price BD (2004) Stable siRNA-mediated silencing of ATM alters the transcriptional profile of HeLa cells. Biochem Biophys Res Commun 317:1037–1044
8. Jung M, Zhang Y, Lee SA, Dritschilo A (1995) Correction of radiation sensitivity in ataxia telangiectasia cells by IκB-α expression corrects. Science 268:1619–1621
9. Albright N (1987) Computer programs for the analysis of cellular survival data. Radiat Res 12:331–340
10. Cheema AK, Timofeeva O, Varghese R, Dimtchev A, Shiekh K, Shulaev V, Suy S, Collins S, Ressom H, Jung M, Dritschilo A (2011) Integrated analysis of ATM mediated gene and protein expression impacting cellular metabolism. J Proteome Res. 10:2651–2657
11. Cheema AK, Varghese RS, Kirilyuk A, Timofeeva O, Lee S, Kirilyuk K, Ressom HW, Jung M, Dritschilo A (2011) Functional proteomics analysis to study ATM dependant signaling in response to ionizing radiation (submitted)
12. Hu ZZ, Huang H, Cheema A, Jung M, Dritschilo A, Wu CH (2008) Integrated bioinformatics for radiation-induced pathway analysis from proteomics and microarray data. J Proteomics Bioinform 1:47–60
13. Wong CM, Cheema AK, Zhang L, Suzuki YJ (2008) Protein carbonylation as a novel mechanism in redox signaling. Circ Res 102:310–318

Malignant Transformation of Human Skin Fibroblasts by Two Alternative Pathways

16

J. Justin McCormick and Veronica M. Maher

Abstract

We developed a telomerase-positive, infinite life span human fibroblast cell strain (MSU-1.0) by transfection of a v-MYC oncogene and spontaneous over-expression of transcription factors SP1/SP3. Loss of expression of p14^{ALT} and enhanced expression of SPRY2 gave rise to the MSU-1.1 cell strain. Unlike MSU-1.0 cells, the MSU-1.1 cells can be malignantly transformed by expression of N-RAS^{LYS61} or H-Ras^{v12} oncoproteins (driven by their original promoters) and expression of a SRC-family protein, v-FES. MSU-1.1 cells can also be malignantly transformed by high expression of these RAS oncogenes or the v-K-RAS oncogene. PDGF-B transformed MSU-1.1 cells give rise to benign tumors (fibromas) in athymic mice. A second route to malignant transformation of the MSU-1.1 cells involves loss of functional TP53 protein by carcinogen treatment and loss of expression of wild type p16^{INK}. These studies indicate 6–8 "hits" are required to activate the oncogenes and inactivate the suppressor genes we identified.

16.1 Brief History of Early Studies on the Growth and Transformation of Fibroblasts in Culture

Fibroblasts are found in all organs and tissues of vertebrates, but are not the dominant cell type of any organ or tissue. Of the more than 200 types of cells that exist, fibroblasts have been the most extensively studied in cell culture. In the 1940s, when cell culture was being developed, investigators found that regardless of which organ or tissue was placed in culture, the cell population soon consisted of spindle-shaped cells identified as

J.J. McCormick (✉)
Department of Microbiology and Molecular Genetics and Department of Biochemistry and Molecular Biology, Carcinogenesis Laboratory – Food Safety and Toxicology Building, Michigan State University, East Lansing, MI 48824-1302, USA
e-mail: mccormi1@msu.edu

J.S. Rhim and R. Kremer (eds.), *Human Cell Transformation: Role of Stem Cells and the Microenvironment*, Advances in Experimental Medicine and Biology 720,
DOI 10.1007/978-1-4614-0254-1_16,

fibroblasts [1, 2]. Some investigators thought that the specialized cells of any organ or tissues dedifferentiated into fibroblasts when placed in culture. Others thought that the culture medium used preferentially selected for the growth of fibroblasts over other cell types, e.g., epithelial cells, which failed to grow and were lost. It was later concluded that this was the case. Many recent papers demonstrating epithelial to mesenchymal cell transition in culture re-opens this question [2].

In 1943, Earle et al. [3] at the National Cancer Institute in Bethesda, MD, discovered that C3H mouse fibroblasts acquire transformed characteristics in culture after a single chemical carcinogen treatment and these cells form sarcomas in C3H mice, leading to the death of the animals. The control cultures (not treated with carcinogen) acquired the same transformed characteristics at a lower frequency and formed sarcomas in 8% of the cases. Since then, many investigators have performed similar experiments with mouse fibroblasts and they also found that spontaneously-transformed malignant cells developed. On rare occasion, there have been reports of the spontaneous malignant transformation of human fibroblasts in culture. However, when we reviewed the literature in 1988 [4], we could not validate any of the reports of spontaneous malignant transformation of human fibroblasts in culture.

Namba et al. [5, 6] reported that repeated carcinogen treatment of human fibroblasts in culture produced cells that caused sarcomas when injected in immunosuppressed mice. We [4] tested the parental cells used in these experiments and their tumor derivatives, and found that they carried the same isozymes. This confirmed the validity of these experiments. Because Namba's tumor-derived cells have multiple chromosomal changes they are not ideal for use in determining the genetic and/or epigenetic changes responsible for malignant transformation.

16.2 Oxygen Exposure to Rodent Fibroblasts in Culture Causes Transformation

A 2009 study by Yoshii and Watanabe [7] examined the question whether the 20% oxygen level (the same as air) typically found in CO_2 incubators is responsible for the spontaneous transformation of rodent fibroblasts in culture. Mouse, Syrian hamster, and human fibroblasts (isolated from embryos at the same stage of development), were exposed to 20% (normoxic), 2% or 0.5% (hypoxic) oxygen continuously. Human fibroblasts did not develop an infinite life span and did not express any transformed properties when exposed to any level of oxygen. Mouse and hamster fibroblasts exposed to the same levels of oxygen developed an infinite life span and expressed anchorage independent growth (a transformed phenotype) especially when exposed to 20% oxygen. Continuous exposure of mouse fibroblasts to 20% oxygen led to cells that formed malignant tumors when injected into syngeneic mice (personal communication, M. Watanabe). Oxidative stress is known to cause DNA damage. Human fibroblasts have been shown to have more robust mechanisms to protect themselves from oxygen stress than rodent fibroblasts [7].

16.3 Strategy Used to Transform Human Fibroblasts

To understand the mechanisms by which human fibroblasts become cancer cells *in vivo*, we transfected genes (oncogenes) that had been shown to be activated or over-expressed in cells from patient-derived sarcomas (fibroblastic or undifferentiated soft tissue sarcomas) into human fibroblastic cell strains. We isolated colonies that were either drug-resistant (plasmid contained a drug-resistant marker) or expressed transformed phenotypes (foci or anchorage independent growth). Independent clonal populations were tested for the level of expression of the oncoprotein of interest and grouped based on high, moderate, or low expression level. These clonal populations were tested for: focus formation, anchorage independent growth, growth in medium with reduced levels of serum, and tumor formation in athymic mice. In addition to studying the role of specific oncogenes, we also evaluated the role of the tumor suppressor genes, *TP53, p14^{ARF}, p16^{INKA}, and LRP12*, in malignant transformation.

Because of space limitations, this paper only describes the details of our research, highlighting the contribution of many of our colleagues. Other researchers have taken a somewhat different approach to reach an understanding of how human fibroblasts become malignant cells. These have been reviewed by Peters and his coworkers [8, 9] as well as by Counter and his coworkers [10].

The process of carcinogenesis involves sequential changes due to enhanced function of an oncogene and/or loss of function of a tumor suppressor gene. These changes typically give the cells a selective growth advantage allowing clonal expansion to a large population. This increases the probability that one of the cells will undergo a spontaneous or carcinogen-induced mutation in another cancer related gene. Each sequential change results in a cell that has acquired a more transformed phenotype, ultimately resulting in a cell that is capable of forming a malignant tumor. Human cancers typically are clonal in origin, i.e., they originate from a single "first cancer cell." Cancers that develop as a result of viral activity may be an exception.

16.4 Development of Infinite Life Span Human Fibroblasts

Many patient-derived cancer cell lines have an infinite life span in culture, while human fibroblasts, like other normal human cells, have a limited life span in culture (~60 population doublings). Infinite life span human fibroblasts, which were chromosomally stable, are ideal to study the genetic and/or epigenetic changes responsible for malignant transformation.

The normal human fibroblast cells used for these experiments (LG1) were obtained from the foreskin tissue of a healthy newborn male. To create infinite life span human cells, we expressed the *v-MYC* gene in LG1 cells [11], a technique that had been reported to cause rat fibroblasts to develop an infinite life span [12]. The *v-MYC* gene (linked to a neomycin-resistant gene) was transfected into human fibroblasts and neomycin-resistant clonal populations were isolated. These v-MYC expressing cells were grown in culture until the cells senesced. A small number of cells continued to grow beyond senescence (more than 150 population doublings). These cells are considered immortal, i.e., having an infinite life span.

The newly established infinite life span cells gave rise to two distinct clonal populations in culture. The cells, designated MSU-1.0, were derived from large diffuse clones that grew relatively slowly (72 h doubling time) and were diploid. The cells, designated MSU-1.1, were derived from small punctual clones that grew more rapidly (18–24 h doubling time) and carried two stable marker chromosomes, 45 chromosomes total. One of the marker chromosomes has a loss of the 11p15 region similar to that observed in embryonal rhabdomyosarcomas [13]. The *H-RAS* gene is located on the lost fragment. We concluded that the diploid MSU-1.0 strain must have arisen first and the MSU-1.1 strain arose from it. The MSU-1.0 and MSU-1.1 cells were both neomycin-resistant, the *v-MYC*-neo vector was integrated in the same location in both, and they carried the same genetic markers as the parental LG1 fibroblasts. Neither cell strain formed tumors in athymic mice. LG1 cells and their many derivatives, MSU-1.0, MSU-1.1, etc., are referred to as the MSU1 lineage of cells.

After the discovery of the telomerase gene, *h-TERT* [14], we tested LG1, MSU-1.0, MSU-1.1 for telomerase activity and found that the MSU-1.0 and MSU-1.1 cell strains had such activity, but the LG1 parental cells did not [15]. It was later reported that the human *h-TERT* expression was controlled by the transcription factors MYC and SP1 [16]. We found that both MSU-1.0 and MSU-1.1 cell strains spontaneously express a significantly higher level of SP1 and SP3 protein than the LG1 cells [17]. Therefore, it seems likely that the spontaneous over-expression of SP1 and SP3 in a subpopulation of the v-MYC-expressing LG1 cells leads to the expression of telomerase and as a result, the infinite life span property of MSU-1.0 and MSU-1.1 cell strains. Studies have shown that the transfection of the human *c-MYC* gene into WI38 human fibroblasts can also lead to infinite life span cells by a similar process [18, 19]. The *c-MYC* gene is over-expressed in the SHAC human fibrosarcoma cell line [20] as well as in other human sarcomas [21, 22].

16.5 INK4a-p14^{ARF}

The MSU-1.1 cells have a shorter doubling time (18–24 h) than the LG1 or MSU-1.0 cells, strongly suggesting that they have an aberration in a growth factor pathway. We found that p14^{ARF}, a protein that controls the cell cycle, is not expressed in the MSU-1.1 cells, but is expressed at the same level in both the MSU-1.0 and LG1 cells [23]. We also found that other normal human fibroblasts express p14^{ARF}, but HT1080 and SHAC patient-derived human fibrosarcoma cell lines do not. The p14^{ARF} protein functions as a growth suppressor by inhibiting MDM2-dependent p53 degradation [24]. Inactivation of the *p14^{ARF}* gene in primary human fibroblasts has been demonstrated to give the cells a strong growth advantage [25], similar to what we have seen in MSU-1.1 cells. The loss of p14^{ARF} expression is found in angiosarcomas [26] and Ewing sarcomas [27].

16.6 Malignant Transformation: The RAS/SRC Pathways

The patient-derived human fibrosarcoma cell line, HT1080, expresses the *N-RAS^{Lys61}* oncogene [28]. HT1080 cells that have a reduction in the amount of the N-RAS^{Lys61} oncoprotein, results in a 1,000-fold reduction in anchorage independent growth. The cells are weakly tumorigenic [29]. Obviously there is a fine balance between the amount of N-RAS and N-RAS^{Lys61} protein that is required for malignant transformation of these cells.

The *K-RAS, N-RAS,* and *H-RAS* genes are highly homologous, but their expression levels vary considerably in different types of normal tissues and their oncogenic forms occur in specific types of human cancer [30]. This led us to carry out studies on the ability of the oncogenic forms of these genes to transform the MSU-1.1 cells. We found that expression of the *v-K-RAS* [31] oncogene caused MSU-1.1 cells to form foci. Cells isolated from the foci grew in an anchorage independent manner and formed spindle cell or myxoid sarcomas in athymic mice. We also studied the ability of the *N-RAS^{Lys61}* oncogene in a high expression vector to malignantly transform MSU-1.1 cells. Cells isolated from resulting foci formed spindle or round cell sarcomas in athymic mice [32]. These cells expressed high amounts of the *N-RAS^{Lys61}* protein, which on Western blot analysis appears as a novel more rapidly migrating band, as we observed in the HT1080 cells.

When we expressed the *N-RAS^{Lys61}* gene (from the HT1080 cells) with its own promoter in MSU-1.1 cells, these cells expressed only a slightly higher level of the N-RAS^{Lys61} protein (the novel more rapidly migrating band) and were more refractive than the parent population. They do not exhibit anchorage independent growth and do not grow in medium with reduced serum [33]. On Western blot analysis, MSU-1.1 cells and HT1080 cells express similar amounts of K-RAS, H-RAS, and N-RAS proteins. It is not surprising that MSU-1.1 cells and HT1080 cells express comparable levels of the three RAS proteins since the sarcoma, which gave rise to the HT1080 cells, developed in part as a result of a mutation at codon 61 (one of the two alleles of the constituent N-RAS genes). We conclude that although high expression of the *N-RAS^{Lys61}* gene malignantly transforms the MSU-1.1 cells, it is likely that in the HT1080 cells, another gene functions to complement it in the malignant transformation process.

We also transfected the MSU-1.1 cells with a vector carrying the *H-RAS^{v12}* oncogene containing either two enhancer sequences, or one enhancer sequence, or no enhancer sequence [34]. Only the cells transfected with the high expression vector (two enhancer sequences) produced foci. Cells isolated from the foci exhibited anchorage independent growth, grew in medium without exogenous growth factors, and produced tumors of 1 cm^3 in 30–34 days. These tumors were found to be poorly differentiated, invasive, spindle cell sarcomas. Chromosomal analysis of the *RAS* oncogene transformed tumor-derived cells indicated that they all carried the two

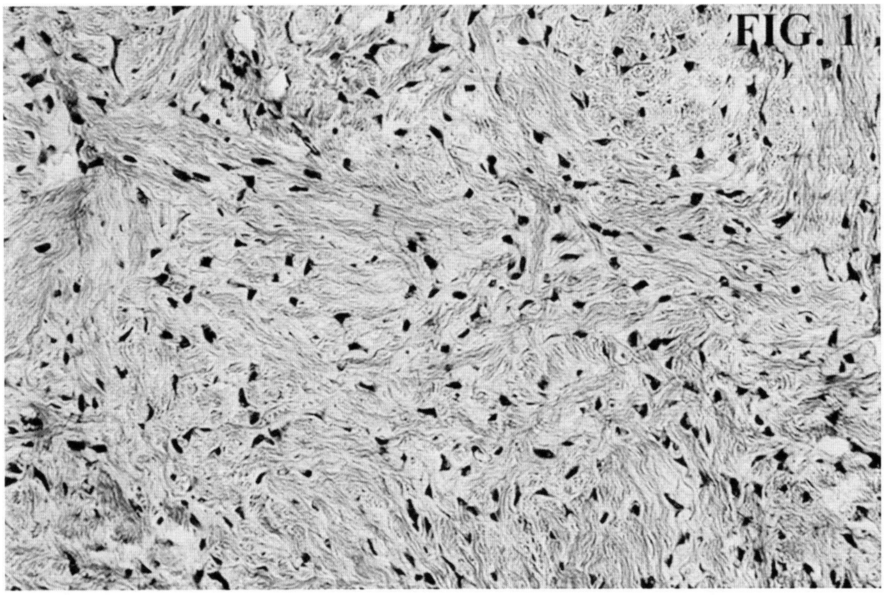
FIG. 1

Fig. 16.1 Representative image of a fibroma generated by the MSU-1-1-PGF-B cell strain

marker chromosomes, which are characteristic of the MSU-1.1 cells and had a modal chromosome number of 45.

However, when we transfected the *H-RAS*^{v12} gene with its own promoter into MSU-1.1 cells, they expressed slightly higher levels of the H-RAS protein and were slightly more refractive than the parental LG1 cells [33]. This cell population exhibited greater anchorage independent growth (small colonies) and grew in medium with reduced growth factors somewhat better than the parental MSU-1.1 cells and is nontumorigenic.

One growth factor that activates the RAS pathway in human fibroblasts is PDGF-B. The simian sarcoma virus that transduces the simian form of the PDGF-B gene causes benign tumors (fibromas) in monkeys [35]. It is also known, unlike normal human fibroblasts, the HT1080 human patient-derived sarcoma cell line expresses PDGF-B protein that drives replication of these cells [36]. We found LG1 human fibroblasts and its derivatives, MSU-1.0 and MSU-1.1, do not express PDGF-B protein [37]. To determine what role PDGF-B gene expression plays in the transformation of human fibroblasts, we transfected the simian sarcoma virus form of the PDGF-B gene (linked to a drug-resistance marker) into the MSU-1.1 cells and selected for drug resistant colonies [37]. Clonal populations that expressed the PDGF-B protein at high levels grow to much higher saturation densities than the MSU-1.1 cells and form large colonies in agarose. As expected, the PDGF-β receptor was activated in the cell strains that produced high levels of PDGF-B. These cells formed small tumors after 72 days, but did not continue to grow and did not regress. They were found to be benign tumors (fibromas) (Fig. 16.1). Tumors left in the animals for 144 days exhibited small islands of sarcomatous cells within the fibromas. When these benign tumors were removed, established in culture, selected for drug resistance, and injected into athymic mice these cells formed predominantly benign tumors with multiple discrete islands of low or high grade spindle, or round cell sarcomas. We concluded that an additional spontaneous change occurred in the benign tumors that gave rise to sarcomatous cells. These fibroma-derived cells had the same chromosome complement as their parental MSU-1.1 cells (45 chromosomes, including the two marker chromosomes).

16.7 Studies on a SRC Family Protein

The *SRC* family of genes consists of nine members that have a close homology and code for nonreceptor tyrosine kinases. All cells express one or more genes from the SRC family. The first oncogene discovered was the viral *SRC* gene (*v-SRC*), the oncogene of the Rous sarcoma virus (RSV) of chickens [38], which is derived from the *c-SRC* gene of chickens. RSV infection of chickens results in a massive viremia, which causes multiclonal spindle cell tumors that kill the animals. RSV also transforms chicken fibroblasts in culture.

To determine whether a SRC-family member plays a role in the malignant transformation of human fibroblasts, we transfected the MSU-1.1 cells with the *v-FES* gene linked to a drug-resistance marker [33]. Clonal populations of drug-resistant cells that expressed the v-FES protein had spindle cell morphology, the same growth factor requirements as their parental MSU-1.1 cells, formed small anchorage independent colonies at an enhanced frequency, and were nontumorigenic. However, v-FES clonal populations kept in culture for 50–60 population doublings acquired the ability to form large colonies in agarose and these populations formed high grade sarcomas of 1 cm^3 in 27–75 days. MSU-1.1 cells not expressing the v-FES protein were nontumorigenic even after extensive culturing. These experiments suggest that the expression of this SRC-family gene (*v-FES*) is required for malignant transformation, but at least one additional genetic and/or epigenetic change is also required. It should be noted that many studies on SRC-family proteins utilize antibodies that cross-react with various members of the family. A 2008 study [39] demonstrates that a SRC-family tyrosine kinase (LYN) plays an essential role in the tumorigenic growth and metastases of Ewing sarcoma. We chose the *v-FES* gene for our studies, because specific FES antibodies were available.

We transfected the *v-FES* oncogene into the H-RAS^{V12} or N-RAS^{Lys61} cells (expressing low levels of these oncoproteins, as described above) to determine whether expression of a SRC-family oncoprotein could complement the RAS oncoprotein and malignantly transform cells. The H-RAS^{V12}v-FES-expressing cells formed large anchorage independent colonies at a high frequency, exhibited enhanced growth in medium with reduced serum, and formed high grade spindle cell sarcomas in athymic mice with a short latency. However, the N-RAS^{Lys61}-v-FES expressing cells did not form anchorage independent colonies, or exhibit enhanced growth in medium with reduced serum, and did not form tumors.

We carried out an additional study transfecting the *v-FES* oncogene into the MSU-1.2 cells [40]. This cell strain is a spontaneous clonal derivative of the MSU-1.1 cells selected for its ability to grow rapidly in medium without exogenous growth factors. It does not form large colonies in agarose and does not form tumors. The v-FES expressing cells formed large colonies in agarose and formed invasive, spindle cell sarcomas with a short latency. This result is consistent with that described above for the H-RAS^{V12}v-FES cells indicating that expression of v-FES provides an essential step in the malignant transformation process.

We also carried out a parallel experiment using MSU-1.1 cells with low levels of expression of the PDGF-B protein [37]. In culture, these cells exhibited minimal characteristics of transformation and were nontumorigenic in athymic mice. Like the H-RAS^{V12}v-FES-expressing cells, the PDGF-B-v-FES-expressing cells formed large anchorage independent colonies at a high frequency, grew in medium with reduced serum, and formed high grade spindle cell sarcomas (Fig. 16.2) in athymic mice with a short latency.

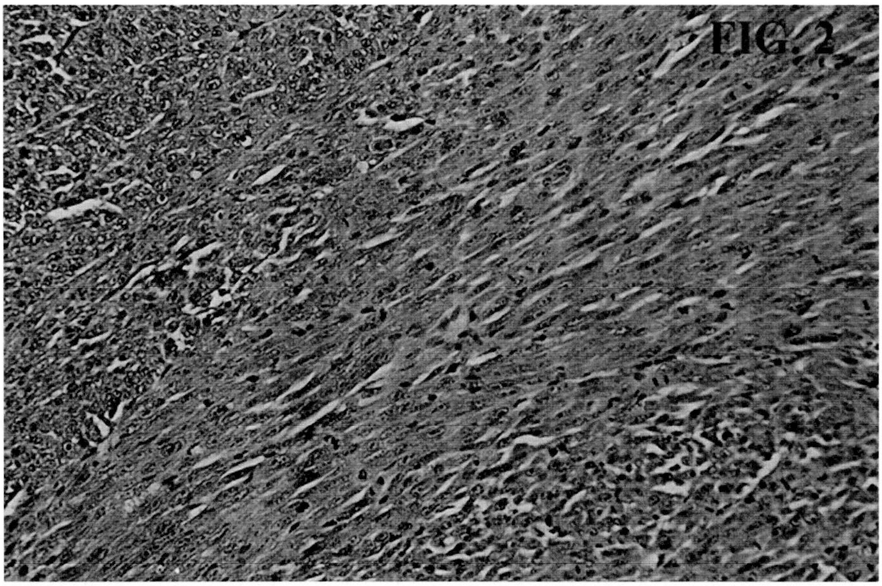

Fig. 16.2 Representative image of a fibrosarcoma generated by the MSU-1.1-H-RAS^{v12}-v-FES cell strain

16.8 Summary of Oncogene Transformation Experiments

The results described above indicate that malignant transformation of human fibroblasts requires specific mutations in the *MYC, SP1/SP3, p14^{ARF}, RAS* (or *PDGF-B*), *v-FES* genes (or homolog's), or their promoter regions. If these mutations were spontaneous or carcinogen-induced, six "hits" would be required for malignant transformation. As a tumor suppressor gene, p14^{ARF} requires a "hit" in each of the *two* alleles and the four oncogenes require a single specific "hit" in *one* allele. Some of the oncogenes require specific mutations in the structural gene such as in *H-RAS^{v12}* or *N*-RAS^{Lys61}. Other oncogenes such as SP1/SP3 require a higher steady state expression level of the wild type gene.

The transfection of oncogenes creates transformed cells that express the same characteristics as cells from patient-derived tumors. However, the oncogene transformed cells are not identical to cells from patient-derived tumors for a number of reasons: (1) the transfected cells typically carry an extra normal copy of the gene of interest; (2) the transfected gene is usually not under the control of its ordinary promoter sequence; (3) the sequence by which the changes were made in the cells is arbitrary except that the infinite life span cells must be an early change; and (4) the site of integration of transfected gene(s) is random. Despite such considerations, the transfection of genes has turned out to be a highly effective way of modeling cancer as demonstrated by our experiments (Fig. 16.3) and that of many investigators.

16.9 Malignant Transformation: The TP53/TP16^{INK4a} Pathway

As a result of the transfection experiments described above, it was clear that the MSU-1.1 cells differed from the normal fibroblasts (LG1) and the MSU-1.0 cells, because, although they expressed some limited transformed characteristics, they did not form tumors. With two additional changes: (1) low expression of H-RAS^{V12} or PDGF-B; and (2) expression of the FES oncogene, the cells could form malignant tumors.

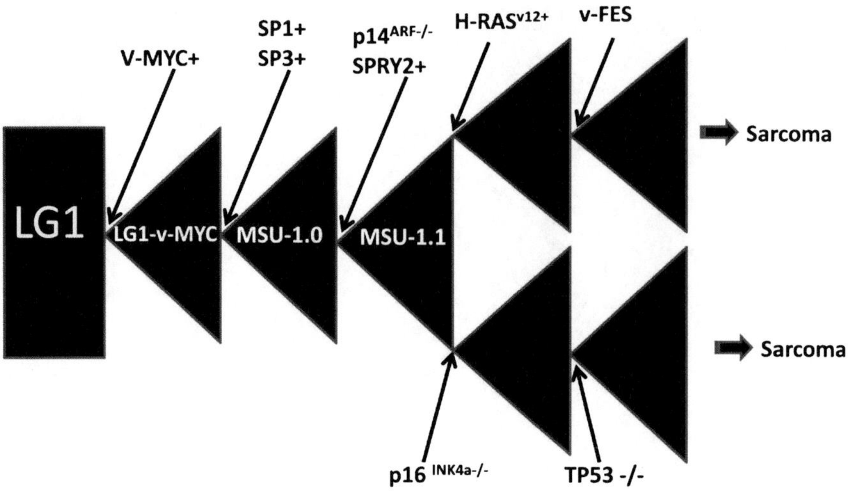

Fig. 16.3 Schematic outline of the main cell strains in the MSU1 lineage. LG1 cells are a population of foreskin-derived human fibroblasts from a newborn. The *triangle* just to the right of the LG1 *box* represents a clonal population of LG1 cells expressing the v-MYC oncoprotein as a result of the transfection of the *v-MYC* oncogene. The next triangle to the right, labeled MSU-1.0, represents a clonal population of LG1-v-MYC cells that spontaneously expressed higher levels of the SP1/SP3 protein. The MSU-1.0 cells express telomerase activity unlike the parental LG1-v-MYC cells. The next triangle to the *right*, labeled, MSU1.1, represents a clonal population of cells that spontaneously expressed SPRY2 and have lost expression of $p14^{ARF}$. The next triangle to the *upper right* represents a clonal population of MSU-1.1 cells that express low levels of the H-Ras^{V12} oncoprotein. The *H-RAS^{V12}* oncogene is driven by its endogenous promoter. The next triangle to the *right* represents a clonal population of the MSU-1.1- H-Ras^{V12} cells that express the *v-FES* oncoprotein. These cells form sarcomas with a short latency. The triangle to the *lower right* of the MSU-1.1 cells represents a clonal population of MSU-1.1 cells that have lost expression of $p16^{INK4a}$ or express a mutant form of the protein. The next triangle to the *right* represents a clonal population of MSU-1.1-$p16^{INK4a-/-}$ that has lost functional TP53 activity. These cells form sarcomas with a short latency. Many other cells strains have been derived from this lineage and are described in the text

Alternatively, high expression of mutant RAS was sufficient to cause the MSU-1.1 cells to form sarcomas. We hypothesized that treatment of MSU-1.1 cells with a carcinogen at a dose known to induce mutations would be sufficient to mutate a proto-oncogene to an oncogene in the MSU-1.1 cells. MSU-1.1 cells that had mutations induced by carcinogen treatment can be expected to express enhanced transformed characteristics; e.g., focus formation, anchorage independent growth, etc.

Therefore we treated MSU-1.1 cells with ionizing radiation at a mutagenic dose and selected for focus formation as we had after transfection of the H-RAS^{V12} or PDGF-B oncogenes. Foci were observed in the irradiated cultures at a high frequency. Only a few foci were observed in the controls. Cells isolated from some of the induced-foci formed sarcomas in athymic mice [41]. We carried out similar experiments with various classic carcinogenic agents: MNU [42], Benzo(a)pyrene diol epoxide [43], UV radiation [44], as well as additional experiments with ionizing radiation [45]. We found that foci occurred in a dose-dependent manner with all agents. The cells isolated from the foci had either a spindle or cobblestone morphology, form a high frequency of large anchorage independent colonies, exhibit enhanced growth in medium with reduced serum, and form high grade spindle cell sarcomas in athymic mice with a short latency. Fifty to ninety per cent of the focus-derived cells formed sarcomas in athymic mice.

The tumor-derived cells obtained from the initial ionizing radiation experiments were karotyped. As expected, the two MSU-1.1 marker chromosomes were observed in all cell strains tested. In addition, all strains exhibited chromosome breaks and had random chromosome numbers. One copy of chromosome 17 lacked the p-arm in one of the irradiated MSU-1.1 cell strains. Because the TP53 gene is located on the p-arm of chromosome 17, it suggests that the other allele of the TP53 gene might be inactivated or lost.

Western blot analysis for TP53 protein expression indicates that the radiation-transformed tumor-derived cell strain lacking one copy of the p-arm of chromosome 17 did not express the TP53 protein. This suggests that the loss of TP53 protein is responsible for focus formation in that strain and its tumorigenicity. We hypothesized that the TP53-expressing focus-derived cells might carry two nonfunctional *TP53* genes resulting from carcinogen-induced, inactivating point mutations in both alleles of the gene. If this were the case, it is difficult to explain why the frequency of induced-foci was similar to the frequency of HPRT-induced mutations [42] given that the parental LG1 cell strain is derived from a male and therefore only carries a single copy of the *HPRT* gene [11].

To resolve this issue, we carried out a yeast assay [46] that measures the transactivating ability of the TP53 protein and can distinguish between cell strains carrying two, one or, no functional copies of the *TP53* gene. We tested 28 focus-derived cell strains resulting from MNU treatment. Fourteen of the 28 cell strains were nonfunctional for p53 protein. Ten of these 14 cell strains were tumorigenic suggesting that the other 4 cell strains still required an additional change to become tumorigenic. The 14 focus-derived cell strains, which were wild type for TP53, did not form tumors indicating that there is an alternative mechanism by which foci form.

To determine whether MNU treatment induces point mutations, we sequenced the *TP53* gene in several of the tumor-derived cell strains that were found to express nonfunctional TP53 protein using the yeast assay [42]. The *TP53* genes in these cell strains had a mutation at codon 215. This finding strongly suggests that there was a mutation at codon 215 in one allele of the *TP53* gene in a subpopulation of MSU-1.1 cells. If so, a recombination event between the mutant and wild type *TP53* genes would result in 50% of the cells carrying a 215 codon mutation in both alleles and 50% of the cells carrying two wild type *TP53* genes. However, we would be unable to identify the cells carrying two wild type alleles because they lack a distinctive phenotype. If this hypothesis is correct, MNU would induce random recombination events on the p-arm of chromosome 17. To determine whether this had occurred, we found 6 informative microsatellite markers on the p-arm of chromosome 17 of the MSU-1.1 cells. Using them as probes, we discovered 7 different patterns of recombination (loss of heterozygosity) on the p-arm. Similar studies on chromosomes 11 and 18 showed no change in heterozygosity indicating that the focus assay was indeed selecting for cells carrying two identical mutant alleles of the *TP53* gene. A smaller parallel study conducted with the same pool of MSU-1.1 cells using ionizing radiation instead of MNU gave similar results [42]. We found that transfection and expression of wild type TP53 protein in an ionizing radiation transformed cell strain (originally lacking expression of any TP53 protein) caused these cells to lose the characteristics of the transformed phenotype, and most importantly, the ability to form tumors [47].

The CDKN2A/p16^{INK4a} locus functions to regulate the cell cycle and is inactivated in many types of human tumors including soft tissue sarcomas [48]. To determine whether p16^{INK4a} protein is expressed in the two patient-derived human fibrosarcoma cell lines (HT1080, SHAC), we carried out Western blotting on these cells and the cells strains of the MSU1 lineage. The HT1080 and SHAC cell lines do not express the p16^{INK4a} protein. The carcinogen-transformed tumor-derived MSU-1.1 cells exhibit a markedly reduced expression of p16^{INK4a} and/or a more rapidly migrating form of the p16^{INK4a} protein. The tumor-derived RAS and PDGF-B transformed MSU-1.1 cells express the p16^{INK4a} protein at the same molecular weight and in an equivalent amount to their parental MSU-1.1 cells.

MSU-1.1 cells express p16^{INK4a} protein at the same level as the MSU-1.0 and LG1 cell strains. Some isolated clonal populations of MSU-1.1

cells expressed reduced amounts of p16^{INK4a} or the more rapidly migrating form of the protein. They did not form tumors [23]. To determine the role of p16^{INK4a} in the transformation of MSU-1.1 cells, populations with and without p16^{INK4a} expression were treated with ionizing radiation at mutagenic doses and assayed for focus-formation. The foci in the p16^{INK4a} negative cells were larger and more distinct than the foci in the p16^{INK4a} positive cells. Most importantly, there was an eightfold increase in the number of foci in the p16^{INK4a} negative cells. The frequency of foci in the nonirradiated p16^{INK4a} positive and p16^{INK4a} negative cells was the same. This is consistent with a report [25] that the loss of p16^{INK4a} expression in human fibroblasts does not affect growth, but in fibroblasts that do not express p53, the loss of p16^{INK4a} provides a growth advantage.

All of the carcinogen-transformed tumorigenic cells derived fromMSU-1.1 cells exhibited loss of functional TP53 and a marked reduction or a more rapidly moving form of the p16^{INK4a} protein. To determine the role of p16^{INK4a} in the transformation of MSU-1.1 cells by transfection of oncogenes, populations with and without p16^{INK4a} expression were transfected by the H-RAS oncogene and assayed for focus formation. As above, the foci in the p16^{INK4a} negative cells were larger and more distinct than the foci in the p16^{INK4a} positive cells. Again, there was an eightfold increase in the number of foci in the p16^{INK4a} negative cells and the frequency of foci in the nontransfected p16^{INK4a} positive and p16^{INK4a} negative cells was the same [49]. However, Western blot analysis of tumor-derived cells from our oncogene transfection experiments showed these cells expressed p16^{INK4a} equivalent to the parental cells, suggesting that p16^{INK4a} is not relevant to the malignant transformation of these cells. We conclude that loss of p16^{INK4a} expression helps focus formation, but is directly involved in malignant transformation only in the case of cells that have lost functional TP53. Other mutagenic carcinogens that we used in parallel studies, i.e., BPDE [43], UV [44] also inactivate p53 and are known to cause recombination. The p16^{INK4a} protein is known to interact negatively with CDK4 and CDK6 proteins. Thus, a reduction in the amount of p16^{INK4a} and/or the loss of functional p16^{INK4a} is the equivalent of amplification of *CDK4* and/or *CDK6* genes, i.e., it accelerates cell cycle progression and cell proliferation [50], which are characteristics of focus-forming cells.

16.10 Summary of Carcinogen Transformation Experiments

The results described above indicate that malignant transformation of human fibroblasts requires specific mutations in the *MYC, SP1/SP3, p14^{ARF}, p16^{INK4a},* and *TP53* genes. If these specific mutations are spontaneous or carcinogen-induced, eight "hits" would be required for malignant transformation. *p14^{ARF}*, *p16^{INK4a},* and *TP53* (tumor suppressor genes), require a "hit" in each of the *two* alleles, and the two oncogenes (*MYC and SP1/SP3*) require a single specific "hit" in *one* allele to enhance expression of these genes (Fig. 16.3).

Note that the cells in the initial strain of the lineage, LG1-vMYC+, did not have a selective growth advantage or a change in cell morphology that we could identify. This situation is similar to that described by Bozic et al., [51] who states, "In the case of solid tumors, the mutation that initiated the process might actually be the second 'hit' in a tumor suppressor gene, the first hit affects one allele, without causing a growth change, whereas the second hit in the opposite allele, leaves the cell without any functional suppressor . . ." In our studies, expression of the *v-MYC* gene did not confer any phenotype that we could observe, but spontaneous over-expression of the genes, e.g., *SP1/SP3* (presumably in a single cell) resulted in a telomerase positive infinite life span cell population (MSU-1.0) that continued to grow when the LG1-vMYC + cells senesced. All subsequent changes that were either initiated or spontaneous resulted in obvious changes in enhanced growth under one or more specific conditions and/or a change in morphology, leading to populations of cells that were distinct from their parental population.

16.11 Effector Genes

As is now understood, each of the oncogenes or suppressor genes (modified in the process of carcinogenesis), function through a network of downstream effector genes. These effector genes exhibit enhanced or reduced activity, or other modifications. We have carried out detailed studies of several important effector genes in the MSU1 lineage. A summary of these studies follows.

16.12 MET and SP1/SP3

MET is reported to be highly expressed in some human sarcoma-derived cell lines [52]. We examined the expression level of MET in carcinogen-transformed, tumor-derived, MSU-1.1 cells [53]. We found enhanced expression of the MET protein in 10/11 of these cell strains and phosphorylated MET (activated form) in 11/11 of these cell strains. The *HGF* gene, the ligand for MET, was expressed in all of the cell strains. We also found that 4/5 patient-derived, human fibrosarcoma cell lines had enhanced expression of MET when compared with normal human fibroblasts.

MET is a tyrosine kinase, which is embedded in the cell membrane, and is the receptor for the extracellular protein, HGF. The binding of HGF to MET causes phosphorylation of the MET protein, which sets up a signaling process that causes cell spreading, migration, and induces blood vessel sprouting. We reduced the expression of MET or a combination of MET and HGF to near normal levels by the use of specific ribozymes. The cell strains with reduced expression failed to form tumors in athymic mice, clearly indicating enhanced MET expression is essential for the tumorigenicity of these cells [53].

The MET gene carries binding sites in its promoter region for the transcription factor SP1/SP3. The level of SP1 in cells that over-expressed the MET protein was determined. We found that the expression of the SP1 protein and the phosphorylated form of MET correlated with each other. Further experiments demonstrated a correlation of the expression levels of SP1 protein and a similar transcription factor, SP3, i.e., the amount of these two proteins changed expression levels in the same proportion in all of the normal and transformed fibroblast strains we studied [17]. We sequenced the promoter region of SP3 and found functional SP1/SP3 binding sites. Downregulation of the SP1/SP3 protein expression levels in normal and transformed cells, using a specific SP3 ribozyme, also downregulated SP1/SP3 expression [54]. These transformed cells with downregulated SP1/SP3 failed to make tumors. Three genes (*uPAR, ERGF* and *VEGF*), known to have multiple SP1/SP3 binding sites in their promoters, exhibited a marked reduction in their protein levels in transformed cells with reduced SP1/SP3 [54]. The proteins of these three genes have been shown to play an important role in carcinogenesis [55–57]. High expression of SP1/SP3 contributes to tumor formation and correlates with the loss of expression of p16^{INK} and/or the loss of functional TP53.

16.13 Urokinase Plasminogen Activator (uPA)

uPA is a serine protease that is involved in tissue remodeling and cell migration. In cancers, uPA has frequently been reported to be over-expressed, causing tumor invasion and metastasis [55]. uPA binds to its own receptor, the urokinase plasminogen activator receptor (u-PAR), focusing proteolytic activity on the external cell membrane of the cancer cell. We analyzed the cells of the MSU1 lineage for receptor-bound uPA (53 kDa) activity [58]. We found that the parental LG1 cells and its two nontumorigenic derivatives, MSU-1.0 and MSU-1.1, have uPA activity levels of 3.8–7.1 PU/24 h/10^6 cells. Nontumorigenic H-RAS and N-RAS MSU-1.1 cell strains expressed very low levels of their RAS oncoproteins and have activity levels of 8 and 10 PU/24 h/10^6 cells, respectively. Tumorigenic derivatives of MSU-1.1 cells (10/10) from either the RAS or p53 pathway (Fig. 16.3) have significantly higher activity levels, 16–40 PU/24 h/10^6 cells. This is the only assay on the tumorigenic variants of the MSU-1.1 cells in which we found a perfect correlation with the tumorigenic potential of the cells. Five/five patient-derived human fibrosarcoma cell lines have activity levels of 23–134 PU/24 h/10^6 cells. We also measured the activity levels of uPA in the cell culture growth medium from all of the above cell lines/strains. Medium from the normal and tumorigenic cell lines exhibited activity levels of 19–226 PU/24 h/10^6 cells. There was no correlation between the uPA activity in the cell culture growth medium and the tumorigenic potential of the cells. A study by Allgayer et al., [59] demonstrates that when the K-RAS oncogene was disrupted in a human colon cancer cell line (HCT116), the expression of u-PAR was reduced by 50–85% and its ability to degrade the protein laminin was reduced 80%. Therefore, the high activity of the RAS oncoprotein in the RAS-transfected MSU-1.1 cells or changes in the expression of the pathways controlled by RAS, could explain the high receptor bound uPA activity. However, this does not explain the high receptor bound uPA activity in the mutant TP53/TP16^{INK4a} tumorigenic MSU-1.1 cells.

16.14 RHO GTPases: RAC1 and CDC42

Recently, we examined the role of two RHO GTPases, RAC1 and CDC42, which act downstream of H-RAS in maintaining the transformed phenotype in the tumor-derived, H-RAS^{V12}–transformed MSU-1.1 cells [60]. By expression of dominant negative RAC1^{N17} and/or CDC42^{N17} mutants in these transformed cells, we found that secreted uPA was reduced 60 and 70%, respectively. Simultaneous expression of both dominant negative genes in the transformed cells resulted in an additive reduction in uPA.

We also found that RAC1 plays a major role in the ability of the tumor-derived, H-RAS^{V12}–transformed MSU-1.1 cells to grow in medium with reduced serum, and in focus formation. CDC42 plays a major role in the ability of these cells to grow in an anchorage independent manner. Expression of both RAC1 and CDC42 was required for tumor formation by the H-RAS^{V12} transformed cells. However, unlike the situation in NIH3T3 mouse fibroblasts, neither expression of RAC1^{V12} or CDC42^{V12} in MSU-1.1 cells was sufficient to cause malignant transformation [61, 62].

16.15 Sprouty 2 (SPRY2)

We carried out a genome-wide study (Affymetrix) on the expression of m-RNA in the MSU-1.0, MSU-1.1, and the H-RAS^{v12}-transformed MSU-1.1 cells. We found that the MSU-1.1 cells exhibited a modest increase in SPRY2 mRNA and protein compared with the MSU-1.0 and LG1 cells [63]. In contrast, the tumor-derived H-RAS^{v12}-transformed MSU-1.1 cells exhibited a very large increase in SPRY2 mRNA and protein. A similar increase in the amount of epidermal growth factor receptor (EGFR) protein was found.

When we stably reduced the expression of SPRY2 by the use of a shRNA, the H-RAS^{v12}-transformed MSU-1.1 cells lost their transformed characteristics in culture and no longer formed tumors in athymic mice. There was no change in expression level of the H-RAS^{v12} oncoprotein. Additional studies [64] indicated that SPRY2, H-RAS^{v12}, c-CBL, and CIN85 proteins formed a complex which results in sustained EGFR levels, a critical factor in maintaining the transformed characteristics of the cells.

Recently, Schaaf et al., [65] found that the silencing of SPRY1 in human embryonal rhabdomyosarcoma-derived cell lines (carrying a mutated *RAS* gene) results in the loss of their ability to form tumors. This study in conjunction with our study illustrates the essential role of RAS oncoproteins in causing higher expression of SPRY1 and SPRY2 proteins, which in turn gives mesenchymal cells transformed characteristics including tumor formation. A recent study [66] indicates that SPRY2 controls c-MET expression and metastasis in colon cancer cells.

16.16 Low Density Lipoprotein Receptor-Related Protein (LRP12)

Using differential display [67], we compared the mRNA of the MSU-1.1 cells with the mRNA of tumor-derived, chemically transformed MSU-1.1 cells. We discovered a novel gene we called *ST7* that showed a sixfold lower expression in the transformed cell strain [68]. Shortly after we published this study, the sequence for the *LRP3* gene was published [69]. Because *LRP3* and *ST7* exhibited a very strong similarity, we renamed *ST7, LRP12.* We demonstrated that LRP12 is a transmembrane protein with an extracellular and a cytoplasmic domain [70]. The latter domain binds SARA, the SMAD anchor for receptor activation. It also binds RACK1, the receptor for activated protein kinase C, and MIBP, Muscle Integrin Binding Protein. LRP12 is strongly expressed in normal human fibroblasts including LG1, but exhibits reduced levels of expression or is not expressed in 5/5 human patient-derived fibrosarcoma cell lines and in 6/6 tumor-derived strains from the MSU1 lineage. Enhanced expression of LRP12 in carcinogen-transformed tumor-derived MSU-1.1 cells markedly reduces the tumor-forming ability of these cells indicating that *LRP12* functions as a tumor suppressor gene [71].

16.17 Fibulin-1D

Using differential display, we compared the mRNA of the MSU-1.1 cells with the mRNA of tumor-derived, chemically transformed MSU-1.1 cells. We discovered that the *fibulin-1D* gene exhibited markedly reduced expression in the transformed cell strains [72]. The *fibulin-1D* gene codes for an extracellular matrix protein, which is expressed in normal human fibroblasts, but not in malignantly transformed fibroblasts of the MSU1 lineage or in the patient-derived fibrosarcomas. Expression of Fibulin-1D protein in SHAC cells or tumor-derived, chemically transformed MSU-1.1 cells resulted in loss of anchorage independent growth and an extended latency in tumor formation.

16.18 Summary

To our knowledge, this analysis is the most detailed description of the genetic/epigenetic changes required for malignant transformation of human skin fibroblasts (Fig. 16.3). The MSU-1.0 cells acquired the infinite life span phenotype by expression of telomerase. The MSU-1.1 cells acquired the ability to replicate rapidly in medium with reduced serum because of their loss of expression of the $p14^{ARF}$ protein. Whether the slightly enhanced expression of SPRY2 plays a role in the transformation of the MSU-1.1 cells is not clear. We conclude that *one route to malignant transformation of the MSU-1.1 cells* involves: [1] the expression of oncogenic RAS at low levels *and* expression of a SRC-family protein (v-FES), or (2) high expression of oncogenic RAS proteins, or (3) expression of PDGF-B, which gives rise to benign tumors that can become malignant tumors. *A second route to malignant transformation of the MSU-1.1 cells* involves the loss of the ability to produce wild type $p16^{INK}$ protein *and* transcriptionally active TP53 protein. The HT1080 human fibrosarcoma-derived cell line follows the first route. It has a mutant N-RAS oncogene and a functional TP53 protein [42]. The SHAC human fibrosarcoma-derived cell line has a nonfunctional TP53 protein [42] and is reported [20] to have a mutated N-RAS^{His61} and over-express c-MYC. Neither of these cell lines express $p14^{ARF}$ or mutant $p16^{INK}$. The sequence of the various genetic changes described here reflects the order in which we carried out the experiments. In our system, like others, the development of infinite life span cells must be an early change. We hypothesize that the other changes could happen in any sequence. We typically relied on finding newly acquired transformed characteristics to determine the role of specific genetic changes.

In the early literature on human fibroblast transformation, most cell lines are designated as positive or negative for tumor formation without specifying the type of tumor that was formed. More recent literature frequently describes the tumors in some detail. We had pathologists examine slides from all of the tumors generated in our studies. Currently, there is an open discussion in the pathology literature regarding the legitimacy of the many categories of soft tissue sarcomas and the appropriate criteria for inclusion or exclusion of tumors in these categories. Therefore, we are reexamining the various MSU-1.1 derived tumors to determine the pattern of gene expression in the fibromas and sarcomas.

Acknowledgements This research was supported by Department of Energy Contract EV04659, Training Grant ES07076 from the National Institute of Environmental Health Sciences, DHHS Grant CA32924 from the National Cancer Institute, DHHS Grant CA21289 from the National Cancer Institute, a grant from the Michigan College Foundation, Dow Chemical Company, Department of Energy Contract 60524, Contract ES65152 from the National Institute of Environmental Health Sciences, DHHS Grant AG11026 from the National Institute of Aging, Baxter Healthcare Corporation, DHHS Grant 60907 from the National Cancer Institute, Training Grant ES07255 from the National Institute of Environmental Health Sciences (Boley), International Atomic Energy Agency Contract 7250JR2/RBITC, DHHS Grant CA21253 from the National Cancer Institute, DHHS grant CA82885 from the National Cancer Institute, Elsa U. Pardee Foundation, Barnett Rosenberg Fellowship (Battle), DHHS grant CA098305 from the National Cancer Institute, and DHHS Grant CA32924 from the National Cancer Institute.

We would like to acknowledge our main collaborators involved in this research: D.M. Appledorn, M.A. Battle, S.E. Boley, J. Coleman, K.-H.T. Dao, J.E. Dillberger, R.W. Dunstan, M. Elkassaby, T.R. Fox, J.D. Fratkin, D.G. Fry, G. Henry, S. Huang, P.J. Hurlin, J. Jankun, A. Jurek, E. Kaplan, A.S. Khan, S. Kleff, S. K. Kohler, H. Liang, C.-C. Lin, P. Lito, Z. Lou, C. Louden, E. Mahoney-Leo, T. P. McManus, B.D. Mets, L. D. Milam, T. L. Morgan, S. O'Reilly, H. Palmer, J. Przybyszewski, J. Qing, D.S. Reinhold, P.A. Ryan, J.M. Scheid, R. Schilz, K.C. Silinskas,S. D. Sleight, M. Walicka, Q. Wang, D. Wei, J.C. Wigle, D. M. Wilson, J. Wray, D. Yang, J. Zhang. We would also like to acknowledge the editorial assistance of: B.A. Heinlen, S.K. Kohler, and S.A. O'Reilly.

References

1. Earle WR (1943) Production of malignancy in vitro IV. The mouse fibroblast cultures and changes seen in living cells. J Natl Cancer Inst 4:165–212
2. Roussos ET, Keckesova Z, Haley JD, Epstein DM, Weinberg RA, Condeelis JS (2010) AACR special conference on epithelial-mesenchymal transition and cancer progression and treatment. Cancer Res 70: 7360–7364
3. Earle WR, Nettleship A, Schilling EL, Stark TH, Straus NR, Brown MF, Shelton E (1943) Production of malignancy in vitro V. Results of injections of cultures into mice. J Natl Cancer Inst 4:213–227
4. McCormick JJ, Maher VM (1988) Towards an understanding of the malignant transformation of diploid human fibroblasts. Mutat Res 199:273–291
5. Namba M, Nishitani K, Fukushima F, Kimoto T, Utsunomiya J, Hayflick L (1981) Neoplastic transformation of human diploid fibroblasts treated with chemical carcinogens and Co-60 Y-rays. Gann Monogr Cancer Res 27:221–230
6. Namba M, Nishitani K, Hyodoh F, Fukushima F, Kimoto T (1985) Neoplastic transformation of human diploid fibroblasts (KMST-6) by treatment with Co-60 gamma rays. Int J Cancer 35:275–280
7. Yoshii H, Watanabe M (2009) Intervention of oxygen-control ability to radiation sensitivity, cell aging and cell transformation. J Radiat Res (Tokyo) 50:127–137
8. Drayton S, Peters G (2002) Immortalisation and transformation revisited. Curr Opin Genet Dev 12:98–104. Review
9. Drayton S, Brookes S, Rowe J, Peters G (2004) The significance of p16INK4a in cell defenses against transformation. Cell Cycle 3:611–615. Review
10. Kendall SD, Adam SJ, Counter CM (2006) Genetically engineered human cancer models utilizing mammalian transgene expression. Cell Cycle 5:1074–1079
11. Morgan TL, Yang D, Fry DG, Hurlin PJ, Kohler SK, Maher VM, McCormick JJ (1991) Characteristics of an infinite life span diploid human fibroblast cell strain and a near-diploid strain arising from a clone of cells expressing a transfected v-myc oncogene. Exp Cell Res 197:125–136
12. Land H, Chen AC, Morgenstern JP, Parada LF, Weinberg RA (1986) Behavior of myc and ras oncogenes in transformation of rat embryo fibroblasts. Mol Cell Biol 6:1917–1925
13. Gallego Melcón S, Sánchez de Toledo Codina J (2007) Molecular biology of rhabdomyosarcoma. Clin Transl Oncol 9:415–419. Review
14. Takakura M, Kyo S, Kanaya T, Hirano H, Takeda J, Yutsudo M, Inoue M (1999) Cloning of human telomerase catalytic subunit (hTERT) gene promoter and identification of proximal core promoter sequences essential for transcriptional activation in immortalized and cancer cells. Cancer Res 59:551–557
15. Wang Q, Maher VM, McCormick JJ (1993) Telomerase activity in human fibroblasts correlates with the rate of cell growth, not with the rate of changes in telomere length. Proceedings-International Symposium on Molecular Mechanisms of Radiation and Chemical Carcinogen-Induced Cell Transformation, sponsored by the Department of Energy and the Commission of European Communities, Mackinac Island, MI, Sep. 19–23, Abstract #26
16. Kyo S, Takakura M, Taira T, Kanaya T, Itoh H, Yutsudo M, Ariga H, Inoue M (2000) Sp1 cooperates

with c-Myc to activate transcription of the human telomerase reverse transcriptase gene (hTERT). Nucleic Acids Res 28:669–677
17. Lou Z, Maher VM, McCormick JJ (2005) Identification of the promoter of human transcription factor Sp3 and characterization of the role of Sp1 and Sp3 in the expression of Sp3 protein. Gene 351:51–59
18. Tao L, Dong Z, Zannis-Hadjopoulos M, Price GB (2001) Immortalization of human WI38 cells is associated with differential activation of the c-myc origins. J Cell Biochem 82:522–534
19. Casillas MA, Brotherton SL, Andrews LG, Ruppert JM, Tollefsbol TO (2003) Induction of endogenous telomerase (hTERT) by c-Myc in WI-38 fibroblasts transformed with specific genetic elements. Gene 316:57–65
20. Andeol Y, Nardeux PC, Daya-Grosjean L, Brison O, Debrian J, Suarez H (1988) Both N-ras and c-myc are activated in the SHAC human stomach fibrosarcoma cell line. Int J Cancer 41:732–737
21. Missiaglia E, Selfe J, Hamdi M, Williamson D, Schaaf G, Fang C, Koster J, Summersgill B, Messahel B, Versteeg R, Pritchard-Jones K, Kool M, Shipley J (2009) Genomic imbalances in rhabdomyosarcoma cell lines affect expression of genes frequently altered in primary tumors: an approach to identify candidate genes involved in tumor development. Genes Chromosomes Cancer 48:455–467
22. Tsiatis AC, Herceg ME, Keedy VL, Halpern JL, Holt GE, Schwartz HS, Cates JM (2009) Prognostic significance of c-Myc expression in soft tissue leiomyosarcoma. Mod Pathol 22:1432–1438
23. Taggie-Hoffman A, McCormick JJ (1999) Decreased p16INK4a expression in tumorigenic fibroblastic cell lines derived from MSU-1.1 cells [unpublished studies]
24. Ozenne P, Eymin B, Brambilla E, Gazzeri S (2010) The ARF tumor suppressor: structure, functions and status in cancer. Int J Cancer 127:2239–2247. Review
25. Voorhoeve PM, Agami R (2003) The tumor-suppressive functions of the human INK4a locus. Cancer Cell 4:311–319
26. Weihrauch M, Markwarth A, Lehnert G, Wittekind C, Wrbitzky R, Tannapfel A (2002) Abnormalities of the ARF-p53 pathway in primary angiosarcomas of the liver. Hum Pathol 33:884–892
27. Obana K, Yang HW, Piao HY, Taki T, Hashizume K, Hanada R, Yamamoto K, Tanaka Y, Toyoda Y, Takita J, Tsuchida Y, Hayashi Y (2003) Aberrations of p16INK4A, p14ARF and p15INK4B genes in pediatric solid tumors. Int J Oncol 23:1151–1157
28. Brown R, Marshall CJ, Pennie SG, Hall A (1984) Mechanism of activation of an N-ras gene in the human fibrosarcoma cell line HT1080. EMBO J 3:1321–1326
29. Paterson H, Reeves B, Brown R, Hall A, Furth M, Bos J, Jones P, Marshall C (1987) Activated N-ras controls the transformed phenotype of HT1080 human fibrosarcoma cells. Cell 51:803–812
30. Hamad NM, Elconin JH, Karnoub AE, Bai W, Rich JN, Abraham RT, Der, CJ, Counter CM (2002) Distinct requirements for Ras oncogenesis in human versus mouse cells. Genes Dev 16:2045–2057. Review
31. Fry DG, Milam LD, Dillberger JE, Maher VM, McCormick JJ (1990) Malignant transformation of an infinite life span human fibroblast cell strain by transfection with v-Ki-ras. Oncogene 5:1415–1418
32. Wilson DM, Yang D, Dillberger JE, Dietrich SE, Maher VM, McCormick JJ (1990) Malignant transformation of human fibroblasts by a transfected N-ras oncogene. Cancer Res 50:5587–5593
33. Lin C, Maher VM, McCormick JJ (1995) Malignant transformation of human fibroblast strain MSU-1.1 by v-fes requires an additional genetic change. Int J Cancer 63:140–147
34. Hurlin PJ, Maher VM, McCormick JJ (1989) Malignant transformation of human fibroblasts caused by expression of a transfected T24 HRAS oncogene. Proc Natl Acad Sci USA 86:187–191
35. Wong-Staal F, Dalla-Favera R, Gelmann EP, Manzari V, Szala S, Josephs SF, Gallo RC (1981) The v-sis transforming gene of simian sarcoma virus is a new onc gene of primate origin. Nature 294:273–275
36. Pantazis P, Pelicci PG, Dalla-Favera R, Antoniades HN (1985) Synthesis and secretion of proteins resembling platelet-derived growth factor by human glioblastoma and fibrosarcoma cells in culture. Proc Natl Acad Sci USA 82:2404–2408
37. Yang D, Kohler SK, Maher VM, McCormick JJ (1994) v-sis oncogene-induced transformation of human fibroblasts into cells capable of forming benign tumors. Carcinogenesis 15:2167–2175
38. Weiss R, Teicg N, Varmus H, Coffin J (eds) (1985) RNA tumor viruses: molecular biology of tumor viruses, 2nd ed. 2/Supplements and appendices. Cold Spring Harbor Laboratory, New York
39. Guan H, Zhou Z, Gallick GE, Jia SF, Morales J, Sood AK, Corey SJ, Kleinerman ES (2008) Targeting Lyn inhibits tumor growth and metastasis in Ewing's sarcoma. Mol Cancer Ther 7:1807–1816
40. Lin C, Wang Q, Maher VM, McCormick JJ (1994) Malignant transformation of a human fibroblast cell strain by transfection of a v-fes oncogene, but not by transfection of a gag-human c-fes construct. Cell Growth Differ 5:1381–1387
41. Reinhold DS, Walicka M, Elkassaby M, Milam LD, Kohler SK, Dunstan RW, McCormick JJ (1996) Malignant transformation of human fibroblasts by ionizing radiation. Intl J Radiat Biol 69:707–715
42. Boley SE, McManus TP, Maher VM, McCormick JJ (2000) Malignant transformation of human fibroblast cell strain MSU-1.1 by N-methyl-N-nitrosourea: evidence of elimination of p53 by homologous recombination. Cancer Res 60:4105–4111
43. Yang D, Louden C, Reinhold DS, Kohler SK, Maher VM, McCormick JJ (1992) Malignant transformation of human fibroblast cell strain MSU-1.1 by (+)-7β,8α-dihydroxy-9α,10α-epoxy-7,8,9,10-tetrahydrobenzo[a]pyrene. Proc Natl Acad Sci USA 89:2237–2241
44. Ashraf-Khan S (1996) Malignant transformation of the human fibroblast cell strain MSU-1.1 by ultraviolet

radiation: correlation of tumorigenicity with anchorage independence and growth factor independence. ProQuest, UMI Dissertations Publishing
45. O'Reilly S, Walicka M, Kohler SK, Dunstan R, Maher VM, McCormick JJ (1998) Dose-dependent transformation of MSU-1.1 human fibroblasts and characterization of the neoplastic characteristics. Radiation Res 150:577–584
46. Schärer E, Iggo R (1992) Mammalian p53 can function as a transcription factor in yeast. Nucleic Acids Res 20:1539–1545
47. McManus TP, McCormick JJ (2000) Restoration of wild-type p53 in focus-derived, tumorigenic human fibroblast strain lacking p53 expression inhibits focus formation and tumorigenicity of the cells without affecting cell growth in culture [unpublished studies]
48. Pérot G, Chibon F, Montero A, Lagarde P, de Thé H, Terrier P, Guillou L, Ranchère D, Coindre JM, Aurias A (2010) Constant p53 pathway inactivation in a large series of soft tissue sarcomas with complex genetics. Am J Pathol 177:2080–2090
49. O'Reilly SA, McCormick JJ (2002) Analysis of genetic changes involved in carcinogen-induced malignant transformation of a human fibroblast cell strain in culture [unpublished studies]
50. Ortega S, Malumbres M, Barbacid M (2002) Cyclin D-dependent kinases, INK4 inhibitors and cancer. Biochim Biophys Acta 1602:73–87
51. Bozic I, Antal T, Ohtsuki H, Carter H, Kim D, Chen S, Karchin R, Kinzler KW, Vogelstein B, Nowak MA (2010) Accumulation of driver and passenger mutations during tumor progression. Proc Natl Acad Sci USA 107:18545–18550
52. Birchmeier C, Birchmeier W, Gherardi E, Vande Woude GF (2003) MET, metastasis, motility and more. Nat Rev Mol Cell Biol 4:915–925
53. Liang H, O'Reilly S, Liu Y, Abounader R, Laterra J, Maher VM, McCormick JJ (2004) Sp1 regulates expression of MET, and ribozyme-induced down-regulation of MET in fibrosarcoma-derived human cells reduces or eliminates their tumorigenicity. Int J Oncol 24:1057–1068
54. Lou Z, O'Reilly S, Liang H, Maher VM, Sleight SD, McCormick JJ (2005) Down-regulation of overexpressed Sp1 protein in human fibrosarcoma cell lines inhibits tumor formation. Cancer Res 65:1007–1017
55. Taubert H, Würl P, Greither T, Kappler M, Bache M, Lautenschläger C, Füssel S, Meye A, Eckert AW, Holzhausen HJ, Magdolen V, Kotzsch M (2010) Co-detection of members of the urokinase plasminogen activator system in tumour tissue and serum correlates with a poor prognosis for soft-tissue sarcoma patients. Br J Cancer 102:731–737
56. Teng HW, Wang HW, Chen WM, Chao TC, Hsieh YY, Hsih CH, Tzeng CH, Chen PC, Yen CC (2011) Prevalence and prognostic influence of genomic changes of EGFR pathway markers in synovial sarcoma. J Surg Oncol 103:773–781
57. Kilvaer TK, Valkov A, Sorbye S, Smeland E, Bremnes RM, Busund LT, Donnem T (2010) Profiling of VEGFs and VEGFRs as prognostic factors in soft tissue sarcoma: VEGFR-3 is an independent predictor of poor prognosis. PLoS One 14:e15368
58. Jankun J, Maher VM, McCormick JJ (1991) Malignant transformation of human fibroblasts correlates with increased activity of receptor-bound plasminogen activator. Cancer Res 51:1221–1226
59. Allgayer H, Heiss MM, Riesenberg R, Grutzner KU, Tarabichi A, Babic R, Schildberg FW (1997) Urokinase plasminogen activator receptor (uPA-R): one potential characteristic of metastatic phenotypes in minimal residual tumor disease. Cancer Res 57:1394–1399
60. Appledorn DM, Dao K, O'Reilly S, Maher VM, McCormick JJ (2010) Rac1 and Cdc42 are regulators of HRasV12-transformation and angiogenic factors in human fibroblasts. BMC Cancer 10:13
61. Qiu RG, Abo A, McCormick F, Symons M (1997) Cdc42 regulates anchorage-independent growth and is necessary for Ras transformation. Mol Cell Biol 17:3449–3458
62. Qiu RG, Chen J, Kirn D, McCormick F, Symons M (1995) An essential role for Rac in Ras transformation. Nature 374:457–459
63. Lito P, Mets BD, Kleff S, O'Reilly S, Maher VM, McCormick JJ (2008) Evidence that Sprouty 2 is necessary for sarcoma formation by H-Ras oncogene-transformed human fibroblasts. J of Biol Chem 283: 2002–2009
64. Lito P, Mets BD, Appledorn DM, O'Reilly S, Maher VM, McCormick JJ (2009) Sprouty 2 regulates DNA Damage-induced apoptosis in Ras-transformed human fibroblasts. J Biol Chem 284:848–854
65. Schaaf G, Hamdi M, Wijnenburg D, Lakeman A, Geerts D, Ersteeg R, Kool M (2010) Silencing of SPRY1 Triggers Complete Regression of Rhabdomyosarcoma Tumors Carrying a Mutated RAS. Gene Cancer Res 70:762–771
66. Yigsaw Y, Dartin L, Pierre S, Scholich K, Patel TB (2001) The C terminus of Sprouty is important for modulation of cellular migration and proliferation. J Biol Chem 276:22742–22747
67. Liang P, Averboukh L, Pardee AB (1993) Distribution and cloning of eukaryotic mRNAs by means of differential display: refinements and optimization. Nucleic Acids Res 21:3269–3275
68. Qing J, Wei D, Maher VM, McCormick JJ (1999) Cloning and characterization of a novel gene encoding a putative transmembrane protein with altered expression in human tumors. Oncogene 18:335–342
69. Ishii H, Kim DH, Fujita T, Endo Y, Saeki S, Yamamoto TT (1998) cDNA cloning of a new low-density lipoprotein receptor-related protein and mapping of its gene (LRP3) to chromosome bands 19q12-q13. 2. Genomics 51:132–135
70. Battle MA, Maher VM, McCormick JJ (2003) ST7 is a novel low-density lipoprotein receptor-related protein (LRP) with a cytoplasmic tail that interacts with

proteins related to signal transduction pathways. Biochemistry 42:7270–7282
71. Battle MA, McCormick JJ (2001) Expression of ST7, a novel low-density lipoprotein receptor-related protein (LRP), inhibits tumorigenicity in human fibrosarcomas [unpublished studies]
72. Qing J, Maher VM, Tran H, Argraves WS, Dunstan RW, McCormick JJ (1997) Suppression of anchorage-independent growth and matrigel invasion and delayed tumor formation by elevated expression of fibulin-1D in human fibrosarcoma-derived cell lines. Oncogene 15:2159–2168

17 A Novel Tumor Suppressor, *REIC/Dkk-3* Gene Identified by Our In Vitro Transformation Model of Normal Human Fibroblasts Works as a Potent Therapeutic Anti-tumor Agent

Masakiyo Sakaguchi, Nam-ho Huh, and Masayoshi Namba

Abstract

Reduced Expression in Immortalized Cell (REIC) was cloned by subtractive hybridization method as a gene whose expression is reduced in many human immortalized and neoplastic tumor cells. The REIC, when overexpressed by an adenovirus (Ad-REIC), exhibited a dramatic therapeutic effect on a wide variety of human cancers through a mechanism triggered by ER-stress-mediated JNK activation. In addition to this direct effect on cancer cells, Ad-REIC exerted another cytotoxicity on human cancers, an indirect host-mediated effect due to overproduction of IL-7 by mis-targeted normal cells. This "one-bullet two-arms" finding may lead to a powerful new therapeutic approach to the treatment of human cancers.

17.1 Introduction

Selective elimination of tumor cells is a key issue in treating human cancers. During malignant conversion and progression, various genetic changes take place in cells that could be potential targets for cancer gene therapy. Among the genes, it is well known that tumor suppressors like p53, PTEN, and MDA7/IL24, whose loss in expression or function are a critical step in the process of carcinogenesis, exert a selective killing effect on cancer cells when they are overexpressed [1–3].

In addition to them, a novel tumor suppressor gene, REIC/Dkk-3 was originally identified from our in vitro transformation model of normal human fibroblasts by subtractive hybridization technique as a gene whose expression is downregulated in human immortalized fibroblast cell lines [4]. This molecule was subsequently recognized to display remarkably decreased expression property in a wide range of cancer species [4–9]. Interestingly, a majority of human cancer cells, but not normal cells, undergo apoptotic cell death when exposed to adenovirus vector carrying REIC/Dkk-3 (Ad-REIC) in vitro as well as in vivo [7–11]. Ad-REIC has also another mechanism of action against human cancer, an indirect host-mediated effect due to overproduction of IL-7 by mis-targeted normal cells [12]. Thus, Ad-REIC shows potent and selective anti-tumor

M. Namba (✉)
Niimi College, 1263-2 Nishigata, Niimi 718-8585, Japan
e-mail: mnamba@md.okayama-u.ac.jp

J.S. Rhim and R. Kremer (eds.), *Human Cell Transformation: Role of Stem Cells and the Microenvironment*, Advances in Experimental Medicine and Biology 720,
DOI 10.1007/978-1-4614-0254-1_17,

function that implies great promise for clinical application. In this manuscript, we mentioned about current understanding of the functions and properties of REIC/Dkk-3 with potential to be a prominent agent for cancer specific gene therapy.

17.2 Establishment of Immortalized Human Cell Lines

Accumulating pieces of evidence indicate that cellular immortalization is a key step for human carcinogenesis, so that we first intended to establish an in vitro transformation model of normal human fibroblasts, and then to analyze what genetic and cytological changes are involved in the transformation process using our model. After many attempts, we succeeded in establishing three immortalized human fibroblast cell lines by treatment with either a chemical carcinogen, 4-nitroquinoline 1-oxide (SUSM-1 and OUMS24F) or Co-60 gamma ray (KMST-6) [13]. In the course of these studies we learned that the immortalization step is essential for the neoplastic transformation of normal cells. Once normal human fibroblasts were immortalized, they were easily transformed into neoplastic ones [14]. Thus, our in vitro transformation model of normal human fibroblasts is good tool for studying the immortalization and following neoplastic transformation processes of normal human fibroblasts.

17.3 Properties of REIC/Dkk-3

17.3.1 Identification of REIC/Dkk-3

The finding that when immortalized human cells were fused with senescent cells, the fused cells stopped growing indicated that the senescent phenotype is dominant, whereas the immortal one is recessive [15]. This phenomenon suggests that tumor suppressor genes are dominantly present in senescent cells and contrary to this the genes are declined in immortalized cells. Based on this hypothesis, we focused on studying the genetic changes in immortalized fibroblast cell lines (KMST-6, 358 the PDL), as compared with their normal senescent counterparts (KMS-6, 48 the PDL) by the representative difference analysis system (subtractive hybridization method). As a result, several cDNA clones whose expression was downregulated in the immortalized cells were isolated. One of them was identified and named REIC (Reduced Expression in Immortalized Cell) [4]. A search of the DDBJ database showed that the amino acid sequence of REIC is identical to that of Dkk-3 protein, which is a novel member of the *Dickkopfs (Dkks)* gene family required for head induction in amphibian embryos and a potent Wnt inhibitor, so that the present common name is REIC/Dkk-3. REIC/Dkk-3 was consistently expressed to a lower degree in three immortalized human cell lines. In addition, the expression was markedly lower in several human tumor-derived cell lines (Hep3B, HuH-6, and HuH-7, hepatocellular carcinoma; HuCCT-1, cholangiocarcinoma; PC3, DU-145, and LNCaP, prostate cancer; Caki-1, Caki-2, ACHN, and KPK-1, renal cancer; NCCIT, testicular germ cell cancer; HeLa, cervical cancer; SaOS-2, osteosarcoma; A-549, lung cancer; MCF-7, breast cancer) [4–9]. These observations support a critical role of REIC/Dkk-3 as a tumor suppressor gene.

17.3.2 Gene Location and Protein Structure of REIC/Dkk-3

The chromosomal localization of *REIC/Dkk-3* gene is 11p15 where LOH of several genetic markers frequently occurs near the REIC/Dkk-3 locus in a variety of human carcinomas [16]. In addition to the LOH, it is also reported that silence of gene expression caused by genomic imprinting occurred on the 11p15 in many cancer species with high frequency [17]. Furthermore, hypermethylation of the REIC/Dkk-3 promoter has been found in a subset of human tumors [18]. Thus, LOH, genomic imprinting, and promoter methylation may prominently influence the reduced expression of REIC/Dkk-3 in cancer cells.

An open reading frame (ORF) of human REIC/Dkk-3 encodes a protein 350 amino acids with an estimated molecular weight of 38,390 and an isoelectric point of 4.3. The protein is a secretory protein possessing N-terminal signal peptide (putative amino acids region: 1–21) and four potential *N*-glycosylation sites (at amino acids 96N, 106N, 121N, 204N, respectively). REIC protein comprises one potential coiled-coil motif (amino acids 40–84), which is predicted to involve in protein–protein interaction, and two Dickkopf cysteine-rich motifs (amino acids 147–195 and 208–284).

17.3.3 Tissue and Subcellular Distributions of REIC/Dkk-3

REIC/Dkk-3 seems to be differently expressed depending on tissue types. Among the human tissues examined, the REIC/Dkk-3 transcript was highly expressed in the heart and brain, which contain a large number of postmitotic cells; moderately expressed in the skeletal muscle, colon, spleen, kidney, liver, small intestine, placenta, and lung; and barely detectable in the thymus and peripheral blood leukocytes [4].

In cultured normal human fibroblasts, REIC/Dkk-3 protein is localized especially in perinuclear region (Endoplasmic Reticulum (ER)) and sparsely distributed in cytoplasmic space as dots (secretory granules), implying a specific distribution of the protein in the typical secretory route through the ER [12].

17.4 Selective Anti-Tumor Function of REIC/Dkk-3

17.4.1 REIC/Dkk-3 and Apoptosis

REIC/Dkk-3 has a growth suppressive function in human cancer cells since forced expression of REIC/Dkk-3 using a plasmid vector inhibited growth of a human Saos-2 osteosarcoma cell line [16]. This finding and expression properties of REIC/Dkk-3 indicate that it may act as a tumor suppressor gene and could provide a new means of treatment for human cancers.

To more efficiently express *REIC/Dkk-3* gene and to study the mechanism by which this gene suppresses growth of tumor cells, a replication incompetent adenovirus carrying human REIC/Dkk-3 cDNA (Ad-REIC) was constructed [7]. Initial studies in the context of prostate cancer cells (PC3, DU-145, and LNCaP lines) demonstrated that over-expression of REIC/Dkk-3 by the in vitro infection with Ad-REIC markedly inhibited cellular proliferation by inducing apoptosis. The same apoptotic events occurred in testicular cancer (NCCIT) and mesothelioma (211H, H28, H2052, H2452), whereas no apoptosis was observed in various normal human cells (OUMS-24 fibroblasts and epidermal keratinocytes), indicating that this effect was specific to cancer cells [7–12]. Furthermore, we showed tumor-specific induction of apoptosis by Ad-REIC in vivo. A single injection of Ad-REIC into tumors formed by subcutaneous transplantation of human prostate cancer cells (PC3) into mice resulted in four out of five mice becoming tumor-free [7]. Thus, the Ad-REIC has dramatic therapeutic effects on human cancers, while no apoptotic effects on normal human cells.

17.4.2 Mechanistic Insights into the Cancer Specific Apoptosis Induced by REIC/Dkk-3

In depth analyses into the mechanism of REIC/Dkk-3 action in eliciting cancer specific apoptosis, we used PC3 prostate cancer cell line, the most sensitive cell line among examined. The amount and intracellular localization of β-catenin were not affected by Ad-REIC infection, excluding the possibility of involvement of the canonical pathway of Wnt signaling (unpublished data). Further studies of the potential mechanism indicated a functional upregulation of the c-jun N-terminal kinase (JNK) that could activate pro-apoptotic molecule Bax (mitochondrial translocation) with a reduction of anti-apoptotic Bcl-2 protein, irrespective of cellular *p53* gene status. The mitochondrial translocation of Bax protein was also linking to the release of cytochrome c into the cytoplasm, and finally leads to apoptotic cell death [7–10].

However, it is not clear how JNK was activated by Ad-REIC infection in a cancer specific manner. In this issue, it should be noted that MDA-7/IL-24, the other secretory tumor suppressor protein, selectively induces apoptosis in many different types of human cancer cells. Adenovirus-mediated over-expression of MDA-7/IL-24 exerts an activation of JNK pathway through the ER-stress sensing system [19]. With this clue, we also found that Ad-REIC actually causes an ER-stress, which is possibly involved in the JNK activation and subsequent cancer specific apoptosis like MDA-7/IL-24 [12, 20].

17.4.3 REIC/Dkk-3 and Metastasis

Tumor metastasis is mediated by multiple genetic and physiological changes in cancer cells that frequently occur in a temporal manner during the process of tumor progression [21]. Key components of the metastatic phenotype are tumor cell migration and invasion. The ability of REIC/Dkk-3 to affect tumor cell migration and invasion has been evaluated in the context of direct intratumoral administration of Ad-REIC to the mouse RM-9 prostate cancer cells using an orthotopic mouse prostate cancer model. We demonstrated that the Ad-REIC treatment inhibited prostate cancer growth and lymph node metastasis, resulting in prolonged mice survival in the model. This therapeutic response was also consistent with the in vitro suppression of cell migration and invasion following to a reduced activity of matrix metalloprotease-2 (MMP-2) [22].

17.4.4 REIC/Dkk-3 and Multidrug Resistance

Multidrug resistance (MDR) is most commonly occurred in drug-resistant cancer and often associated with an over-expression of the *MDR1* gene encoding *P*-glycoprotein [23, 24]. This phenomenon is a major problem in the chemotherapy of progressive cancers, since the protein extrudes anti-cancer drugs. Hence, the in vitro effects of Ad-REIC were also investigated in terms of the sensitivity of multidrug-resistant MCF7/ADR cells, a human breast cancer MCF7 subline, to doxorubicin and of the *P*-glycoprotein expression. Infection of Ad-REIC to the MCF7/ADR cells in culture induced remarkable downregulation of *P*-glycoprotein expression through the JNK activation, and sensitized its drug resistance against doxorubicin. These results imply that Ad-REIC might be an attractive agent against drug-resistant cancers in combination with conventional anti-neoplastic agents [25].

17.4.5 Overcoming a Tolerance to Ad-REIC

The induction of apoptosis in cancer cells by Ad-REIC is mainly due to ER stress caused by overproduction of REIC/Dkk-3 in the ER. However, it has also become evident that some human cancers are resistant to Ad-REIC-induced ER-stress. Therefore, it is important to understand the resistant mechanism to overcome the tolerance to Ad-REIC.

In this issue, we isolated resistant clones from a human prostate cancer cell line, PC3, that is sensitive to Ad-REIC and found that over-expression of BiP/GRP78 (BiP), an ER-residing chaperone protein, is a major determinant of the acquired and inherent resistance to Ad-REIC. Downregulation of BiP using siRNA restored sensitivity of tumors formed by transplantation of a resistant clone to therapeutic application of Ad-REIC in vivo [20]. Thus, BiP is useful for diagnosis of inherent and acquired resistance of cancers and also as a target molecule to overcome resistance to the gene therapeutic Ad-REIC.

17.4.6 Indirect Anti-Tumor Effect of Ad-REIC

As described above, we showed that a single injection of Ad-REIC into tumors formed by transplantation of PC3 cells into mice, resulted in 4 out of 5 mice becoming tumor-free [7].

This was more efficient than had been expected considering that the infection efficiency of an adenovirus vector cannot be 100% in solid tumors in vivo. The efficient therapeutic effect, however, is conceivable when Ad-REIC has an additional noncell autonomous mechanism for tumor suppression. We also showed that intratumoral injection of Ad-REIC resulted in suppression of local metastasis of a prostate cancer cell line in an orthotopic model [22]. This may be at least in part due to an indirect effect via Ad-REIC-infected normal cells in the environment surrounding cancer cells.

Then, we investigated the effect of Ad-REIC infection in normal human fibroblasts, which is insensitive to Ad-REIC-mediated apoptosis. When cultured normal human fibroblasts were infected with Ad-REIC, secreted Interleukin-7 (IL-7) level was elevated. After the transplantation of Ad-REIC-infected NHF with untreated PC3 cells in vivo, the tumor growth was significantly suppressed. This effect was partially abrogated by the injection of an IL-7 antibody. Furthermore, the effect was not observed in an in vitro co-culture system. These results show that the infection of normal cells by Ad-REIC activates the immune system through production of IL-7, while Ad-REIC infection of cancer cells results in a potent selective cell-autonomous killing function without immune suppression [12].

We further examined the mechanisms of the induction of IL-7 by Ad-REIC in normal human fibroblasts. Application of REIC/Dkk-3 recombinant protein did not show any appreciable effect on the expression and secretion of IL-7. IRE1α, an ER stress sensor protein, was activated on infection of normal human fibroblasts with Ad-REIC. Activation of IRE1α resulted in the induction of IRF-1 via ASK1, p38, and STAT1. Finally, STAT1-mediated up-regulation of IRF-1 caused the transcriptional activation of the *IL-7* gene. These results indicate that ER stress exerted by overproduction of REIC/Dkk-3 leads to induction of IL-7, a nonredundant pleiotropic regulator for both innate and acquired immune systems, including the development of NK cells [12].

17.5 Conclusion

In our proceedings of REIC/Dkk-3 study, the molecule has progressed from cloning as a novel gene, whose expression is reduced in various immortalized and neoplastic tumor cells [4–9], to elucidation of selective anti-cancer properties in vitro (in culture) and in vivo (in animal model) [7–12]. When REIC/Dkk-3 cDNA was used as therapeutic adenovirus (Ad-REIC), Ad-REIC has revealed a number of interesting and unique anti-cancer properties, including direct cancer-killing property without any harmful effect to normal cells [7–11], metastatic inhibition property [22], suppressive function of multidrug resistance [25], and immune-modulating activity [12]. These selective multitargeting anti-cancer functions of Ad-REIC show a great promise for clinical application.

17.6 Future Direction

Downregulation of BiP sensitized resistant cancer cells not only to Ad-REIC as shown in the present study, but also to many chemotherapeutic agents [20, 26]. Transfection with siRNA in vivo was effective in reducing target proteins in the present experimental settings [20]. However, since it is unlikely that a similar method can be applied to human cancers, safer and more effective methods for the gene delivery are to be developed. Substances to inhibit BiP have been intensively screened and some candidate molecules have been reported [27, 28]. Based on the fact that normal human fibroblasts were not sensitized to Ad-REIC by downregulation of BiP [20], combination of a BiP inhibitor with Ad-REIC probably leads to a potent and specific therapeutic approach against inherent or acquired resistant cancers in the near future.

Adenovirus vector is one of the most widely used vectors for gene therapies, including those for cancers. One of the serious problems associated with adenovirus vector-based gene therapies is that expression of Coxsackievirus and Adenovirus Receptor (CAR) [29, 30], a receptor for the adenovirus vector, is often reduced among advanced cancers. Recently, we succeeded in enhancing expression level of a cargo gene by structural modification of vector. A newly designed vector showed 10–100-fold higher expression than that attained by existing high-expression vectors (unpublished data). This vector will certainly improve the therapeutic effect of Ad-REIC for a wide variety of human cancers.

References

1. Nielsen LL, Maneval DC (1998) P53 tumor suppressor gene therapy for cancer. Cancer Gene Ther 5: 52–63
2. Saito Y, Swanson X, Mhashilkar AM, Oida Y, Schrock R, Branch CD, Chada S, Zumstein L, Ramesh R (2003) Adenovirus-mediated transfer of the PTEN gene inhibits human colorectal cancer growth in vitro and in vivo. Gene Ther 10:1961–1969
3. Fisher PB (2005) Is mda-7/IL-24 a "magic bullet" for cancer? Cancer Res 65:10128–10138
4. Tsuji T, Miyazaki M, Sakaguchi M, Inoue Y, Namba M (2000) A REIC gene shows down-regulation in human immortalized cells and human tumor-derived cell lines. Biochem Biophys Res Commun 268: 20–24
5. Nozaki I, Tsuji T, Iijima O, Ohmura Y, Andou A, Miyazaki M, Shimizu N, Namba M (2001) Reduced expression of REIC/Dkk-3 gene in non-small cell lung cancer. Int J Oncol 19:117–121
6. Kurose K, Sakaguchi M, Nasu Y, Ebara S, Kaku H, Kariyama R, Arao Y, Miyazaki M, Tsushima T, Namba M, Kumon H, Huh NH (2004) Decreased expression of REIC/Dkk-3 in human renal clear cell carcinoma. J Urol 171:1314–1318
7. Abarzua F, Sakaguchi M, Takaishi M, Nasu Y, Kurose K, Ebara S, Miyazaki M, Namba M, Kumon H, Huh NH (2005) Adenovirus-mediated overexpression of REIC/Dkk-3 selectively induces apoptosis in human prostate cancer cells through activation of c-Jun-NH2-kinase. Cancer Res 65:9617–9622
8. Tanimoto R, Abarzua F, Sakaguchi M, Takaishi M, Nasu Y, Kumon H, Huh NH (2007) REIC/Dkk-3 as a potential gene therapeutic agent against human testicular cancer. Int J Mol Med 19:363–368
9. Kobayashi T, Sakaguchi M, Tanimoto R, Abarzua F, Takaishi M, Kaku H, Kataoka K, Saika T, Nasu Y, Miyazaki M, Kumon H, Huh NH (2008) Mechanistic analysis of resistance to REIC/Dkk-3-induced apoptosis in human bladder cancer cells. Acta Med Okayama 62:393–401
10. Abarzua F, Sakaguchi M, Tanimoto R, Sonegawa H, Li DW, Edamura K, Kobayashi T, Watanabe M, Kashiwakura Y, Kaku H, Saika T, Nakamura K, Nasu Y, Kumon H, Huh NH (2007) Heat shock proteins play a crucial role in tumor-specific apoptosis by REIC/Dkk-3. Int J Mol Med 20:37–43
11. Kashiwakura Y, Ochiai K, Watanabe M, Abarzua F, Sakaguchi M, Takaoka M, Tanimoto R, Nasu Y, Huh NH, Kumon H (2008) Down-regulation of inhibition of differentiation-1 via activation of activating transcription factor 3 and Smad regulates REIC/Dickkopf-3-induced apoptosis. Cancer Res 68: 8333–8341
12. Sakaguchi M, Kataoka K, Abarzua F, Tanimoto R, Watanabe M, Murata H, Than SS, Kurose K, Kashiwakura Y, Ochiai K, Nasu Y, Kumon H, Huh NH (2009) Overexpression of REIC/Dkk-3 in normal fibroblasts suppresses tumor growth via induction of interleukin-7. J Biol Chem 284:14236–14244
13. Namba M, Mihara K, Fushimi K (1996) Immortalization of human cells and its mechanisms. Crit Rev Oncog 7:19–31
14. Namba M, Nishitani K, Fukushima F, Kimoto T (1988) Multistep carcinogenesis of normal human fibroblasts. Human fibroblasts immortalized by repeated treatment with Co-60 gamma rays were transformed into tumorigenic cells with Ha-ras oncogenes. Anticancer Res 8:947–958
15. Stein GH, Namba M, Corsaro CM (1985) Relationship of finite proliferative lifespan, senescence, and quiescence in human cells. J Cell Physiol 122:343–349
16. Tsuji T, Nozaki I, Miyazaki M, Sakaguchi M, Pu H, Hamazaki Y, Iijima O, Namba M (2001) Antiproliferative activity of REIC/Dkk-3 and its significant down-regulation in non-small-cell lung carcinomas. Biochem Biophys Res Commun 289:257–263
17. Feinberg AP (1999) Imprinting of a genomic domain of 11p15 and loss of imprinting in cancer: an introduction. Cancer Res 59:1743–1746
18. Kobayashi K, Ouchida M, Tsuji T, Hanafusa H, Miyazaki M, Namba M, Shimizu N, Shimizu K (2002) Reduced expression of the REIC/Dkk-3 gene by promoter-hypermethylation in human tumor cells. Gene 282:151–158
19. Pataer A, Hu W, Xiaolin L, Chada S, Roth JA, Hunt KK, Swisher SG (2008) Adenoviral endoplasmic reticulum-targeted mda-7/interleukin-24 vector enhances human cancer cell killing. Mol Cancer Ther 7:2528–2535
20. Tanimoto R, Sakaguchi M, Abarzua F, Kataoka K, Kurose K, Murata H, Nasu Y, Kumon H, Huh NH (2010) Down-regulation of BiP/GRP78 sensitizes resistant prostate cancer cells to gene-therapeutic overexpression of REIC/Dkk-3. Int J Cancer 126: 1562–1569
21. Guarino M, Rubino B, Ballabio G (2007) The role of epithelial-mesenchymal transition in cancer pathology. Pathology 39:305–318
22. Edamura K, Nasu Y, Takaishi M, Kobayashi T, Abarzua F, Sakaguchi M, Kashiwakura Y, Ebara S, Saika T, Watanabe M, Huh NH, Kumon H (2007) Adenovirus-mediated REIC/Dkk-3 gene transfer inhibits tumor growth and metastasis in an orthotopic

prostate cancer model. Cancer Gene Ther 14: 765–772
23. Gottesman MM, Pastan I, Ambudkar SV (1996) P-glycoprotein and multidrug resistance. Curr Opin Genet Dev 6:610–617
24. Gottesman MM, Ling V (2006) The molecular basis of multidrug resistance in cancer: the early years of P-glycoprotein research. FEBS Lett 580:998–1009
25. Kawasaki K, Watanabe M, Sakaguchi M, Ogasawara Y, Ochiai K, Nasu Y, Doihara H, Kashiwakura Y, Huh NH, Kumon H, Date H (2009) REIC/Dkk-3 overexpression downregulates P-glycoprotein in multidrug-resistant MCF7/ADR cells and induces apoptosis in breast cancer. Cancer Gene Ther 16:65–72
26. Lee AS (2007) GRP78 induction in cancer: therapeutic and prognostic implications. Cancer Res 67:3496–3499
27. Park HR, Tomida A, Sato S, Tsukumo Y, Yun J, Yamori T, Hayakawa Y, Tsuruo T, Shin-ya K (2004) Effect on tumor cells of blocking survival response to glucose deprivation. J Natl Cancer Inst 96: 1300–1310
28. Ermakova SP, Kang BS, Choi BY, Choi HS, Schuster TF, Ma WY, Bode AM, Dong Z (2006) (-)-Epigallocatechin gallate overcomes resistance to etoposide-induced cell death by targeting the molecular chaperone glucose-regulated protein 78. Cancer Res 66:9260–9269
29. Meier O, Greber UF (2003) Adenovirus endocytosis. J Gene Med 5:451–462
30. Freimuth P, Philipson L, Carson SD (2008) The coxsackievirus and adenovirus receptor. Curr Top Microbiol Immunol 323:67–87

Index

J.S. Rhim and R. Kremer (eds.), *Human Cell Transformation: Role of Stem Cells and the Microenvironment*, Advances in Experimental Medicine and Biology 720, DOI 10.1007/978-1-4614-0254-1, © Springer Science+Business Media, LLC 2011

U

V

W